DENTAL SCIENCE, MATERIALS AND TECHNOLOGY

BIOMECHANICS OF DENTAL IMPLANTS

HANDBOOK FOR RESEARCHERS

DENTAL SCIENCE, MATERIALS AND TECHNOLOGY

Additional books in this series can be found on Nova's website under the Series tab.

Additional e-books in this series can be found on Nova's website under the e-book tab.

DENTAL SCIENCE, MATERIALS AND TECHNOLOGY

BIOMECHANICS OF DENTAL IMPLANTS

HANDBOOK FOR RESEARCHERS

MURAT ÇEHRELI
EDITOR

Nova Science Publishers, Inc.
New York

LIBRARY OF CONGRESS CATALOGING-IN-PUBLICATION DATA

Biomechanics of dental implants : handbook of researchers / editor, Murat Gehreli.
p. ; cm.
Includes bibliographical references and index.
ISBN 978-1-62100-780-7 (hardcover)
I. Gehreli, Murat.
[DNLM: 1. Dental Implants. 2. Biomechanics. WU 640]

617.6'93--dc23

2011038319

Published by Nova Science Publishers, Inc. † New York

This book is dedicated to Lara and Onat

CONTENTS

PREFACE

Biomechanics is applied to many fields of dentistry including treatments involving the use of dental implants to provide a means for predicting the clinical outcome. Several experimental and theoretical approaches are being used to accomplish this goal. In writing this humble book, our sincere hope was to provide a brief guide to the apprentice, although any reader interested in dental implant biomechanics can benefit this book to get the most complete picture of clinical/animal/experimental biomechanical studies related to state-of-art dental implants. I also wished to include a brief section related to post and core restorations, since these devices as well as orthodontic implants are assigned to the same biomechanical analyses for predicting their clinical outcome.

I would like to thank deeply to all the contributors for their unrequited efforts and the job well-done. Without their work, this book would be incomplete.

CONTRIBUTORS

Serdar Arıkan
Department of Restorative Dentistry
Ordu University, Ordu, Turkey

Lyndon F. Cooper
Bone Biology and Implant Therapy Laboratory,
Department of Prosthodontics, University of North Carolina at Chapel Hill, Chapel Hill, NC, USA

Murat Çehreli
Department of Prosthetic Dentistry
Ordu University, Ordu, Turkey

Marco Degidi
Private Practice,
Bologna, Italy

Joke Duyck
Department of Prosthetic Dentistry
Catlolic University of Leuven
Leuven, Belgium

Steven Eckert
Mayo Clinic,
Rochester, MN, USA

Giovanna Iezzi
Dental School
University of Chieti-Pescara, Italy

Siegfried V. N. Jaecques
Leuven Medical Technology Centre (LMTC) and Division of Biomechanics and Engineering Design (BMGO), Catholic University of Leuven, Leuven, Belgium

Maureen E. Lynch
Department of Biomedical Engineering Weill Hall, Cornell University
Ithaca, NY, USA

Carlo Mangano
Private Practice,
Gravedona, Italy

Daniela B.S. Mendonça
Bone Biology and Implant Therapy Laboratory,
Department of Prosthodontics, University of North Carolina at Chapel Hill, Chapel Hill, NC, USA

Gustavo Mendonça
Bone Biology and Implant Therapy Laboratory,
Department of Prosthodontics, University of North Carolina at Chapel Hill, Chapel Hill, NC, USA

Marjolein C. H. van der Meulen
Sibley School of Mechanical andAerospace Engineering Upson Hall and Research Division Hospital for Special Surgery
Cornell University, Ithaca, NY, USA

Ignace Naert
Department of Prosthetic Dentistry
Catlolic University of Leuven, Leuven, Belgium

Burak Sadık
Department of Endodontics
Ordu University, Ordu, Turkey

Antonio Scarano
Dental School
University of Chieti-Pescara, Italy

Jamil Awad Shibli
Department of Periodontology
Dental Research Division, Guarulhos University Guarulhos, SP, Brazil

Regina Mericske-Stern
Department of Prosthodontics
University of Bern, Switzerland

Vittoria Perrotti
Dental School,
University of Chieti-Pescara, Italy

Roberto S. Pessoa
Department of Diagnostic and Surgery, Division of Periodontics, UNESP – São Paulo State University, Araraquara, Brazil

Adriano Piattelli
Dental School
University of Chieti-Pescara, Italy

Ergin Tönük
Department of Mechanical Engineering
Middle East Technical University, Ankara, Turkey

Katleen Vandamme
Department of Prosthetic Dentistry
Catlolic University of Leuven, Leuven, Belgium

In: Biomechanics of Dental Implants
Editor: Murat Çehreli
ISBN 978-1-62100-780-7

Chapter 1

MECHANICAL PROPERTIES OF BONE TISSUE

Maureen E. Lynch and Marjolein C. H. van der Meulen

1. INTRODUCTION

The mechanical performance of any object depends on the forces applied to the structure, the resulting burdens (stresses, strains) placed on the materials that comprise the structure, and the ability of the materials to withstand and sustain the burdens (material properties). As a living material, bone is uniquely capable of responding and adapting to mechanical cues that influence its size and shape. For example, with the loss of teeth, individuals experience decreased mastication and biting functionality, and the strength and stiffness of the mandible decreases accordingly [1, 2]. Conversely, as a result of increased exercise, long bones increase their size and shape in response to large incident forces, which increases their ability to withstand loads [3, 4]. Therefore, understanding the mechanical performance of bone can help to understand the factors that determine skeletal structure and predict changes that occur with aging, prosthetic implantation, fracture and other clinical conditions.

1.1. Structural vs. Material Behavior

When a force is applied to a structure, such as a whole bone, the structure deforms, and strains and stresses develop throughout the material. In the skeleton, structural behavior refers to the mechanical behavior of a whole bone, such as an entire tibia or mandible, and is a function of the material properties and geometry or architecture of the structure. In contrast, material behavior refers to the behavior under load of the bone tissue. Material properties are intrinsic characteristics and independent of geometry or architecture.

To further illustrate the differences between structural and material behaviors, consider a cylinder undergoing pure compression (Figure 1a). As increasing force is applied, the cylinder deforms and shortens. The resulting behavior can be characterized by the force-deformation curve that describes the structural behavior of the cylinder (Figure 1b). This behavior is analogous to the behavior when loading a whole bone in vivo. The slope of the initial linear portion is defined as the structural stiffness, *k*:

$$k = AE / L \tag{Eq.1}$$

and is proportional to the cross-sectional area of the cylinder (*A*) and Young's modulus (*E*) of the material, and inversely proportional to the length (*L*). Therefore, if two different cylinders are compressed, each of the same material and length yet different cross-sectional areas, each cylinder will exhibit a different force-deformation curve. Consequently, each cylinder will have a different structural stiffness and strength. Similarly, if two whole bones consisting of the same material but different geometries, such as the tibia and femur, undergo compression, each structure will have a different force-deformation curve and structural stiffness.

To eliminate geometric effects and isolate material behavior, the force-deformation curve can be normalized by dividing force by cross-sectional area and deformation by the original length of the cylinder for structures consisting of a single material and uniform geometry. The resulting behavior is the stress-strain curve (Figure 1c). Material behavior is independent of specimen geometry and describes the intrinsic behavior of the material. The linear portion of the stress-strain curve is described by Hooke's Law relating stress (σ) to strain (ε):

$$\sigma = E\varepsilon \tag{Eq.2}$$

The slope is defined as the Young's modulus or elastic modulus, *E*, which describes the intrinsic stiffness of the material. The elastic modulus describes ability of the material to resist permanent deformation (elasticity). In this chapter our focus will be on these material properties for bone tissue.

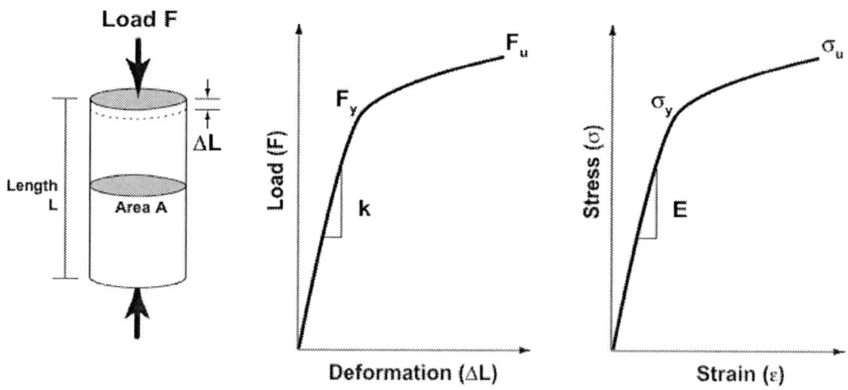

Figure 1. a) Cylinder of length L and cross-sectional area A subjected to uniaxial compression by force F. The change in height is ΔL. b) Structural behavior: The force-deformation curve describes the structural behavior of the cylinder as a function of the applied load F and resulting deformation ΔL. The slope of the linear portion is the structural stiffness, k. The behavior transitions from linear to nonlinear at the yield point, F_y. Failure occurs at the ultimate load, F_u. c) Material behavior: Corresponding stress-strain curve, the normalized force-deformation curve, describes the material behavior of the cylinder and is independent of geometry. The slope of the linear portion is the material stiffness or Young's modulus, E. The behavior transitions from elastic to plastic or permanent deformation at the yield stress, σ_y. Failure occurs at the ultimate stress, σ_u.

1.2. Stress and Strain

Stress is the measure of the internal force intensities that have developed within a material due to an applied force. Strain is the relative deformation, either normal or shear, that occurs in response to an applied force (Figure 2). The applied force can be mathematically resolved into normal (perpendicular to surface) and shear components (parallel to surface). Tension and compression are pure normal forces and torsion is a pure shear force. Normal stress (σ) is calculated as the force perpendicular to the surface divided by cross-sectional area. Normal strain (ε) refers to lengthening or shortening and is measured as change in length divided by original length. Shear stress (τ) is the force parallel to the surface divided by the cross-sectional area. Shear strain (γ) refers to parallel movement of one plane section relative to an adjacent section and is measured as the angular change measured in radians. For torsional loading, the resulting torque-twist curve (analogous to the force-deformation curve) is normalized to generate the shear stress-strain curve, from which the shear modulus, G, (analogous to E) is determined.

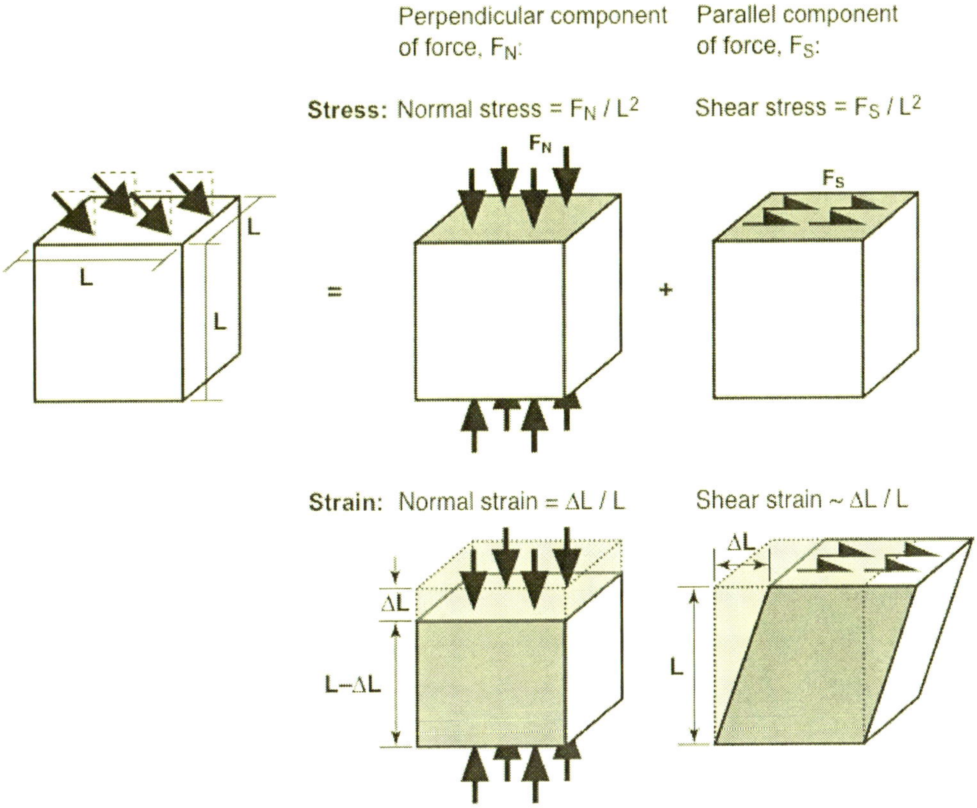

Figure 2. Cube of material (side length L) with applied load F. The load can be decomposed into normal (F_N) and shear (F_S) components. (top) The normal and shear stresses are calculated as the respective force per area (L2). (bottom) The normal strain is calculated as the change in length (ΔL) per original length (L) and the shear strain is the angular change that is approximated as the linear angle (ΔL) divided by the original length (L).

1.3. Material Properties

Material properties can be determined from mechanical tests using standardized specimens under controlled conditions. The simplest mechanical tests are uniaxial with loading applied in compression, tension or torsion. In a uniaxial test, the specimen is progressively loaded in a single direction until failure. The magnitude of the stresses and strains that develop under applied forces, as well as the ability of the object to withstand and sustain the stresses and strains, characterize the material's overall mechanical performance. To characterize more complex mechanical behaviors, multiaxial tests can be used where forces are applied in multiple directions. Data obtained from such tests are valid only for a material with the same microstructure and environmental conditions as the test specimens.

For bone, elastic modulus (material stiffness) and strength are generally considered the most important material mechanical properties, but the parameter of interest will depend on the research question. These parameters are derived from the stress-strain data for uniaxial, monotonic tests (Figure 3). The stress-strain behavior is generally divided into two regions: elastic and plastic. The elastic portion is the linear region of the curve and describes the ability of the material to return to its original configuration upon removal of applied forces. As discussed previously, this linear behavior is described by Equation 2, where the slope is E. As the applied forces increase, the material undergoes permanent deformation, known as yielding, after which point the specimen will not return to its original configuration. In this case, the stresses have increased beyond the elastic limit, known as the yield stress or yield strength, σ_Y. The region beyond the yield stress is the plastic, or post-yield, region. The ultimate stress or ultimate strength, σ_U, is the maximum stress the material can withstand and the fracture point, σ_f, is the stress at which the material breaks. Toughness is the area under the stress-strain curve and is the total energy the material absorbed before fracture.

In addition to mechanical tests of physical specimens, ultrasonic velocity transmission is another method to determine the elastic behavior of a material that is especially useful for bone. The velocity at which an ultrasonic wave (v) passes through a material depends on its density (ρ) and intrinsic elasticity (c):

$$c = \rho v^2 \qquad \text{(Eq.3)}$$

The exact form of this relationship depends on whether the wave velocity is measured parallel (longitudinal mode) or perpendicular (transverse mode) to its propagation direction and the geometric dimensions of the specimen and the wave frequency. Cortical bone is typically tested using frequencies between 2-10 MHz while cancellous bone tissue requires lower frequencies, between 50-100 kHz, due to its porous structure [5]. Elastic moduli of both cortical and cancellous bone determined using ultrasonic methods compare well to moduli measured from traditional mechanical tests [5, 6].

One advantage of the ultrasonic testing method is the ability to test smaller specimens. A second advantage is the ability to determine to the elastic constants in different directions. For an isotropic material, the mechanical behavior is independent of the orientation of the material relative to the loading direction.

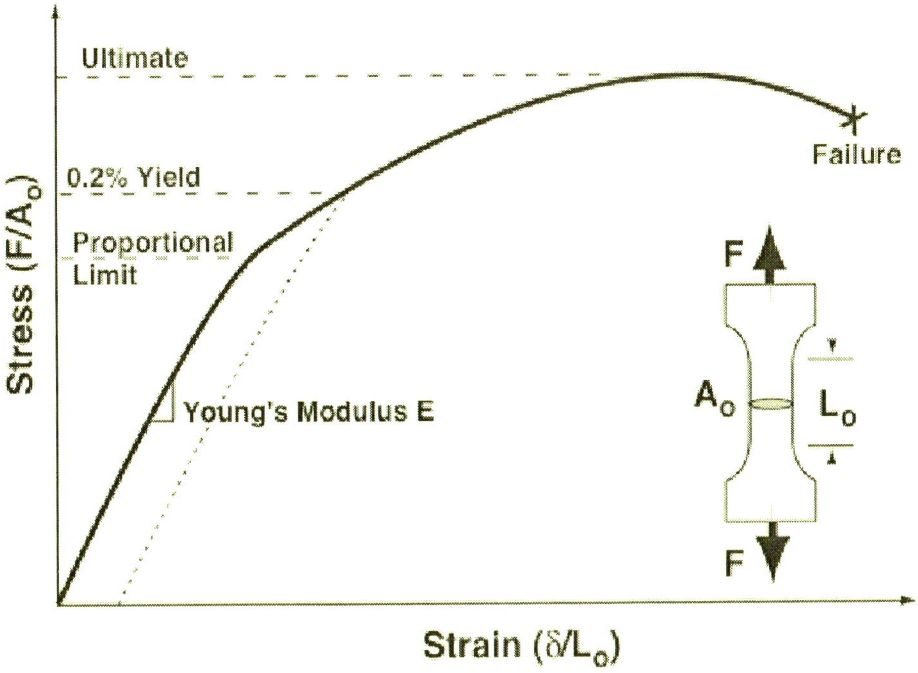

Figure 3. Stress-strain behavior for a materials test. These measurements depend on the loading mode (tension, compression, bending, or shear) and are independent of specimen size and shape. The material stiffness or Young's modulus (E) is determined from the initial linear elastic region. Yield stress (σ_y) is the transition from linear to nonlinear behavior and is calculated based on 0.2% offset of the linear region (dashed line); and ultimate strength (σ_u) is the maximum stress or material strength. For the tensile specimen shown, F is the tensile load, A_o is the original area, and L_o is the original specimen length.

However, the material properties of anisotropic materials depend on the orientation of the material relative to the loading direction (Figure 4). Isotropic materials are uniquely characterized by two material constants: the elastic modulus (Young's modulus) and Poisson's ratio, ν. The Poisson's ratio describes the ability of the material to resist deformation perpendicular to the direction of the applied forces, or how much the material will bulge when compressed or contract when stretched. The Poisson's ratio is defined as the strain perpendicular to the loading direction divided by the strain in the loading direction. Anisotropic materials are characterized by more than two and as many as twenty one material constants.

Bone tissue is an anisotropic material. Anisotropic materials may exhibit a degree of symmetry with respect to properties in a particular orientation. Cortical bone is generally considered to be transversely isotropic, where the material properties within a transverse section are isotropic and are different from the material properties in the longitudinal direction (parallel to the longitudinal axis of osteons) [7-11]. Similarly, in the human mandible, maximal stiffness is generally aligned with osteonal orientation (Figure 4b) [11, 12]. Cancellous bone also exhibits anisotropy and has at least orthotropic symmetry [13, 14]. The mechanical properties of cancellous bone in the human mandible are also anisotropic [15]. In some locations, transverse isotropy has been reported [16, 17]. Transversely isotropic materials require two moduli (transverse and longitudinal) and three Poisson's ratios to uniquely characterize the elastic material properties.

Additional material property considerations include the time-dependency of material behavior. Materials whose elastic behavior depends on loading rate are known as viscoelastic. Similarly, time-dependent permanent deformation is known as viscoplasticity. The properties discussed above are for uniaxial behavior, but most normal daily skeletal loading is dynamic and induces stresses well below the yield stress. When a material is subjected to cyclic stresses or strains at subfailure levels, microscopic fatigue damage accumulates [18-20], causing gradual and progressive loss of mechanical integrity.

Ultimately, the material may fail at stresses that are lower than the yield stress. To determine the fatigue properties of a material, standardized specimens are subjected to specified cyclic stresses and loaded repetitively until failure, producing an a stress-life cycle, or S-N, curve. From this curve, the maximum number of cycles a material can withstand at a given stress. In addition, some materials exhibit an endurance limit, a stress plateau below which failure will never occur for an infinite number of loading cycles. Mechanistically, fatigue failure involves the initiation of microcracks that propagate and coalesce into large cracks until failure occurs.

2. Bone Tissue Morphology

Two types of bone material are found in the skeleton: cortical and cancellous bone. Cortical bone forms the diaphyses of long bones and external shell of short bones such as the vertebrae and mandible. Microstructurally, cortical bone consists of cylindrical osteons that surround a central Haversian canal. In the cortex, osteons are aligned longitudinally. Cancellous bone is found in the ends of long bones and in the interior of irregular bones such as the vertebrae and mandible. The tissue microstructure consists of trabecular rods and plates that form a cellular or lattice structure. The solid phase of both tissues is a composite of an organic phase that is primarily Type I collagen and an inorganic phase consisting of calcium carbonate mineral. Combined, the weak polymer and brittle mineral give bone tissue unique strength and toughness behaviors that are different from the individual constituents. Cortical and cancellous bone tissue can be distinguished primarily based on porosity, the volume occupied by non-mineralized tissue relative to total tissue volume [21]. For a bulk sample, porosity is often characterized by apparent density. Apparent density is calculated as mass of mineralized tissue divided by total volume (mineralized and non-mineralized) of the specimen. Cortical bone is fairly dense with a porosity of 5-10% and apparent density of 1.5-1.8 g/cm^3. Cancellous bone is comprised of individual trabeculae that form of a lattice of rod- or plate-like struts creating a very porous material (50-90%) with low apparent density (0.5-1.0 g/ cm^3). In the human mandible, the porosity of cortical bone is 3.5% and condylar cancellous bone is 79% [22]. The porosity within bone affects the mechanical properties of bone tissue (Figure 5). With increasing apparent density, both modulus and strength increase, such that cortical bone is significantly stiffer and stronger than cancellous bone. The wide range of apparent densities found in cancellous bone produce a correspondingly wide range of material properties.

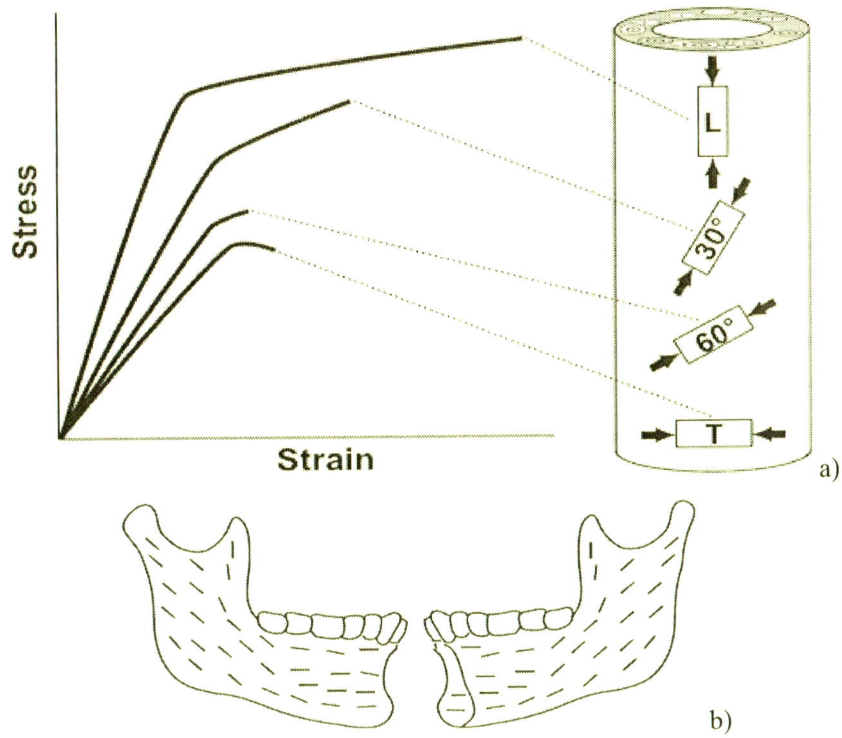

Figure 4. a)Anisotropic mechanical behavior of femoral cortical bone as a function of osteonal orientation. Adapted from [7]. b) Schematic of osteonal orientation in the human mandible indicated. Lines indicate osteonal direction, which is coincident with direction of maximum stiffness. Adapted from [11, 12, 27].

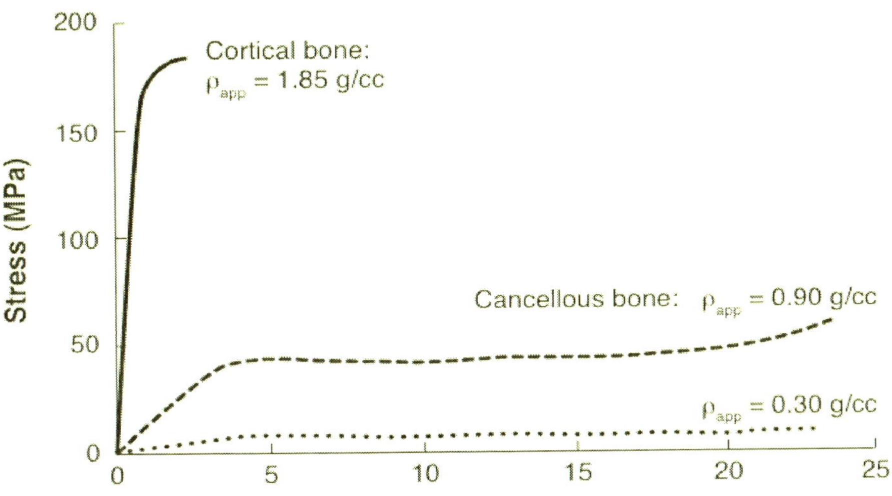

Figure 5. Stress and strain behavior of cortical and cancellous bone tissue of different apparent densities (ρ_{app}) in compression. Adapted from [63, 64].

3. Cortical Bone Mechanical Properties

3.1. Material Properties

Cortical bone is characterized by osteonal microstructure, which gives the material a preferred orientation relative to the longitudinal axes of the osteons. The elastic behavior of cortical bone is anisotropic, with different moduli measured in the longitudinal and transverse directions (Figure 6). Cortical bone is stiffest in the longitudinal direction. The longitudinal modulus of the cortex of long bones ranges from 10–20 GPa and is approximately 50% greater than the transverse modulus and five times greater than the shear modulus (Table 1) [7, 23]. In the mandible, the longitudinal modulus of cortical bone ranges between 17 – 25 GPa and is 1 – 2x greater than the transverse modulus, depending on anatomical location (Table 2) [24-27]. In addition, the moduli of cortical bone are similar under tension and compression (Figure 7).

The ultimate strength of cortical bone is different in the longitudinal and transverse directions, but also depends on the loading modality (Figure 6), therefore strength is asymmetric. Cortical bone is strongest in compression for both longitudinal (~190 MPa compressive strength vs. ~130 MPa tensile strength) and transverse directions (~130 MPa compressive strength vs. ~50 MPa tensile strength) [7]. Considered as a whole, the properties of cortical bone are well adapted to withstand compressive forces aligned with the orientation of its osteons.

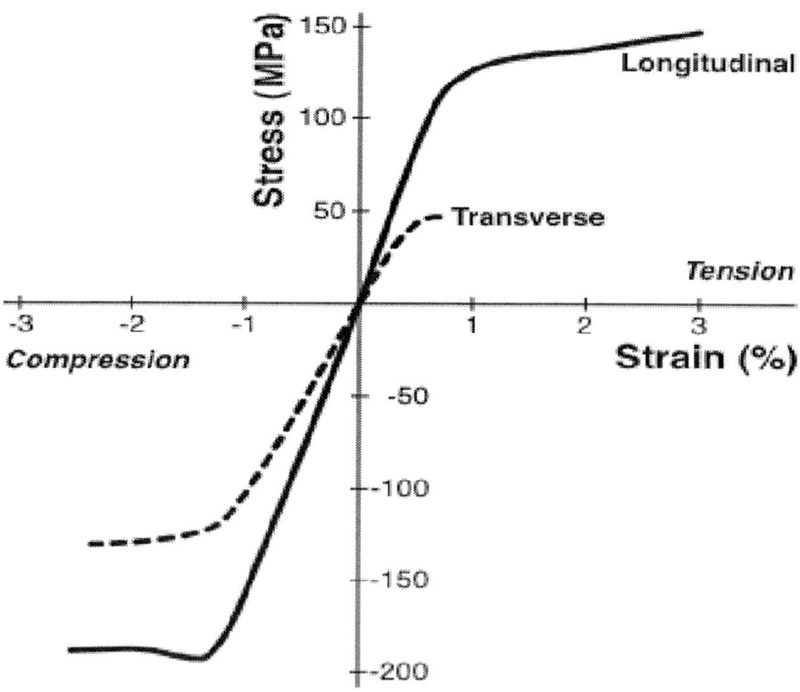

Figure 6. Typical stress-strain curves for human cortical bone samples loaded to failure in tension and compression in the longitudinal and transverse directions. Adapted from [85].

Table 1. Elastic moduli (E) for human cortical bone determined by mechanical testing (MT) or ultrasonic (US) measurement. Anatomic location indicated. E_{rad} = modulus in radial direction of transverse section; E_{tan} = modulus in tangential direction of transverse section; E_{long} = modulus in longitudinal direction

Bone	Femur	Femur	Tibia	Tibia	Mandible	Mandible
	(MT)	(US)	(MT)	(US)	(MT)	(US)
	[7]	[10]	[84]	[6]	[25, 26]	[24]
E_{rad} (GPa)	11.5	18.8	6.9	12.0	6.9	11.3
E_{tan} (GPa)	11.5	18.8	8.5	12.4	8.2	13.8
E_{long} (GPa)	17.0	27.4	18.4	20.0	17.3	19.4

Table 2. Site specific variation of elastic moduli for cortical bone in the human mandible [27]. E_{long} = modulus in longitudinal direction; E_{trans} = modulus in transverse direction

Site	E_{long} (GPa)	E_{trans} (GPa)
Lingual alveolor process	25.2	12.7
Facial inferior border	22.5	13.6
Facial inferior border	23.0	16.0
Lingual symphysis	21.1	16.8
Lingual midbody	22.9	20.6

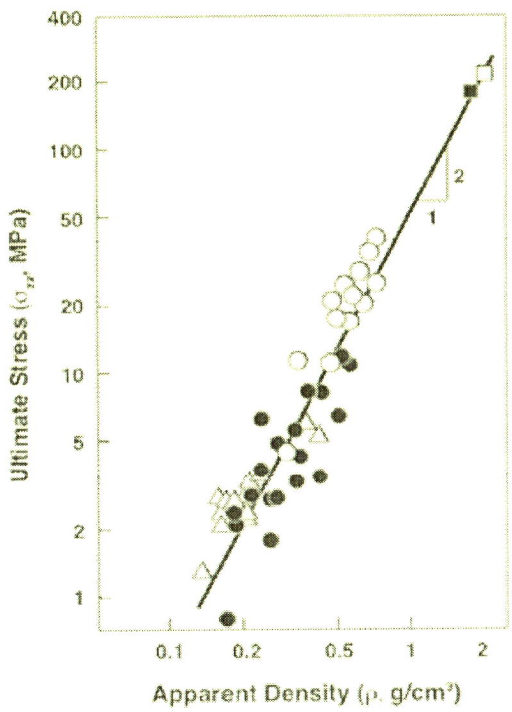

Figure 7. The effects of apparent density on cancellous bone strength of human and bovine tissue. Adapted from [63].

When loaded in the longitudinal direction, cortical bone is a relatively tough, ductile material, which may be important for withstanding traumatic loading. The strains at fracture are greater than those at yielding, therefore cortical bone can undergo substantial deformation and absorb a large amount of energy prior to fracture. In contrast, cortical is relatively brittle in the transverse direction, reflected by the fact that the ultimate strain is close to the yield strain.

3.2. Viscoelasticity

The mechanical properties of cortical bone change with strain rate ($\dot{\varepsilon}$) [28-30] (Figure 8). Generally, as the strain rate increases, both stiffness and strength increase, although ultimate strength is more sensitive to strain rate than modulus [28]. For typical activities, such as walking and running (strain rates between 0.001 and 0.03 sec^{-1}), the modulus is relatively constant [31-35]. In this physiological range, cortical bone also becomes stronger with more strenuous activities. At higher strain rates, representing traumatic loading, cortical bone transitions from a ductile to a more brittle material [28, 29, 36]. The compressive strength of mandibular cortical bone is sensitive to strain rate, but within a lower strain range (0.0001 – 0.1 sec^{-1}) [35], which may reflect that the jaw is not adapted for the same magnitude of high-impact forces as the appendicular skeleton. For example, the ductile-brittle transition is ~0.0024 sec^{-1} for mandibular bone whereas the transition occurs between 1-100 sec^{-1} for the femur [28, 35].

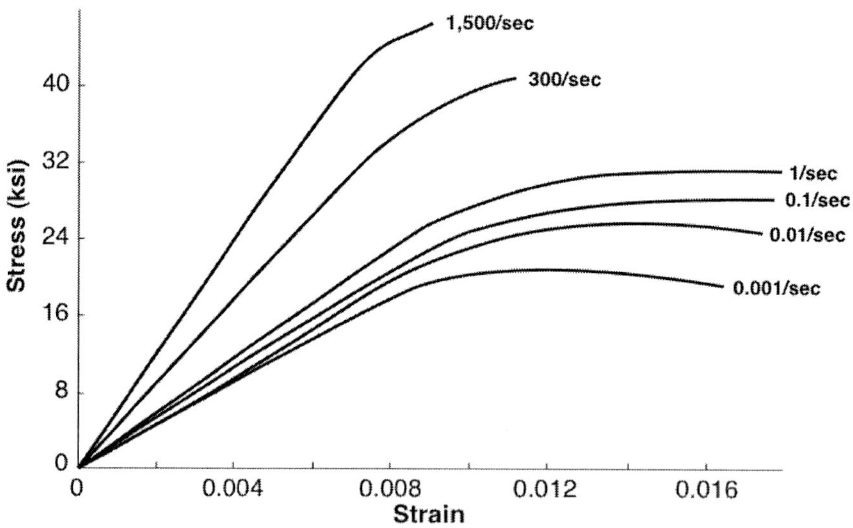

Figure 8. Strain rate dependency of stress-strain behavior for cortical samples taken from the human femur and aligned with the longitudinal axis. Strain rates indicated. Adapted from [28].

3.3. Fatigue

During activities of daily living, the skeleton is continually exposed to cyclic loading at low loads. Fatigue behavior is determined by microcrack initiation and propagation. Materials generally either resist crack initiation then propagate cracks easily once they form, or conversely, initiate cracks easily then inhibit crack propagation effectively. Unlike conventional materials, bone is constantly turning over through remodeling. One function of remodeling is believed to be the repair of fatigue microdamage. This mechanism is consistent with experimental observations that cracks form readily in bone, but do not propagate easily due to a variety of microstructural and material features. When damage accumulation exceeds the natural repair processes, stress fractures occur in vivo. These fractures are common in ballet dancers, military recruits and runners [37, 38].

In vitro tests may underestimate in vivo fatigue behavior due to the absence of these repair processes; however, useful information is obtained regarding the overall behavior and key characteristics. Bone tissue fatigue behavior is similar to that of conventional fiber-reinforced composite materials. In laboratory tests of cortical bone, fatigue failure correlates better with strain range than with the applied stress range (Figure 9) [39, 40]. Cortical bone is more susceptible to fatigue damage in tension than in compression reflecting that the material is weaker under tensile loads [41]. At physiological strains, the fatigue life of bone cortical bone is twice as long for compressive than tensile loading.

3.4. Additional Effects on Cortical Material Properties

A variety of individual characteristics affect the mechanical properties of cortical bone, including age, disease and diet. Elastic modulus and strength do not change with age [42, 43]. However, postyield strain reduces with age in tension and is accompanied by a reduction in energy absorption. These changes may result from increased fatigue damage present in the tissue with increasing age [44]. Age-related changes in cortical bone mechanical properties are similar in males and females [43]; however, microdamage accumulation occurs more rapidly in women than men [45].

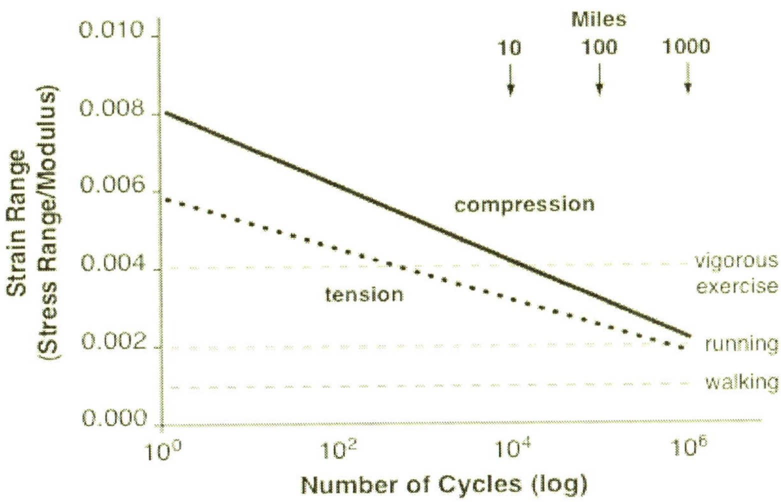

Figure 9. Fatigue life expressed as strain range for cortical bone samples from the human femur and aligned with the longitudinal axis. Fatigue performance is greater for compressive than tensile loading. Strain ranges for typical activities (walking, running and exercise) are indicated. Typical numbers of cycles for specific mileages also indicated. Adapted from [40].

4. Cancellous Bone Mechanical Properties

4.1. Material Properties

Cancellous bone exhibits substantial heterogeneity across and within anatomical sites due to variations in apparent density and architecture, which have a strong effect on the mechanical behavior of cancellous bone. Furthermore, apparent density and architecture change with aging and disease. Material properties can vary by up to 2 orders of magnitude depending on anatomical location [46-50]. For example, modulus of human vertebral cancellous bone can be up to an order of magnitude lower than that in the proximal tibia or femur [49, 51-53]. Properties can vary substantially within a single location as well [46, 49, 53, 54]. Within the mandibular condyle, stiffness values range from 200-1400 MPa depending on the apparent density [54]. As expected, elastic moduli increase as apparent density increases (Figure 10) [2, 54]. As in cortical bone, cancellous bone exhibits asymmetry with respect to loading modality. Both the yield strength (5.7 – 32 MPa) and ultimate strength (7.0 – 43 MPa) are 30-40% higher in compression than in tension [55-57] and are lowest in shear [58].

Cancellous bone absorbs a considerable amount of energy post-yield under compressive loading because the mechanism of failure is based on the aggregate failure of individual trabeculae [59]. When cancellous bone is loaded to failure in compression, individual trabeculae bend and compress first, then individual trabeculae begin to fail [59-62]. Ultimately, all trabeculae fail and the pores are compressed to form a bulk material that then bears the load. Therefore, two key properties affect the mechanical behavior of cancellous bone: apparent density and trabecular architecture.

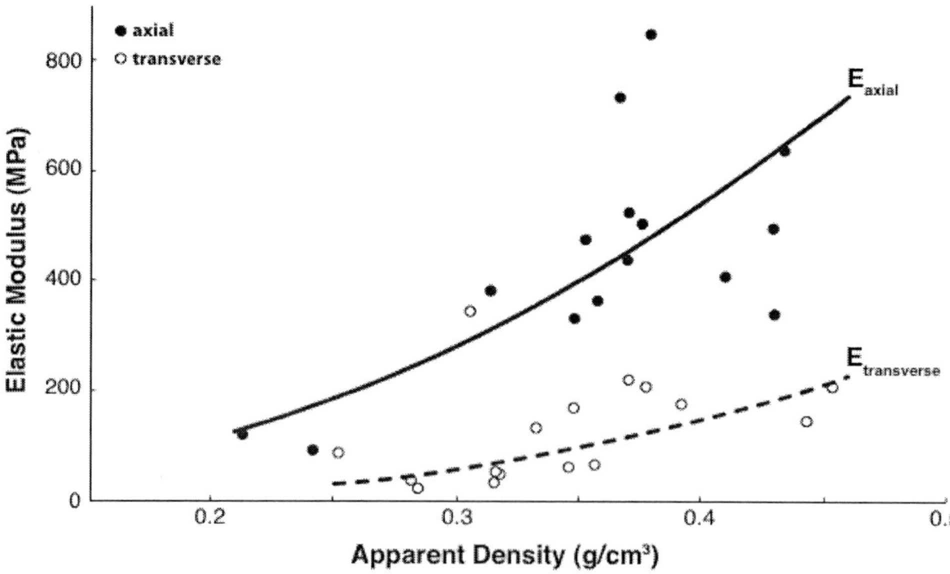

Figure 10. Stiffness increases with increasing apparent density for tissue samples from the human mandibular condyles. Anisotropy is evident; the elastic modulus in the axial direction (E_{axial}) is significantly greater than the modulus measured in the transverse direction ($E_{transverse}$). Adapted from [15].

Cancellous bone exhibits a wide range of apparent densities. The effects of apparent density on both the modulus and ultimate strength of cancellous bone follow a power-law relationship [46, 63-67] (Figure 7), where γ is the material property, ρ is apparent density, and A and B are experimentally derived constants:

$$\gamma = A\rho^{B} \quad \text{(Eq.4)}$$

The most important parameter is the exponent B, which describes how quickly modulus and strength change with apparent density. B ranges from 1-4, depending on site, orientation, and the specific material property being examined [55, 63, 65]. Empirically, the exponent for modulus of cancellous bone ranges from 1-3 [57, 63, 65] and strength is generally of a lower order than modulus. For example, the ultimate compressive strength of human tibial cancellous bone changes with apparent density following an exponent of 2, while the modulus follows an exponent of 3 [63]. However, a linear modulus-density relationship, rather than power law, may be sufficient for cancellous tissue loaded in the direction of the predominant trabecular orientation [57]. In the mandibular condyle, B ranges between 2.2 and 3.2 for modulus, depending on the loading direction [15]. For ultimate strength of cancellous bone, B ranges from 1.5 to 2.2 [57, 68].

In addition to apparent density, the trabecular architecture of cancellous bone affects its mechanical properties. Architecture includes the shape, spacing, and number of the individual trabeculae and their general orientation at a given anatomic site. The modulus of cancellous bone exhibits anisotropy and depends on the loading direction relative to the orientation of the trabeculae (Figure 10) [2, 16]. Cancellous bone is stronger (2.8x) and stiffer (3.4x) in the longitudinal direction relative to the transverse direction [52]. In general, cancellous bone is stiffer and stronger in the predominant trabecular direction [17]. In the mandibular condyle, orientation explained 50% of the mechanical behavior of cancellous bone [66]. Indeed, the modulus of mandibular cancellous bone is greater in the direction of predominant trabecular alignment [2, 15], likely an adaptation to resist typical compressive forces during mastication [69]. Consequently, condylar mechanical properties in edentate individuals decline relative to dentate individuals due to loss of function [1, 2].

4.2. Viscoelasticity

Cancellous bone, like cortical bone, is also sensitive to strain rate when subjected to loading [60, 63, 67]. The effects of strain rate also follow a power law relationship, similar to Equation 4, and can be combined with apparent density for both compressive modulus and ultimate strength:

$$\gamma = A\dot{\varepsilon}^{0.06}\rho^{B} \quad \text{(Eq.5)}$$

where γ is the material property, strain rate ($\dot{\varepsilon}$) is raised to the 0.06 power, ρ is apparent density, and A and B are experimentally derived constants [63]. Because the exponent for apparent density ($B > 1$) is greater than that for strain rate, the influence of strain rate on mechanical properties is weaker than density (Figure 10). Furthermore, at traumatic strain rates (~10 sec^{-1}), the presence of bone marrow hydraulically increases the strength of cancellous bone and increases the amount of energy absorbed.

4.3. Fatigue

Less data are available for cancellous bone fatigue behavior than cortical bone. These studies are complicated by the wide range of apparent densities of the tissue and variations in architecture. Furthermore, experimental tests are mostly conducted in compression due to limitations in the ability to grip cancellous specimens during tensile testing. As in cortical bone, fatigue behavior of cancellous bone is better characterized by applied strain range rather than stress range (Figure 11) [70, 71].

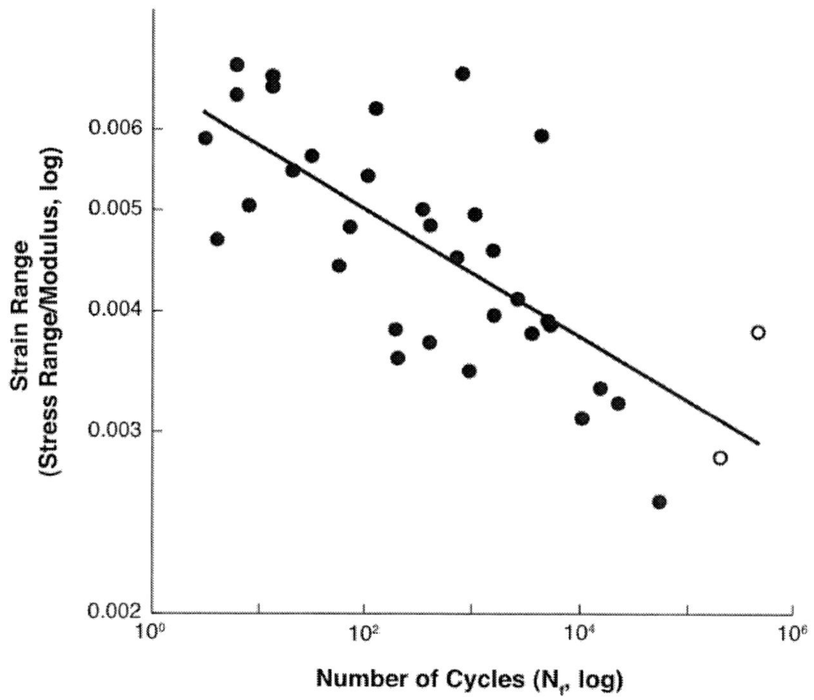

Figure 11. Fatigue life expressed as strain range for cancellous bone samples from human lumbar vertebrae and aligned with the inferior-superior direction. Cyclic loading applied in compression. Two specimens (open circles) did not fail after one week of loading. Adapted from [77].

Failure is driven by microcrack initiation and accumulation within the tissue of trabeculae, rather than fracture of entire individual trabeculae [72, 73]. Cyclic loading results in large reductions in apparent stiffness [71, 74, 75] and cumulative creep deformations [76, 77]. Experimentally, fatigue behavior is relatively constant among species and across anatomical sites, suggesting that cyclic failure occurs at the ultrastructural level rather than at the level of an individual trabeculae [77]. However, cancellous bone is significantly weaker when cyclic loading is applied off-axis from the predominant trabecular orientation [71]. For example, human vertebrae resist fatigue failure much longer when they are compressed in the direction of the spinal column than when they are compressed in the direction perpendicular to the spinal column.

4.4. Additional Effects on Cancellous Material Properties

In general, cancellous apparent density declines with aging, correlating with a decline in mechanical properties in the human vertebra, hip, and femur, and mandibular condyle [47, 48, 52, 78-80]. The apparent density changes are similar in both sexes. As in the cortex, microdamage accumulation in cancellous tissue increases with age [81]. Even when the changes in mechanical properties are normalized for density, mechanical properties deteriorate with aging, suggesting other factors, such as trabecular lattice connectivity, may play an important role [52]. In addition, aging also increases general trabecular alignment in human cancellous bone, attributed to a reorientation of trabeculae in the primary direction and a loss of trabeculae in other directions [82]. Resistance to fatigue decreases and sensitivity to off-axis loading increases with aging, making cancellous bone more susceptible to fatigue damage and subsequent fracture and more sensitive to 'off-axis' loading [71, 83].

Cancellous bone is often disproportionately affected by disease, likely reflecting the greater metabolic activity of cancellous bone than cortical bone per volume. Bone remodeling occurs on surfaces and the trabecular microstructure increases surface area/volume. With osteoporosis cancellous bone volume decreases and microarchitecture deteriorates, thereby reducing both strength and stiffness.

Conclusion

The mechanical performance of bone tissue governs skeletal function and is complex. The material properties depend not only on the volume fraction of tissue, but also on the orientation of the loading relative to the microstructure. In addition, the material properties differ with loading mode and loading rates. A key consideration that is not addressed here is that bone is mechanosensitive, and the tissue properties are strongly shaped by the loads experienced by the skeleton in vivo, particularly during growth and development. Therefore, as we consider the mechanical behavior of the tissue, we must also remember that the tissue structure and whole bone architecture are dynamic properties that reflect the loading history experienced by the skeleton.

References

[1] Giesen, E.B., Ding, M., Dalstra, M. and van Eijden, T.M. (2003) Reduced mechanical load decreases the density, stiffness, and strength of cancellous bone of the mandibular condyle. *Clinical Biomechanics (Bristol, Avon)*,18,358-363.

[2] van Eijden, T.M., van Ruijven, L.J. and Giesen, E.B. (2004) Bone tissue stiffness in the mandibular condyle is dependent on the direction and density of the cancellous structure. *Calcified Tissue International*,75,502-508.

[3] Jones, H.H., Priest, J.D., Hayes, W.C., Tichenor, C.C. and Nagel, D.A. (1977) Humeral hypertrophy in response to exercise. *Journal of Bone and Joint Surgery America*,59,204-208.

[4] Kontulainen, S., Sievanen, H., Kannus, P., Pasanen, M. and Vuori, I. (2003) Effect of long-term impact-loading on mass, size, and estimated strength of humerus and radius of female racquet-sports players: a peripheral quantitative computed tomography study between young and old starters and controls. *Journal of Bone and Mineral Research*,18,:352-359.

[5] Ashman, R.B., Corin, J.D. and Turner, C.H. (1987) Elastic properties of cancellous bone: measurement by an ultrasonic technique. *Journal of Biomechanics,*20,979-986.

[6] Ashman, R.B., Cowin, S.C., Van Buskirk, W.C. and Rice, J.C. (1984) A continuous wave technique for the measurement of the elastic properties of cortical bone. *Journal of Biomechanics,*17,349-361.

[7] Reilly, D.T. and Burstein, A.H. (1975) The elastic and ultimate properties of compact bone tissue. *Journal of Biomechanics,*8,393-405.

[8] Reilly, D.T., Burstein, A.H. and Frankel, V.H. (1974) The elastic modulus for bone. *Journal of Biomechanics,*7, 271-275.

[9] Lang, S.B. (1970) Ultrasonic method for measuring elastic coefficients of bone and results on fresh and dried bovine bones. *IEEE Transactions of Biomedical Engineering*,17,101-105.

[10] Yoon, H.S. 6 Katz, J.L. (1976) Ultrasonic wave propagation in human cortical bone--II. Measurements of elastic properties and microhardness. *Journal of Biomechanics* 9,459-464.

[11] Nomura, T., Gold, E., Powers, M.P., Shingaki, S. and Katz, J.L. (2003) Micromechanics/structure relationships in the human mandible. *Dental Materials*,19,167-173.

[12] Schwartz-Dabney, C.L. and Dechow, P.C. (2002) Edentulation alters material properties of cortical bone in the human mandible. *Journal of Dental Research*,81,613-617.

[13] Zysset, P.K., Goulet, R.W. and Hollister, S.J. (1998) A global relationship between trabecular bone morphology and homogenized elastic properties. *Journal of Biomechanical Engineering*,120,640-646.

[14] Yang, G., Kabel, J., van Rietbergen, B., Odgaard, A., Huiskes, R. and Cowin, S.C. (1998) The anisotropic Hooke's law for cancellous bone and wood. *Journal of Elasticity*,53,125-146.

[15] Giesen, E.B., Ding, M., Dalstra, M. and van Eijden, T.M. (2001) Mechanical properties of cancellous bone in the human mandibular condyle are anisotropic. *Journal of Biomechanics,*34,799-803.

[16] Williams, J.L. 6 Lewis, J.L. (1982) Properties and an anisotropic model of cancellous bone from the proximal tibial epiphysis. *Journal of Biomechanical Engineering*,104,50-56.

[17] Odgaard, A., Kabel, J., van Rietbergen, B., Dalstra, M. and Huiskes, R. (1997) Fabric and elastic principal directions of cancellous bone are closely related. *Journal of Biomechanics,*30,487-495.

[18] Carter, D.R., Caler, W.E., Spengler, D.M. and Frankel, V.H. (1981) Uniaxial fatigue of human cortical bone. The influence of tissue physical characteristics. *Journal of Biomechanics,*14,461-470.

[19] Carter, D.R. and Hayes, W.C. (1977) Compact bone fatigue damage: a microscopic examination. *Clinical Orthopedics and Related Research*,127,265-274.

[20] Schaffler, M.B., Radin, E.L. and Burr, D.B. (1989) Mechanical and morphological effects of strain rate on fatigue of compact bone. *Bone*,10,207-214.

[21] Carter, D.R. and Spengler, D.M. (1978) Mechanical properties and composition of cortical bone. *Clinical Orthopedics and Related Research*,135,192-217.

[22] Renders, G.A., Mulder, L., van Ruijven, L.J. and van Eijden, T.M. (2007) Porosity of human mandibular condylar bone. *Journal of Anatomy*,210,239-248.

[23] Cowin, S. (1989) Bone Mechanics. Boca Raton: CRC Press, Inc.

[24] Dechow, P.C., Nail, G.A., Schwartz-Dabney, C.L. and Ashman, R.B. (1993) Elastic properties of human supraorbital and mandibular bone. *American Journal of Physical Anthropology*,90,291-306.

[25] Arendts, F.J. and Sigolotto, C. (1989) [Standard measurements, elasticity values and tensile strength behavior of the human mandible, a contribution to the biomechanics of the mandible--I]. *Biomedical Technology (Berl).*,34,248-255.

[26] Arendts, F.J. and Sigolotto, C. (1990) [Mechanical characteristics of the human mandible and study of in vivo behavior of compact bone tissue, a contribution to the description of biomechanics of the mandible--II]. *Biomedical Technology (Berl).*, 35,123-130.

[27] Schwartz-Dabney, C.L. and Dechow, P.C. (2003) Variations in cortical material properties throughout the human dentate mandible. *American Journal of Physical Anthropology*,120,252-277.

[28] McElhaney, J.H. (1966) Dynamic response of bone and muscle tissue. *Journal of Applied Physiology*,21,1231-1236.

[29] Saha, S. and Hayes, W.C. (1976) Tensile impact properties of human compact bone. *Journal of Biomechanics*,9,243-251.

[30] Wright, T.M. and Hayes, W.C. (1976) Tensile testing of bone over a wide range of strain rates: effects of strain rate, microstructure and density. *Medical and Biological Engineering*, 14,671-680.

[31] Lanyon, L.E., Goodship, A.E., Pye, C.J. and MacFie, J.H. (1982) Mechanically adaptive bone remodelling. *Journal of Biomechanics*,15,141-154.

[32] Lanyon, L.E., Hampson, W.G., Goodship, A.E. and Shah, J.S. (1975) Bone deformation recorded in vivo from strain gauges attached to the human tibial shaft. A*cta Orthopedica Scandinavia*,46,256-268.

[33] Lanyon, L.E., Paul, I.L., Rubin, C.T., Thrasher, E.L., DeLaura, R., Rose, R.M. and Radin, E.L. (1981) In vivo strain measurements from bone and prosthesis following total hip replacement. An experimental study in sheep. *Journal of Bone and Joint Surgery America*,63,989-1001.

[34] O'Connor, J.A., Lanyon, L.E. and MacFie, H. (1982) The influence of strain rate on adaptive bone remodelling. *Journal of Biomechanics*,15,767-781.

[35] Robertson, D.M. and Smith, D.C. (1978) Compressive strength of mandibular bone as a function of microstructure and strain rate. *Journal of Biomechanics*,10-12, 455-471.

[36] Carter, D.R. and Caler, W.E. (1985) A cumulative damage model for bone fracture. *Journal of Orthopedic Research*,3,84-90.

[37] Milgrom, C., Giladi, M., Stein, M., Kashtan, H., Margulies, J.Y., Chisin, R., Steinberg, R. and Aharonson, Z. (1985) Stress fractures in military recruits. A prospective study showing an unusually high incidence. *Journal of Bone and Joint Surgery Britian,* 67,732-735.

[38] Fredericson, M., Jennings, F., Beaulieu, C. and Matheson, G.O. (2006) Stress fractures in athletes. *Topics in Magnetic Resonance Imaging*,17,309-325.

[39] Currey, J.D. (1970) The mechanical properies of bone. *Clinical Orthopedics*,73,210-231.

[40] Carter, D.R., Caler, W.E., Spengler, D.M. and Frankel, V.H. (1981) Fatigue behavior of adult cortical bone: the influence of mean strain and strain range. *Acta Orthopedica Scandinavia*,52,481-490.

[41] Caler, W.E. and Carter, D.R. (1989) Bone creep-fatigue damage accumulation. *Journal of Biomechanics*,22,625-635.

[42] Rose, E.C., Hagenmuller, M., Jonas, I.E. and Rahn, B.A. (2005) Validation of speed of sound for the assessment of cortical bone maturity. *European Journal of Orthodontics*,27,190-195.

[43] Burstein, A.H., Reilly, D.T. and Martens, M. (1976) Aging of bone tissue: mechanical properties. *Journal of Bone and Joint Surgery America*,58-A,82-86.

[44] Schaffler, M.B., Choi, K. and Milgrom, C. (1995) Aging and matrix microdamage accumulation in human compact bone. *Bone*,17,521-525.

[45] Burr, D.B., Forwood, M.R., Fyhrie, D.P., Martin, R.B., Schaffler, M.B. and Turner, C.H. (1997) Bone microdamage and skeletal fragility in osteoporotic and stress fractures. *Journal of Bone and Mineral Research*,12,6-15.

[46] Goldstein, S.A., Wilson, D.L., Sonstegard, D.A. and Matthews, L.S. (1983) The mechanical properties of human tibial trabecular bone as a function of metaphyseal location. *Journal of Biomechanics*,16,965-969.

[47] Mosekilde, L. and Mosekilde, L. (1988) Iliac crest trabecular bone volume as predictor for vertebral compressive strength, ash density and trabecular bone volume in normal individuals. *Bone*,9,195-199.

[48] Mosekilde, L., Viidik, A. and Mosekilde, L. (1985) Correlation between the compressive strength of iliac and vertebral trabecular bone in normal individuals. *Bone*,6,291-295.

[49] Hvid, I. and Hansen, S.L. (1985) Trabecular bone strength patterns at the proximal tibial epiphysis. *Journal of Orthopedic Research*,3,464-472.

[50] Keaveny, T.M., Morgan, E.F., Niebur, G.L. and Yeh, O.C. (2001) Biomechanics of trabecular bone. *Annual Reviews of Biomedical Engineering*,3,307-333.

[51] Linde, F. and Hvid, I. (1989) The effect of constraint on the mechanical behaviour of trabecular bone specimens. *Journal of Biomechanics*, 1989;22(5):485-490.

[52] Mosekilde, L., Mosekilde, L. and Danielsen, C.C. (1987) Biomechanical competence of vertebral trabecular bone in relation to ash density and age in normal individuals. *Bone*,8,79-85.

[53] Rohlmann, A., Zilch, H., Bergmann, G. and Kolbel, R. (1980) Material properties of femoral cancellous bone in axial loading. Part I: Time independent properties. *Archives of Orthopedic Trauma Surgery*,97,95-102.

[54] van Eijden, T.M., van der Helm, P.N., van Ruijven, L.J. and Mulder, L. (2006)Structural and mechanical properties of mandibular condylar bone. *Journal of Dental Research*,85,33-37.

[55] Keaveny, T.M. and Hayes, W.C. (1993) Mechanical Properties of Cortical and Trabecular Bone. In: B. Hall, editor. Bone, Volume VII: A Treatise. Boca Raton: CRC Press, Inc.;p. 285-344.

[56] Keaveny, T.M., Wachtel, E.F., Ford, C.M. and Hayes, W.C. (1994) Differences between the tensile and compressive strengths of bovine tibial trabecular bone depend on modulus. *Journal of Biomechanics*,27,1137-1146.

[57] Kopperdahl, D.L. and Keaveny, T.M. (1998) Yield strain behavior of trabecular bone. *Journal of Biomechanics*,31,601-608.

[58] Ford, C.M. and Keaveny, T.M. (1996) The dependence of shear failure properties of trabecular bone on apparent density and trabecular orientation. *Journal of Biomechanics*,29,1309-1317.

[59] Hayes, W.C. and Carter, D.R. (1976) Postyield behavior of subchondral trabecular bone. *Journal of Biomedical Materials Research. Biomedical Materials Symposium*,7,537-544.

[60] Ducheyne, P., Heymans, L., Martens, M., Aernoudt, E., de Meester, P. and Mulier, J.C. (1977) The mechanical behaviour of intracondylar cancellous bone of the femur at different loading rates. *Journal of Biomechanics*,10,747-762.

[61] Townsend, P.R., Raux, P., Rose, R.M., Miegel, R.E. and Radin, E.L. (1975) The distribution and anisotropy of the stiffness of cancellous bone in the human patella. *Journal of Biomechanics*,75,363-367.

[62] Turner, C.H. (1989) Yield behavior of bovine cancellous bone. *Journal of Biomechanical Engineering*,111,256-260.

[63] Carter, D.R. and Hayes, W.C. (1977) The compressive behavior of bone as a two-phase porous structure. *Journal of Bone and Joint Surgery America*,59,954-962.

[64] Gibson, L.J. (1985) The mechanical behavior of cancellous bone. *Journal of Biomechanics,*18,317-328.

[65] Rice, J.C., Cowin, S.C. and Bowman, J.A. (1988) On the dependence of the elasticity and strength of cancellous bone on apparent density. *Journal of Biomechanics,*21,155-168.

[66] Giesen, E.B., Ding, M., Dalstra, M. 6 van Eijden, T.M. (2003) Architectural measures of the cancellous bone of the mandibular condyle identified by principal components analysis. *Calcified Tissue International*,73,225-231.

[67] Linde, F., Norgaard, P., Hvid, I., Odgaard, A. and Soballe, K. (1991) Mechanical properties of trabecular bone. Dependency on strain rate. *Journal of Biomechanics,* 24,803-809.

[68] Hansson, T.H., Keller, T.S. and Panjabi, M.M. (1987) A study of the compressive properties of lumbar vertebral trabeculae: effects of tissue characteristics. *Spine,*12,56-62.

[69] van Ruijven, L.J., Giesen, E.B. and van Eijden, T.M. (2002) Mechanical significance of the trabecular microstructure of the human mandibular condyle. *Journal of Dental Research*,81,706-710.

[70] Bowman, S.M., Guo, X.E., Cheng, D.W., Keaveny, T.M., Gibson, L.J., Hayes, W.C. and McMahon, T.A. (1998) Creep contributes to the fatigue behavior of bovine trabecular bone. *Journal of Biomechanical Engineering*,120,647-654.
[71] Dendorfer, S., Maier, H.J., Taylor, D. and Hammer, J. (2008) Anisotropy of the fatigue behaviour of cancellous bone. *Journal of Biomechanics*,41,636-641.
[72] Fyhrie, D.P. and Schaffler, M.B. (1994) Failure mechanisms in human vertebral cancellous bone. *Bone*,15,105-109.
[73] Dendorfer, S., Maier, H.J. and Hammer, J. (2009) Fatigue damage in cancellous bone: an experimental approach from continuum to micro scale. *Journal of Mechanical Behavior of Biomedical Materials*,2,113-119.
[74] Wachtel, E.F. and Keaveny, T.M. (1997) Dependence of trabecular damage on mechanical strain. *Journal of Orthopedic Research*,15,781-787.
[75] Moore, T.L. 6 Gibson, L.J. (2003) Fatigue of bovine trabecular bone. *Journal of Biomechanical Engineering*,125,761-768.
[76] Michel, M.C., Guo, X.D., Gibson, L.J., McMahon, T.A. and Hayes, W.C. (1993) Compressive fatigue behavior of bovine trabecular bone. *Journal of Biomechanics,* 26,453-463.
[77] Haddock, S.M., Yeh, O.C., Mummaneni, P.V., Rosenberg, W.S. and Keaveny, T.M. (2004) Similarity in the fatigue behavior of trabecular bone across site and species. *Journal of Biomechanics,* 37,181-187.
[78] McCalden, R.W., McGeough, J.A. and Court-Brown, C.M. (1997) Age-related changes in the compressive strength of cancellous bone. The relative importance of changes in density and trabecular architecture. *Journal of Bone and Joint Surgery America*,79,421-427.
[79] Mosekilde, L. and Mosekilde, L. (1986) Normal vertebral body size and compressive strength: relations to age and to vertebral and iliac trabecular bone compressive strength. *Bone*,7,207-212.
[80] Ding, M., Dalstra, M., Danielsen, C.C., Kabel, J., Hvid, I. and Linde, F. (1997) Age variations in the properties of human tibial trabecular bone. *Journal of Bone and Joint Surgery Britian*,79,995-1002.
[81] Mori, S., Harruff, R., Ambrosius, W. and Burr, D.B. (1997) Trabecular bone volume and microdamage accumulation in the femoral heads of women with and without femoral neck fractures. *Bone*,21,521-526.
[82] Ding, M., Odgaard, A., Linde, F. and Hvid, I. (2002) Age-related variations in the microstructure of human tibial cancellous bone. *Journal of Orthopedic Research*,20,615-621.
[83] Dendorfer, S., Maier, H.J. and Hammer, J. (2008) How do anisotropy and age affect fatigue and damage in cancellous bone? *Studies in Health Technology and Informatics*, 133,68-74.
[84] Gibson, L.J. and Ashby, M.F. (1988) Cellular Solids: Structure and Properties. 1st ed: Pergamon Press, Ltd.

In: Biomechanics of Dental Implants
Editor: Murat Çehreli
ISBN 978-1-62100-780-7

Chapter 2

ANİMAL MODELS FOR BİOLOGİCAL/BİOMECHANİCAL ASSESSMENT OF ENDOSSEOUS IMPLANTS

Murat Çehreli

1. INTRODUCTION

Animal experiments or animal research, and in vivo testing refers to use of non-human animals in experiments that do not cause extreme pain or disability to the animals. Animal experiments bridges the gap between in vitro and human studies. The earliest attempt to use animals in experiments could be found in Greek writings in 2nd-4th centuries BC. Today, it is estimated that more than 100 million vertabrate animals are used for experiments annually [1]. The use of animals for experimental research is deemed necessary in the event the results in vitro studies cannot be extrapolation to the in vivo situation and the scientific evidence obtained would benefit the health and quality of life of humans and animals. The use of animals in bone and bone biomaterials research has become a scientific as well as an ethical issue. The protocol of the experiments should be approved by the Institutional Animal Care and Use Committee for apropriateness of the content of the project and that the experiments comply with the Animal Welfare Act.

Experimental and clinical hypotheses in the field of tissue engineering which deals with bioactive materials such as dental implants and bone biomaterials could only be adequately studied in living organisms. Not all animal species are useful for biomedical/biomechanical research and therefore, the limitations of each unique model should be kept in mind during planning of the study. The most effective animal model for bone regeneration with or without implantable biomaterials and biomechanics should meet several criteria including [2]:

1. The ideal model for a human is another human. A model is a representation of a real or actual object and, therefore, it is not expected to be identical to the original subject. The candidate model should provide a biological and clinical environment that matches to the greatest extent possible, the actual environment that material will

be used. Consequently, the pathogenesis, disease progression, and histopathologic tissue reaction to implantable biomaterials should match those observed in humans. In addition, the animal model should allow replication of the clinical surgical technique and preferably the design of the implants with possibly the matching clinical dimensions. In the context of biomechanical experiments, the following selection criteria should also be included [3]:

a. Normal activity level of the animal and the functional load-bearing of the region of interest.
b. Available information on skeletal morphology of the animal model including site-specific mechanical properties. Bone tissue micro-architecture, modeling and remodeling properties (to extrapolate the results to the human)
c. Fusion observation intervals
d. Thoroughbred or mongrels
e. Study focus on material or structural properties
f. Mechanical testing equipment
g. Specimen size
h. Type of biomechanical experiment(s)

In biomechanical experiments, larger animals closely match the size and bony anatomy of the human should be considered to ensure that the appropriate size and number of implants could be placed. In general the use of large animals provide several advantages including Haversian and trabecular bone remodeling, greater surface areas and bone volumes allowing placement of orthopedic/oral endosseous implants, and similar postmenopausal bone loss in most animals. Use of large animals, on the other hand, requires a larger space for housing, large animals are difficult to handle, have long life spans, increase the cost of the experiment, and in case of dog and primates might arise for example ethical concerns.

Guidelines for species-specific and site-specific dimensions of implants have been established to avoid pathological fracture of the test site. It is recommended to place Ø 2mm X6 mm cylindrical implants into rabbit tibial and femoral diaphyseal bone. For larger animals such as sheep, goats and dogs the ISO recommends Ø 4 mm X12 mm implants for the same regions. dimensions of cylindrical In addition, the breed of animal used must be considered as it may have an influence on implants dimensions, i.e., large breeds of sheep may allow the use of larger diameter implants in certain locations [4]. Large animal models could also be used to harbor titanium bone chambers, that allow repeated sampling. Despite a number of limitations, bone chambers could be used to testing particular scaffold materials, with or without growth factors, or study skeletal tissue regeneration or response around loaded implants in a well-controlled environment with a high level of reproducibility.

2. The animal model should provide objective and quantifiable parameters to assess the survival and biomechanical performance of the implant or the regenerated tissue. Nevertheless, one should keep in mind that within a field of study, no single animal model will be suitable for all purposes, nor can a model be dismissed as unsuitable for all purposes [5]. The animal model should also be easily available to many researchers for validation and further investigations on a specific topic. Indeed, according to a review by the Animal Procedures Committee [2003], utilization of a readily available animal model might be a legitimate factor in validation.

3. The appropriate animal model should detect and predict clinically relevant differences in biomechanical and biological performance between methods or implantable materials.
4. The cost of animal(s) purchasing, transportation and time for quarantine, if needed, housing, surgical/experimental equipment needs to be considered at the outset of the project. Larger animals such as the pig, dog, goat, and sheep require more space for housing, are difficult to handle, and are more expensive in comparison with small animals such as rats and rabbits.

According to international standards, dogs, sheep, goats, pigs, and rabbits are suitable for testing endosseous implants. It is important to perform power calculations prior to the study to determine the suitable number of test animals that meet the desired power. The number of animals used in the experiments depend on factors including the reproducibility of the surgical procedure, and assessments techniques, evaluation methods, interindividual variations between animals, results of pilot study, if available, and the statistical methods for assessment of data. In addition, it is important to use animals of the same strain, age, and weight to control experimental conditions. Because Sprague-Dawley rats and New Zealand white rabbits have identical genetic traits, and canines and cats are rather heterogeneous, more animals are needed for experiments including latter species. Long-term observations periods for dogs, sheep, goats, pigs, and rabbits are given as 12, 26, 52 and 78 weeks and in certain instances, with the exception of rabbits, 104 weeks [4,6].

2. Animals for Biological/Biomechanical Testing of Endosseous Implants

A variety of species have been used for biological and biomechanical testing of endosseous implants. The sheep and the goat have been extensively used in the context of orthopedic implants. The long bones, maxillary sinus, and recently the iliac crest have been used to place dental implants, particularly by Flanders, As its efficacy in the context of dental implants seems questionable, the sheep and the goat model will not be included in this chapter.

2.1. Rabbit

The Rabbit (White New Zelland- Oryctolagus cuniculus) has been extensively used to test the biological response of endosseous implants with or without biomechanical tests (Table 1). Rabbits have appropriate size for surgical interventions and analysis. Indeed, the most apparent reasons for use of this animal are its ease in handling, moderate size, which alows placement of regular size implants, and relatively rapid skeletal maturity at around 6 months [56]. The femur and the tibia have been the most used bones for implantation, whereas the mandible or the maxilla (Figures 1 and 2) have been used infrequently. Although placement of Ø 2mm X 6 mm cylindrical implants into rabbit tibial and femoral diaphyseal bone has been recommended [6], many studies have included larger implants (Ø 4 mm X 8-

10 mm). Rabbits are generally used for experiments up to 6 months, although longer obervations could also be performed. Most of the studies on the rabbit focus on histomorphometric bone-implant interactions at the unloaded state, and mechanical tests such as removal torque and pull-out tests have been also employed.

Rodents and rabbits have been frequently used in bone biomechanics experiments due to several factors such as ease in standardization of test conditions by acquiring genetically specific or mutant strains, short life spans, abundance of documentation on mechanically-mediated bone response, similarity in lamellar bone structure, remodeling rate, ease in housing, and cost-effectiveness. In studies related to biologic/biomechanical bone-implant interactions, a good understanding of the species-specific structural, mechanical as well as modeling/remodeling bone properties are of utmost importance for correct extrapolating (rather than predicting alone) the data to the human situation [4].

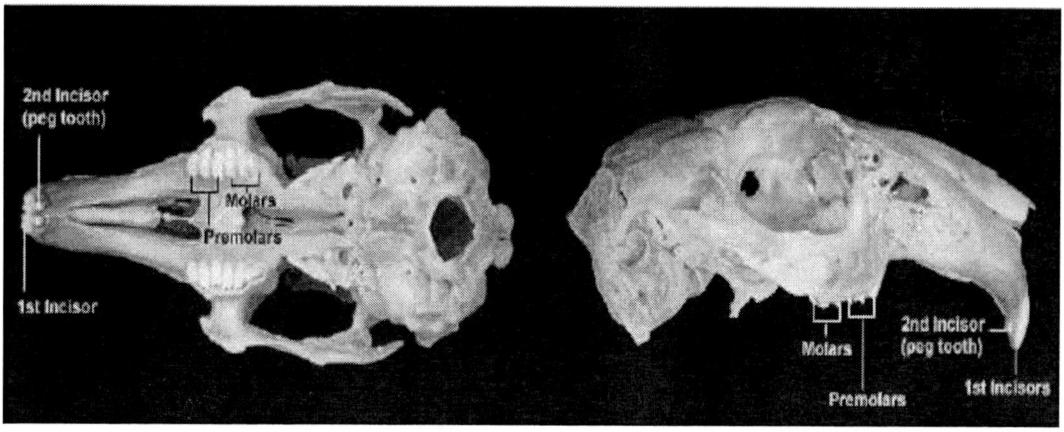

Figure 1. The Rabbit maxilla (with permission from Richard. A. Bowen Department of Biomedical Sciences, ARBL Colorado State University).

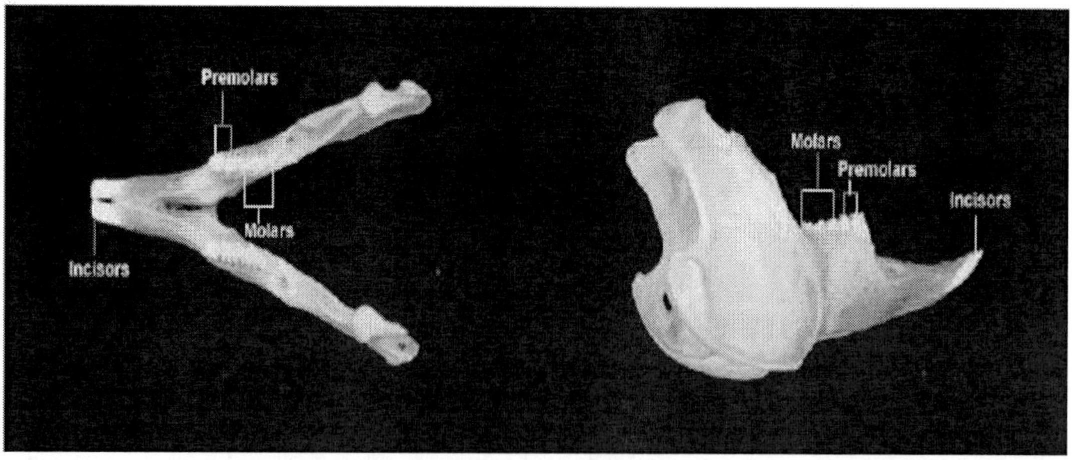

Figure 2. The Rabbit mandible (with permission from Richard. A. Bowen Department of Biomedical Sciences, ARBL Colorado State University).

Table 1. Examples of studies related to endosseous implants on rabbit that perform bone-implant contact measurements and additional assessments

Study	Number of Animals	Region	Number of Implants	Experimental set up	Implant Brand	Healing period (weeks)	Additional Analysis to BIC measurements
Gottlander [7]	12	Tibia	48	Implant surface	Exp	6 weeks and 52 weeks	NA
Gottlander [8]	9	Femur	18	Implant surface	IMZ	36 weeks	NA
Jansen [9]	9	Tibia	51	Implant material	Exp	6 and 16 weeks	Radiographics
Jansen [10]	8	Tibia	48	Implant surface	Dyna (modified)	3 and 12 weeks	XRD
Piattelli [11]	10	Femur	20	Implant surface	Lifecore	24 weeks	Laser scanning microscopy
Gotfredson [12]	26	Tibia	72	Implant surface	Exp	3 and 12 weeks	SEM; Surface profilometry; Radiographics; RT
Dean [13]	55	Femur	110	Implant surface	Exp	24 weeks	XRD, Pull out test
Hulshoff [14]	18	Femur	72	Implant surface	Exp	3, 6, and 9 weeks	XRD; Infrared spectroscopy; Rutherford backscattering spectrometry
Wennerberg [15]	10	Tibia and femur	60	Implant surface and RT	Exp	12 weeks	RT
Wennerberg [16]	9	Tibia	36	Implant surface	Exp	4 weeks	Optical profilometry; RT
Piattelli [17]	45	Femur	90	Implant surface	Lifecore	Up to 8 weeks	NA
Johansson [18]	30	Tibia	132	Implant material	Exp	Up to 52 weeks	3-D optical laser scanning; RT
Cordioli [19]	12	Tibia	48	Implant surface	Exp	5 weeks	Scanning interference microscopy; RT
Johnsson [20]	43	Tibia and femur	136	Osseointegration in irradiated bone	Exp	12 weeks and 52 weeks	RT
Hallgren [21]	10	Tibia	40	Implant surface	Exp	12 weeks	XPS; AFM; Pull out test
Halgren [22]	10	Tibia and femur	60	Implant surface	Nobel Biocare (modified)	12 weeks	Auger electron spectroscopy; Confocal laser scanning profilometry; RFA; RT
Jung [23]	36		144	Implant surface and osteoporosis	Dong Myeong Company	12 weeks	Cell culture test; Surface roughness measurements; RT
Sawase [24]	7	Tibia	28	Implant surface	Exp	4 weeks	AES;AFM; Glancing angle XRD; Potentiodynamic measurements

Table 1. (Continued)

Study	Number of Animals	Region	Number of Implants	Experimental set up	Implant Brand	Healing period (weeks)	Additional Analysis to BIC measurements
Stenport [25]	16	Tibia	32	Systemic human growth hormone administration and implant stability	Exp	2,4, and 8 weeks	RFA DEXA; IGF-1 measurements; Weight measurements
London [26]	11	Tibia	44	Implant surface	3i	Up to 8 weeks	3-D profilometry
Morales [27]	24	Femur	24	Implant surface and fibrin application	Exp	7 weeks	Radiographics
Salata [28]	14	Mandible	28	Autogenous bone graft w/wo e-PTFE membrane	Exp	6 and 24 weeks	NA
Sul [29]	10	Tibia and femur	80	Implant surface	Exp	6 weeks	SEM; XPS; Confocal laser scanning profilometry; Auger electron spectroscopy; Thin film XRD; RT
Sul [30]	10	Tibia and femur	56	Implant surface	Exp	6 weeks	XPS; RT
Sul [31]	12	Tibia	48	Implant surface	Exp	6 weeks	NA
Stefani [32]	32	Tibia	128	Nicotine on osseointegration	Exp	42 days	
Mohammadi [33]	60	Tibia and femur	480	Implant surface	Exp	1, 3, and 6 weeks	FTIR; XRD; ICP-OEC; stylus profilometry
Study	Number of Animals	Region	Number of Implants	Experimental set up	Implant Brand	Healing period (weeks)	Additional Analysis to BIC measurements
Stenport [34]	6	Femur and Tibia	36	Proplyene gycol alginate with or without EMD on osseointegration	Exp	6 weeks	RFA; RT
Ramires [35]	10	Tibia and femur	60	Implant surface	Exp	12 weeks	Laser profilometry; Cell culture test; RT
Duyck [36]	NA	Tibia	14	Controlled loading	Exp	4-18 weeks	NA
Ellingensen [37]	20	Tibia	80	Osseointegration of flouride-modified implants	Exp	Up to 12 weeks	Optical profilometry; RT
Keller [38]	40	Tibia	40	Osseointegration in osteporosis	Exp	4 weeks (postimplantation)	Immunohistochemistry; Pull out test
Goransson [39]	9	Femur	18	Implant surface	Nobel Biocare	12 weeks	RFA
Park [40]	10	Tibia	40	Implant surface	Osstem	6 weeks	Surface profilometry; micro CT; RT
Choi [41]	30	Tibia	152	Implant surface	Exp	4 and 12 weeks	SEM; RT
Hayakawa [42]	16	Femur	32	Implant surface	Exp (press fit)	Up to 12 weeks	Surface scanning
Park [43]	10	Tibia	40	Implant surface	Exp	12 weeks	RT
Park [44]	7	Femur	14	Effect of tetracell	MegaGen	8 weeks	NA

Study	Number of Animals	Region	Number of Implants	Experimental set up	Implant Brand	Healing period (weeks)	Additional Analysis to BIC measurements
				adhesion molecule on osseointegration			
Park [45]	20	Tibia	80	Implant surface	Exp	6 weeks	EDS; SEM; XRD; RT
Scarano [46]	22	Tibia	44	RT of TPS implants w/wo bone defects filled w/wo HA 250-450 μm	Exp	Up to 8 weeks	RT
Vandamme [47]	10	Tibia (Bone chamber model)	30	Controlled displacement for 9 weeks	Exp	9 weeks	NA
Fröjd [48]	10	Tibia and femur	37	Implant surface	Exp	12 weeks	Optical interferometry
Kim [49]	NA	Tibia	NA	Implant surface	Dentium	4 weeks	EDX, SEM, Clinical study
Meirelles [50]	10	Tibia	20	Implant surface	Exp	4 weeks	Optical interferometry and AFM
Meirelles [51]	10	Tibia	40	Implant surface	Astra Tech	4 weeks	Interferometry; SEM; XPS; RT
Susin [52]	16	Tibia and femur	96	Implant surface	NobelBiocare	6 weeks	RT
Vasconcellos [53]	20	Tibia	60	Implant surface	Exp	4 and 8 weeks	Profilometry, SEM, Pushout
He [54]	12	Femur	24	Implant surface	Exp	2 and 4 weeks	SEM, AFM
Park [55]	21	Tibia	126	Implant surface	Exp	6 weeks	SEM; XRD; XPS; RT

Exp: experimental design; XRD: x-ray diffraction; BIC: Bone implant contact; RT: Removal torque test; SEM: scanning electron microscopy; XPS: X-ray photoelectron spectroscopy; AFM: Atomic force microscopy; RFA: resonance frequency analysis; DEXA: dual energy x-ray analysis; FTIR: Fourier transform infrared spectroscopy; ICP-OES: inductively coupled plasma optical emission spectroscopy; EDS: energy dispersion spectroscopy.

Table 2. Mechanical properties of cortical bone in human and rabbit femur and tibia (adapted from[3])

Species	Bone	Test	Mineral content	E (GPa)	Strength (MPa)	ε_f	References
Human	Femur	Tension	0.496	16.7	166	0.029	Currey [57]
	Femur	Tension	11.4	19.7	107	140	Reilly[58]
	Femur	Tension	3.9	11.9	90.6	116	Ascenzi [59]
	Tibia	Tension	18.9	29.2	145	170	Burnstein [60]
	Tibia	Compression	24.5	34.3	183	213	Burnstein [60]
Rabbit	Femur	Bending		13.6±0.4	137±6		An [61]
	Femur	Bending		10.7±2.5	88±20		Ayers [62]
	Tibia	Bending		21.3±0.7	195±6		An [61]
	Tibia	Bending		23.3±7.0	192±47		Ayers [62]

E: Young's modulus; ε_f: ultimate strain; mineral content: mineral weight/wet bone weight.

It is certain that each species demonstrates unique advantages and disadvantages regarding their suitability as a candidate model for the bioloci/biomechanical response of bone tissue to an endosseous implant. In this context, one should keep in mind that the size, anatomy and consequent mechanical properties of rabbit femur and tibia are different from human (Table 2). Although the tibia of the animal has been the most used site, its medulla is composed of low density trabecular bone that does not replicate the structure of human jaw bones. In addition, the medulla of the tibia does not provide any mechanical support for the freshly placed implant, which makes the site inherently weak for dynamic loading experiments. The trabecular bone mass of Rabbits exhibits Haversian modeling, although its highly fatty marrow has different physical properties than that found in humans, making this model unsuitable for autogenous bone and marrow harvesting. Besides, Rabbits and rodents have different loading patterns and minimal intracortical remodeling.

2.2. Canine

There is a growing ethical issue toward use of dogs in eperimental medical studies due to the social/companion status of these animals. The adult dog (mongrel, beagle, foxhound, terrier, and bulldog) model has been, by far, one of the most used animal models for studying tissue changes around implants in loaded and unloaded conditions (Table 3). In addition to use of extremities for placement of implants, the mandible and the maxilla has also been used extensively. Owing to the dimensions of the available bone, the premolar-molar region of the maxilla and mandible may allow use of medium size (Ø 4 mm X 10 mm) implants. Nevertheless, depending on the size and the breed of the dog, there may be a need to fabricate custom implants. The eruption of primary teeth starts at three months after birth up to five weeks, resulting in 28 teeth (Figure 3 and 4) .

The anterior teeth occlude and there are diestamas between canines and the premolar teeth. The unilateral adult dog mandible has three incisors, a canine, four premolars, and two molars. The extraction procedure of teeth in the dog mandible or maxilla requires sectioning of teeth, as the teeth may easily break. Unlike humans, the use of several sutures will not enhance tissue healing in the dog or pig model (personal communication with Dr. Michel Dard, Institute Straumann).

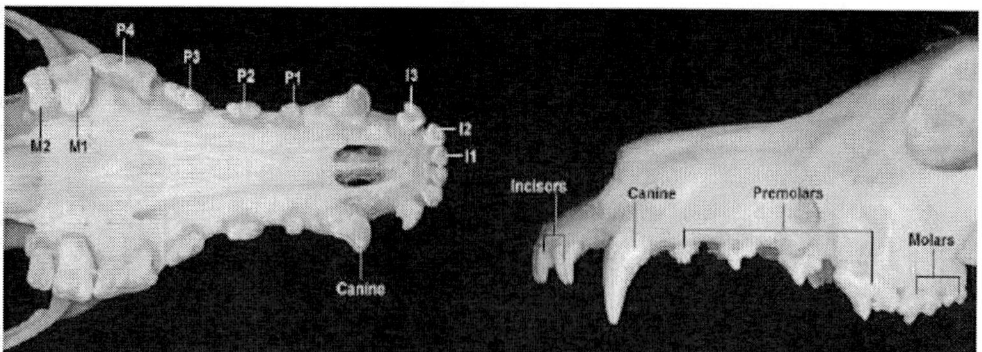

Figure 3. The canine maxilla (with permission from Richard. A. Bowen Department of Biomedical Sciences, ARBL Colorado State University).

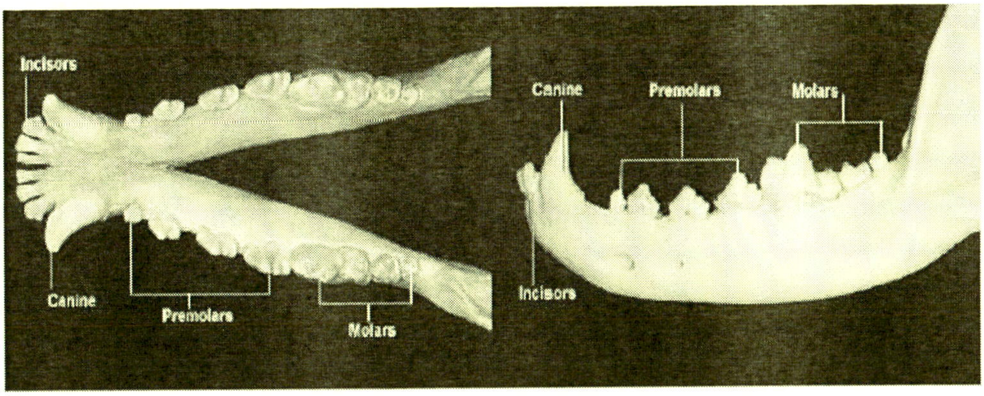

Figure 4. The canine mandible (with permission from Richard. A. Bowen Department of Biomedical Sciences, ARBL Colorado State University).

A study on a fracture properties and their correlations with compositional and microstructural properties on different species showed that while the adult human bone has a secondary osteonal structure, the canine has been found to have a mixed microstructure (secondary osteonal bone and plexiform bone) [96]. The plexiform bone appears in fast growing animals and provides mechanical support higher than woven bone. Studies show that the apparent density and ash density may remarkably vary between sites, although it is almost constant accross species (Table 4). Indeed, the findings by Kuhn et al. [97] suggest a high level of similarity between human and canine distal femur. The dog bone has a lower modulus, but higher ultimate (compressive) strains in comparison to human bone [97,100,101].

Table 3. Examples of studies related to endosseous implants on canine that perform bone-implant contact measurements and additional assessments

Study	Number of Animals	Region	Number of Implants	Experimental set up	Implant Brand	Loaded (L)/Unloaded(UL)	Healing period (weeks)	Additional Analysis
Xue [63]	6	Femur	48	Implant surface and push out test	Exp	UL	Up to 12 weeks	Push out; SEM
Xue [64]	6	Femur	NA	Implant surface	Exp	UL	Up to 12 weeks	SEM
Abrahamsson [65]	6	mandible (premolar)	36	Implant thread	Astra Tech	L (6 months-fixed partial denture)	64 weeks	Standardized intraoral radiographs
Coelho [66]	4	Tibia	38	Implant surface	Exp	UL	2 and 4 weeks	SEM; XPS + ion beam milling, thin-film mode X-ray diffraction,

Table 3. (Continued)

Study	Number of Animals	Region	Number of Implants	Experimental set up	Implant Brand	Loaded (L)/Unloaded(UL)	Healing period (weeks)	Additional Analysis
Coelho [67]	6	Tibia	36	Implant thread	Intraloack international and Bicon	UL	2 and 4 weeks	NA
Granato [68]	6	Tibia	72	Implant surface and length (6 and 11 mm)	Bicon	UL	2 and 4 weeks	RT
Hure [69]	2	Mandible	17	Implant design, immediate placement and early loading	Euroteknika	IL	12 weeks	RFA
Persson [70]	4	Mandible	24	Treatment of periimplantitis	Straumann	UL	24 weeks	NA
Rasmusson [71]	6	Mandible	36	Implant design and surface at bone defect sites	Branemark and AstraTech	UL	16 weeks	RFA
Tehemar [72]	10	Mandible and Maxilla	80	Immediate placement, implant surface, periodontitis, GBR	Sterioss	UL?	12 weeks and 24 weeks	Radiogra-phics
Wikesjö [73]	6	Mandible	36?	GBR around implants	NobelBiocare	UL	16 weeks	Radiogra-phics
Wikesjö [74]	4	Mandible	24?	Vertical bone regeneration around implants	Exp (Implant Innovations)	UL	8 weeks	Radiogra-phics
de Vincente [75]	3	Mandible	20	Treatment of peri-implant bone defects	Straumann	UL	36 weeks	
De Mazetzu [76]	12	Mandible	36	Implant surface	Zimmer Dental	UL	12 and 24 weeks	SEM
De Pauw [77]	5	Zygomatic arch	29	Osseointegration and static loading	Branemark	L (5 N static nonaxial loading after 8 weeks9	16 weeks	
Dubuille [78]	9	Mandible	24	Implant surface and GBR	Exp	UL	64 weeks including biomaterial application, 40 weeks postimplan-tation	SEM, radiographics
Kim [79]	10	Iliac crest	30	GBR around implants	Avana	UL	6 weeks and 12 weeks	NA
Kohal [80]	3	Mandibula	12	Submerged vs nonsubmerged healing	Paragon	UL	24 weeks	Clinical parameters; Periotest
Nociti [81]	5	Mandible	30	GBR in treatment of Periimplantitis	Napio system	UL	16 weeks after treatment	NA

Study	Number of Animals	Region	Number of Implants	Experimental set up	Implant Brand	Loaded (L)/Unloaded(UL)	Healing period (weeks)	Additional Analysis
Perry [82]	5	Mandible	40	Osseointegration in distraction versus autogenous onlay graft sites	Strau mann	UL	8 weeks graft application or distraction	NA
Quahash [83]	8	Mandible	48	Implant surface and GBR	Biomet and Osteotite	UL	8 weeks	
Rothamel [84]	6	Mandible	48	Osseointegration in vertical bone augmentation	Camlog	UL	24 weeks	NA
Todescan [85]	4	Mandible	24	Implant placement depth	Nobel Biocare	UL	24 weeks (3 months submerged and 3 months with healing abutments)	Morphometric measurements
You [86]	6	Tibia	18	Bone regeneration adjacent to implants	Osstem	UL	6 weeks	NA
Berglundh [87]	6	Mandible	48	Functional loading and implant design	Astra Tech and Branemark	L (by fixed prosthesis in occlusal contact) and Control UL	40 weeks	NA
Blanco [88]	5	Mandible	20	Immediate implant placement in flapless versus flap surgery	Straumann	UL	12 weeks	NA
Kim [89]	10	Iliac crest	30	GBR around implants	Avana	UL	6 and 12 weeks	
Kohri [90]	5	Mandible	30	Implant surface and functional loading	Apaceram and Osseodent	L (by fixed prosthesis in occlusal contact) and Control UL	4 or 8 weeks after functional loading	Clinical assessment; Radiographics
Lew [91]	17	Iliac crest	34	GBR around implants	Branemark	UL	Up to 12 weeks	Radiographics
Suzuki [92]	6	Tibia	24	Implant surface	Bicon	UL	2 and 4 weeks	SEM
Schwarz [93]	4	Maxilla and mandible	24	Implant socket preparation using ErYAG laser	Camlog, Oraltronics, and Straumann	UL	2 and 12 weeks	Radiographics
Granato [94]	6	Tibia	36	Implant surface	Bicon	UL	2 and 4 weeks	SEM; RT
Hethe-rington [95]	6	Femur	36	Implant material and surface	Exp	UL	8 weeks	Dektak profilometry; XRD; Pushout test

There seems to be a similarity between bone composition (ash weight, hydroxyproline, extractable proteins and IGF-1 content, water fraction, organic fraction, volatile inorganic fraction and ash fraction) between canine and human bones [102,103]. The bone density in canine and pig model is close to that of humans. However, the trabecular and cortical bone remodelling rates between dog and humans are not similar, which might influence interpretation of time-dependent peri-implant changes [4,104]. Although data on canine trabecular bone turnover suggest a high level of variability between sites, the turnover rate of canine model is higher than humans, in which the remodelling of total bone mass per year ranges between 5-15%, with estimates of the whole body trabecular bone turnover rate ranging from 10-15% per year to 40-55% [4, 105,106].

In the canine model, the impact of age, does not only influence bone turnover, but might have an influence of bone-implant contact percentage of implantable materials [4]. An early study suggest that higher bone-implant contact could be achieved in young greyhounds compared to older ones, which might depend on deterioration of remodelling rate [107].

Another important issue regarding species- and site-specific mechanical properties of bone tissue is the magnitude and quality (tensile vs. compressive) strains engendered during function (Table 5 and 6). An endosseous implant placed in tibia of a canine should not be considered as "unloaded" when a direct external load is not applied, because the entire bone tissue will experience functional strains during gait. Therefore, many studies that have evaluted bone-implant interactions in the unloaded state failed to isolate the mechanical environment. (this is one of the main rasons to use bone chambers).

Table 4. Apparent and ash densities of trabecular bone in human and dog (adapted from[3])

Species	Bone	Apparent density (g/cm^3)	Ash density (g/cm^3)	References
Human	Distal femur	0.43±0.15	0.26±0.08	Kuhn [97]
		0.46	-	Odgaard [98]
Dog	Distal femur	0.44±0.16	0.26±0.08	Kuhn [97]
		0.69±0.98	0.40±0.56[a]	Kang [99]
	Proximal tibia	0.41±0.83[a]	0.22±0.44[a]	Kang [99]
	Humeral	0.84±0.17	0.43±0.06	Kang [99]

[a] Range of average values from different locations.

Table 5. Mechanical properties of trabecular bone in human and canine (adapted from[3])

Species	Anatomic Site	Mineral content	E (GPa)	Strength (MPa)	References
Human	Femoral head		900±710	9.3±4.5	Martens [108]
	Distal femur	0.26±0.08	298±224	5.6±3.8	Kuhn [97]
	Proximal tibia		445±257	5.3±2.9	Linde [109]
Canine	Femoral head		435	12±5.8	Vahey [110]
	Distal femur	0.26±0.08	209±140	7.1±4.6	Kuhn [97]
		0.40-0.56	210-394	13-28	Kang [99]
	Proximal tibia	0.22–0.44	106–426	2–24	Kang [99]

E: Young's modulus; mineral content: mineral weight/wet bone weight.

Table 6. In vivo strain-gauge measurements of human and dog long bones (adapted from[3])

Species	Bone	Aspect	Activity	Principal or maximum strain (με)	Maximum strain rate (με /s)	Reference
Human	Tibia	Anteromedial	Walking	850 T	13,000	Lanyon [112]
			Running	400 C	-4000	
	Tibia	Medial	Walking	540 C, 440 T, 870 S	-7200; 11,000	Burr [113]
			Running	1150 C, 710 T, 1870 S	-30,000; 15,700	
	Femur	Lateral	Two-legged stance	400 T, 115 C		Aamodt [114]
			Stair climbing	1230 T, 570 C		
Canine	Femur	Medial	Lateral Walking	460 C, 240 T		Manley [115]
	Femur	Medial	Walking	310 C, 150 T		Szivek [116]
	Radius	Caudal	Fast walk	1500 C		Rubin and Lanyon [117]

C: compressive strain; T: tensile strain.

Table 7. Mechanical properties of cortical bone in human and pig (adapted from[3])

Species	Anatomic Site	Mineral content	Apparent Density	E (GPa)	Strength (MPa)	References
Human	Femur		0.50±0.16	389±270	7.36±4.00	Rohlmann [129]
	Vertebral body		0.24±0.07	67±44	2.45±1.52	Mosekilde[130]
	Vertebral body		0.17±0.04	291±113	2.23±0.95	Kopperdahl [131]
Pig	Femur			5900±4300		Ko [132]
	Vertebral body	0.46±0.04		1080±470	27.5±3.4	Mosekilde [133]

E: Young's modulus; mineral content: mineral weight/wet bone weight.

The functional strain levels are different at different locations throughout the skeleton. However, peak strains in load-bearing bones are reported to be similar [111]. An early study by Lanyon and coworkers [112] suggested that the strains and strain rates in a human were similar to those in animals. Bone strains and strain rates are fairly similar across different species, despite variations in animal size and bone function. The peak functional strains in long bones, ranging between -2000 and -3000 με during vigorous activity, are similar for a range of animals. Because the functional and parafunctional strains in the peri-implant zone in the jaw bones of man have not been measured so far and strain measurements are rather limited to a few species and long bones, the mechanical environment of an implant placed in the jaw is not known.

According to Frost [118] bone mass will change when induced absolute peak strains in "a" bone fall either below or above the "physiological window" (approximately ranging between 200–1500 με). In this context, two questions arise with regard to implant treatment. First, if the physiological window is really the same for all bones, do these absolute strain magnitudes rule or control bone modelling/remodelling mechanisms in bone supporting implants and if so, is it applicable for various conditions such as differences in bone density, loading history of the implant, and hormonal status and age of the patient? Second, if not applicable, how can mechanically-induced bone loss around dental implants be explained? Surprisingly, little notice has been taken of the fact that strains induced in bone under differing loadings can be a factor leading to implant failure, as high strain concentrations causing incapacity to repair bone breakdown (microfracture) and resorption. However, extensive research is indeed indicated to define bone response to implants in terms of strains induced in bone. This actually implies that a "strain-map" of bone response to unloaded, conventionally-, immediately-, and artificially-loaded implants should be established in various experimental conditions to determine the real thresholds of bone response to implants in the jaws of man [119].

2.3. Pig

On the basic problem of rising public sentiment against using companion animals such as the dog in laboratory testing, the use of pigs might be considered. Commercial breeds of pig are often considered unsuitable for orthopedic/dental research due to their large growth rates and excessive final body weight, and short and thick bones [4]. Although the development of miniature and micropigs has overcome this problem to some extent, the size and behavioral nature of pigs are not favorable as the sheep and the goat [120,121]. The adult pig mandible and maxilla (including the maxillary sinus) have extensively used to assess reactions to implants and bone replacement materials remodelling. Indeed, Aerssens and co-workers [102] found that the bone mineral content and density of pig had similarities with humans, although most of the available data are related to long bones and the vertebra (Table 7). In addition, Mosekilde and co-workers [133] found that the pig had a denser trabecular network and a similar lamellar bone structure to humans and that the bone remodelling processes were also similar [133,134]. The bone regeneration rate of pigs (1.2-1.5mm per day) has been also found to be closer to humans (1.0-1.5 mm per day) than dogs (1.5-2.0 mm/day) [135].

The joint between the skull of the pig and the mandible is formed by a condyle, which allows considerable lateral movement in the horizontal plane. The premolars perform a combination of crushing and grinding action, while the molars have complex crushing mounds. This is one of the main morphologic fatures of omnivore animals such as the pig and the sheep whose mastication depend on griding action. This should also be accounted in experiments with oral implants that integument the oral cavity, because the implants might experience excessive lateral loading even without prosthetic superstuctures. Swine teeth are more similar to human teeth than the teeth of other mammals in relation to shape, morphology and period of development [136,137], although their size is slightly larger. The permanent dentition of the domestic farm pig completes between 4-22 months (incisors: 8-12 months; canines: 8-18 months; premolars: 12-16 months; molars: 4-22 months) resulting in 44 teeth (Table 8) (Figures 5 and 6). The roots of the teeth are deeply implated in bone and the premolars and molars have 2-6 roots [138,139]. Therefore, the tooth extraction, like the dog model, needs sectioning and elevation of the roots with utmost care particularly in the mandible to avoid loss of the lingual plate. The long roots of the mandibular teeth are often aimed toward the lingual and luxation of the teeth might result in fracture of the lingual bone plate. Reflection of flaps may be required to close the wounds. The wounds should be closed by absorbable sutures. Similar to the dog model the dimensions of the available bone in the premolar-molar region of the maxilla and mandible may allow use of medium size (Ø 4 mm X 10 mm) implants. Nevertheless, depending on the size and the breed of the animal, there may be a need to fabricate custom implants.

Table 8. Distribution of permanent teeth in humans and different species

Species	Incisor	Canine	Premolar	Molar	Total (two jaws)
Human	4	2	4	6	32
Dog	6	2	8	5	42
Sheep	4	0	6	6	32
Rabbit	3	0	5	6	28
Pig	6	2	8	6	44

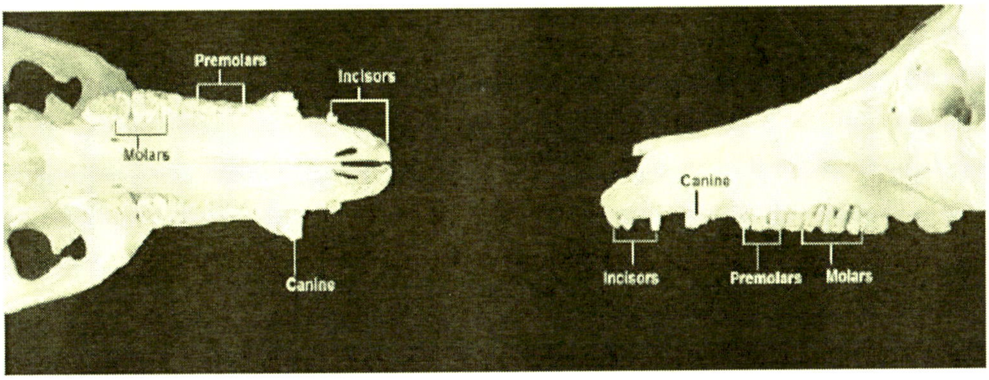

Figure 5. The pig maxilla (with permission from Richard. A. Bowen Department of Biomedical Sciences, ARBL Colorado State University).

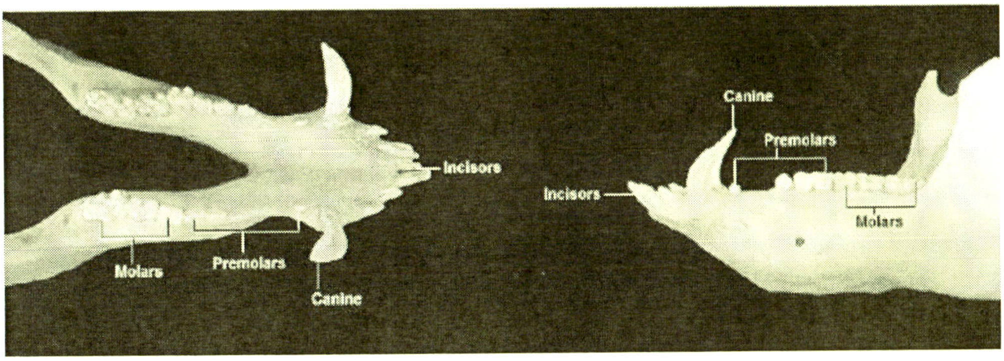

Figure 6. The pig mandible (with permission from Richard. A. Bowen Department of Biomedical Sciences, ARBL Colorado State University).

The structure as well as the size of the craniofacial complex is a concern in biomechanical experiments relating to oral implants. The mice and the rat being monophyodonts have a sinlge set of molar and incisors and their relatively small size have limited application of implants in the maxilla and mandible. Relatively larger animals such as the rabbit and the guinea pig have contuniously erupting incisors and diestameas between incisors and premolars (Figures 3 and 4), and their small oral cavities have limited use in endosseous implant experiments. Although the goat, sheep, dog and the swine have larger oral cavities, the dog and swine oral cavities have been mainly used to test oral implants, whereas the sheep seem as model to study sinus augmentation procedures [140].

2.4. Primate

Nonhuman primates are categorized as New World (marmosets, tamarins, squirrel, owl, and spider monkeys) and Old World (rhesus monkey (Macaca mulatta), cynomolgus monkey (Macaca fascicularis): crab-eating or long-tailed macaque, baboons (Papio hamadryas) or African green (vervet) monkeys (Cercopithecus aethiops)) monkeys that separated in evolution about 60 million years ago. Old World monkeys are the most frequently utilized. Indeed, Old World monkeys accounts for 78% of all primates used in 2002; New World monkeys and prosimians both account for 11%.

It is very difficult to house primates in a naturalistic way in laboratory conditions. Primates can be housed individual cages or in larger cages to form socialy groups, depending on the available facility and research experience. Poor housing may impair the immune system of primates and compromise the outcome of the study. Single cages in size of double the height of the animal should be preferred and used during experiments. Big cages allow sufficient space and pyschosocial activities, although their tendency toward creating social groups may result in continuous harrasment and stres for the subordinates and sometimes death for a monkey introduced to the preexisting group. In addition, most primates used for research purposes are relatively large wild-captured animals difficult to handle and heterogeneity in age, body weight, and oral and general health conditions might impair or not allow controlled test conditions. Monkeys other than baboons have small size. The adult male Rhesus monkey weighs in range of 6.5–12 kg , and has a femur 16.5 cm in length and 1.25 cm in mid-shaft diameter (2). The small size and ease in handling of squirrel monkeys and marmosets might seem as advantage. However, they do not exhibit an inflammatory profile characteristic of human periodontal disease and have not been extensively studied in bone metabolism in part because of a generalized resistance to the actions of glucocorticoids.

Nonhuman primates are phylogenetically closer to humans than other mammalian groups, which may seem as a great advantage with regard to extrapolation of the experimental findings to the human situation. Monkeys have been a good candidate to study bone metabolism due to having mensturul cycles very similar to that of women [141] and the presence of cortical bone osteons and osteonal (Haversian) remodeling that does not normally occur in rodents. (Rodents also continue to model their bones throughout life, their growth plates remain open, hemopoietic bone marrow at most skeletal sites increasing bone turnover rate, their trabecular bone content even in metaphysic long bones are scarce, which makes rodents unsuitable to simulate bone biology of humans.) Osteonal remodeling plays and important role in microdamage repair and bone response to mechanical stimuli. For example, osteonal remodeling in primate mandible is influenced by the stiffness of the diet consumed and simultaneous force exterted [142,143]. Nonhuman primates have been used in studies relating to bone repair, periodontal surgery and regeneration periodontitis and/or gingivitis models, and bone response to oral implants under unloaded and loaded conditions [144-146].

REFERENCES

[1] Cohn, M. (2010) Alternatives to animal testing gaining ground. *The Baltimore Sun*, 26 August.

[2] Muschler, G.F., Raut, V.P., Patterson, T.E., Wenke, J.C. and Hollinger, J.O. (2010) The design and use of animal models for translational research in bone tissue engineering and regenerative medicine. *Tissue Engineering Part B*, 16, 123-45.

[3] Liebschner, M.A. (2004) Biomechanical considerations of animal models used in tissue engineering of bone. *Biomaterials*, 25,1697-714.

[4] Pearce, A.I., Richards, R.G., Milz, S., Schneider, E. and Pearce, S.G. (2007) Animal models for implant biomaterial research in bone: a review. *European Cell Materials*,2,1-10.

[5] Hazzard, D.G., Bronson, R.T., McClearn, G.E. and Strong, R. (1992) Selection of an appropriate animal model to study aging processes with special emphasis on the use of rat strains. *Journal of Gerontology*,47,B63-B64.

[6] International Standard ISO 10993-6. (1994) *Biological evaluation of medical devices* - Part 6. 1-11.

[7] Gottlander, M. and Albrektsson, T. (1991) Histomorphometric studies of hydroxylapatite-coated and uncoated CP titanium threaded implants in bone. *International Journal of Oral and Maxillofacial Implants*,6,399-404.

[8] Gottlander, M., Albrektsson, T. and Carlsson, L.V. (1992) Histomorphometric study of unthreaded hydroxyapatite-coated and titanium-coated implants in rabbit bone. *International Journal of Oral and Maxillofacial Implants*,7,485-90.

[9] Jansen, J.A., van der Waerden, J.P. and Wolke, J.G. (1993) Histological and histomorphometrical evaluation of the bone reaction to three different titanium alloy and hydroxyapatite coated implants. *Journal of Applied Biomaterials*,4,213-9.

[10] Jansen, J.A., van der Waerden, J.P. and Wolke, J.G. (1993) Histologic investigation of the biologic behavior of different *hydroxyapatite plasma-sprayed coatings in rabbits. Journal* of *Biomedical Materials Research*,27,603-10.

[11] Piattelli, A., Piattelli, M., Romasco, N. and Trisi, P. (1994) Histochemical and laser scanning microscopy characterization of the hydroxyapatite-bone ınterface: an experimental study in rabbits. *International Journal of Oral and Maxillofacial Implants*,9,163-168.

[12] Gotfredsen, K., Wennerberg, A., Johansson, C., Skovgaard, L.T. and Hjørting-Hansen, E. (1995) Anchorage of TiO2-blasted, HA-coated, and machined implants: an experimental study with rabbits. *Journal of Biomedical Materials Research*,29,1223-31.

[13] Dean, J.C., Tisdel, C.L., Goldberg, V.M., Parr, J., Davy, D. and Stevenson, S. (1995) Effects of hydroxyapatite tricalcium phosphate coating and intracancellous placement on bone ingrowth in titanium fibermetal implants. *Journal of Arthroplasty*,10,830-8.

[14] Hulshoff, J.E., van Dijk, K., van der Waerden, J.P., Wolke, J.G., Kalk, W. and Jansen, J.A. (1996) Evaluation of plasma-spray and magnetron-sputter Ca-P-coated implants: an in vivo experiment using rabbits. *Journal of Biomedical Materials Research*,31,329-37.

[15] Wennerberg, A., Albrektsson, T., Johansson, C. and Andersson, B. (1996) Experimental study of turned and grit-blasted screw-shaped implants with special emphasis on effects of blasting material and surface topography. *Biomaterials*,17,15-22.

[16] Wennerberg, A., Albrektsson, T., and Andersson, B. (1996) Bone tissue response to commercially pure titanium implants blasted with fine and coarse particles of aluminum oxide. *International Journal of Oral and Maxillofacial Implants*,11,38-45.

[17] Piattelli, A., Manzon, L., Scarano, A., Paolantonio, M. and Piattelli, M. (1998) Histologic and histomorphometric analysis of the bone response to machined and sandblasted titanium ımplants: an experimental study in rabbits. *International Journal of Oral and Maxillofacial Implants*,13,805-810.

[18] Johansson, C.B., Han, C.H., Wennerberg, A. and Albrektsson, T. (1998) A quantitative comparison of machined commercially pure titanium and titanium-aluminum-vanadium implants in rabbit bone. *International Journal of Oral and Maxillofacial Implants*,13,315-21.

[19] Cordioli, G., Majzoub, Z., Piattelli, A. and Scarano, A. (2000) Removal torque and histomorphometric investigation of 4 different titanium surfaces: an experimental study in the rabbit tibia. *International Journal of Oral and Maxillofacial Implants*,15,668-74.

[20] Johnsson, A.A., Sawaii, T., Jacobsson, M., Granström, G. and Turesson, I.(2000) A histomorphometric and biomechanical study of the effect of delayed titanium implant placement in irradiated rabbit bone. *Clinical Implant Dentistry and Related Research*,2,42-9.

[21] Hallgren, C., Sawase, T., Ortengren, U. and Wennerberg, A. (2001) Histomorphometric and mechanical evaluation of the bone-tissue response to implants prepared with different orientation of surface topography. *Clinical Implant Dentistry and Related Research*,3,194-203.

[22] Hallgren, C., Reimers, H., Gold, J. and Wennerberg, A. (2001) The importance of surface texture for bone integration of screw shaped implants: an in vivo study of implants patterned by photolithography. *Journal of Biomedical Materials Research*,15,485-96.

[23] Jung, Y.C., Han, C.H., Lee, I.S. and Kim, H.E. (2001) Effects of ion beam-assisted deposition of hydroxyapatite on the osseointegration of endosseous implants in rabbit tibiae. *International Journal of Oral and Maxillofacial Implants*,16,809-18.

[24] Sawase, T., Wennerberg, A., Baba, K., Tsuboi, Y., Sennerby, L., Johansson, C.B. and Albrektsson, T. (2001) Application of oxygen ion implantation to titanium surfaces: effects on surface characteristics, corrosion resistance, and bone response. *Clinical Implant Dentistry and Related Research*,3,221-9.

[25] Stenport, V.F., Olsson, B., Morberg, P., Törnell, J. and Johansson, C.B. (2001) Systemically administered human growth hormone improves initial implant stability: an experimental study in the rabbit. *Clinical Implant Dentistry and Related Research*,3,135-41.

[26] London, R.M., Roberts, F.A., Baker, D.A., Rohrer, M.D. and O'Neal, R.B. (2002) Histologic comparison of a thermal dual-etched implant surface to machined, TPS, and

HA surfaces: bone contact in vivo in rabbits. *International Journal of Oral and Maxillofacial Implants*,17,369-76.

[27] Morales, M Navarro, R., Almenara, M., Medina, J.M., Melian, M., and Gutierrez, C. (2002) Effects of fibrin on the integration hydroxyapatite coating implants: experimental study in a rabbit model. *Journal of Experimental Animal Science*,42,102-112.

[28] Salata, L.Z., Rasmusson, L. and Kahnberg, K.E. (2002) Effects of a mechanical barrier on the integration of cortical onlay bone grafts placed simultaneously with endosseous implant. *Clinical Implant Dentistry and Related Research*,4,60-8.

[29] Sul,Y.T., Johansson, C.B. and Albrektsson, T. (2002) Oxidized titanium screws coated with calcium ions and their performance in rabbit bone. *International Journal of Oral and Maxillofacial Implants*,17,625-34.

[30] Sul, Y.T., Johansson, C.B., Kang, Y., Jeon, D.G. and Albrektsson, T. (2002) Bone reactions to oxidized titanium implants with electrochemical anion sulphuric acid and phosphoric acid incorporation. *Clinical Implant Dentistry and Related Research*,4,78-87.

[31] Sul, Y.T., Johansson, C.B., Röser, K. and Albrektsson, T. (2002) Qualitative and quantitative observations of bone tissue reactions to anodised implants. *Biomaterials*, 23,1809-17.

[32] Stefani, C.M., Nogueira, F., Sallum, E.A., de Toledo, S., Sallum, A.W. and Nociti, F.H. Jr. (2002) Influence of nicotine administration on different implant surfaces: a histometric study in rabbits. *Journal of Periodontology*,73,206-12.

[33] Mohammadi, S., Esposito, M., Hall, J., Emanuelsson, L., Krozer, A. and Thomsen, P. (2003) Short-term bone response to titanium implants coated with thin radiofrequent magnetron-sputtered hydroxyapatite in rabbits. *Clinical Implant Dentistry and Related Research*,5,241-53.

[34] Franke Stenport, V. and Johansson, C.B. (2003) Enamel matrix derivative and titanium implants. An experimental pilot study in the rabbit. *Journal of Clinical Periodontology,*30,359-363.

[35] Ramires, P.A., Wennerberg, A., Johansson, C.B., Cosentino, F., Tundo, S. and Milella, E. (2003) Biological behavior of sol-gel coated dental implants. *Journal of Materials Science Materials in Medicine,*14,539-45.

[36] Duyck, J., Cooman, M.D., Puers, R., Van Oosterwyck, H., Sloten, J.V. and Naert, I. (2004) A repeated sampling bone chamber methodology for the evaluation of tissue differentiation and bone adaptation around titanium implants under controlled mechanical conditions. *Journal of Biomechanics*,37,1819-22.

[37] Ellingsen, J.E., Johansson, C.B., Wennerberg, A. and Holmén, A. (2004) Improved retention and bone-tolmplant contact with fluoride-modified titanium implants. *International Journal of Oral and Maxillofacial Implants,19*,659-66.

[38] Keller, J.C., Stewart, M., Roehm, M. and Schneider, G.B. (2004) Osteoporosis-like bone conditions affect osseointegration of implants. *International Journal of Oral and Maxillofacial Implants,*19,687-94.

[39] Göransson, A. and Wennerberg, A. (2005) Bone formation at titanium implants prepared with iso- and anisotropic surfaces of similar roughness: an in vivo study. *Clinical Implant Dentistry and Related Research*,7,17-23.

[40] Park, Y.S., Yi, K.Y., Lee, I.S., Han, C.H., Jung, Y.C. (2005) The effects of ion beam-assisted deposition of hydroxyapatite on the grit-blasted surface of endosseous implants in rabbit tibiae. *International Journal of Oral and Maxillofacial Implants*,20,31-8.

[41] Choi, J.W., Heo, S.J., Koak, J.Y., Kim, S.K., Lim, Y.J., Kim, S.H. and Lee, J.B. (2006) Biological responses of anodized titanium implants under different current voltages. *Journal of Oral Rehabilitation,*33,889-97.

[42] Hayakawa, T., Takahashi, K., Yoshinari, M., Okada, H., Yamamoto, H., Sato, M. and Nemoto, K. (2006) Trabecular bone response to titanium implants with a thin carbonate-containing apatite coating applied using the molecular precursor method. *International Journal of Oral and Maxillofacial Implants,*21,851-8.

[43] Park, J.M., Koak, J.Y., Jang, J.H., Han, C.H., Kim, S.K., Heo, S.J. (2006) Osseointegration of anodized titanium implants coated with fibroblast growth factor-fibronectin (FGF-FN) fusion protein. *International Journal of Oral and Maxillofacial Implants*,21,859-66.

[44] Park, J.W., Lee, S.G., Choi, B.J. and Suh, J.Y. (2007) Effects of a cell adhesion molecule coating on the blasted surface of titanium implants on bone healing in the rabbit femur. *International Journal of Oral and Maxillofacial Implants*,22,533-41.

[45] Park, K.H., Heo, S.J., Koak, J.Y., Kim, S.K., Lee, J.B., Kim, S.H. and Lim, Y.J. (2007) Osseointegration of anodized titanium implants under different current voltages: a rabbit study. *Journal of Oral Rehabilitation*,34,517-27.

[46] Scarano, A., Carinci, F., Mangano, C., Quaranta, A. and Piattelli, A. (2007) Removal torque values of titanium implants inserted into bone defects filled with hydroxyapatite: a histologic and histomorphometric analysis in rabbit. *International Journal of Immunopathology and Pharmacology*,20,49-53.

[47] Vandamme, K., Naert, I., Geris, L., Vander Sloten, J., Puers, R. and Duyck, J. (2007) The effect of micro-motion on the tissue response around immediately loaded roughened titanium implants in the rabbit. *European Journal of Oral Science*,115,21-9.

[48] Fröjd, V., Franke-Stenport, V., Meirelles, L. and Wennerberg, A. (2008) Increased bone contact to a calcium-incorporated oxidized commercially pure titanium implant: an in-vivo study in rabbits. *International Journal of Oral and Maxillofacial Surgery*,37,561-6.

[49] Kim, H., Choi, S.H., Ryu, J.J., Koh, S.Y., Park, J.H. and Lee, I.S. (2008) The biocompatibility of SLA-treated titanium implants. *Biomedical Materials*,3,025011.

[50] Meirelles, L., Melin, L., Peltola, T., Kjellin, P., Kangasniemi, I., Currie, F., Andersson, M., Albrektsson, T. and Wennerberg, A. (2008) Effect of hydroxyapatite and titania nanostructures on early in vivo bone response. *Clinical Implant Dentistry and Related Research*,10,245-54.

[51] Meirelles. L,, Currie, F., Jacobsson, M., Albrektsson, T. and Wennerberg, A. (2008) The effect of chemical and nanotopographical modifications on the early stages of osseointegration. *International Journal of Oral and Maxillofacial Implants*,23,641-7.

[52] Susin, C., Qahash, M., Hall, J., Sennerby, L. and Wikesjö, U.M. (2008) Histological and biomechanical evaluation of phosphorylcholine-coated titanium implants. *Journal of Clinical Periodontology,*35,270-5.

[53] Vasconcellos, L.M., Oliveira, M.V., Graça, M.L., Vasconcellos, L.G., Cairo, C.A., Carvalho, Y.R. (2008) Design of dental implants, influence on the osteogenesis and fixation. *Journal of Materials Science Materials in Medicine,*19,2851-7.

[54] He, F.M., Yang, G.L., Li, Y.N., Wang, X.X. and Zhao, S.F. (2009) Early bone response to sandblasted, dual acid-etched and H2O2/HCl treated titanium implants: an experimental study in the rabbit. *International Journal of Oral and Maxillofacial Surgery*,38,677-81.

[55] Park, J.W., Jang, J.H., Lee, C.S. and Hanawa, T. (2009) Osteoconductivity of hydrophilic microstructured titanium implants with phosphate ion chemistry. *Acta Biomaterialia*, 5,2311-2321.

[56] Gilsanz, V., Roe, T.F., Gibbens, D.T., Schulz, E.E., Carlson, M.E., Gonzalez, O. and Boechat, M.I. (1988) Effect of sex steroids on peak bone density of growing rabbits. *American Journal of Physiology,* 255,E416-E421.

[57] Currey, J.D. (1998) Mechanic al properties of vertebrate hard tissues. *Proceedings of the Institute of Mechanical Engineers [H]*,212,:399–411.

[58] Reilly, D.T., Burstein, A.H. and Frankel, V.H. (1974) The elastic modulus for bone. *Journal of Biomechanics*,7,271–5.

[59] Ascenzi, A. and Bonucci, E. (1968) The compressive properties of single osteons as a problem of molecular biology. *Calcified Tissue Research*,(Suppl: 44-44a).

[60] Burstein, A.H., Reilly, D.T. and Martens, M. (1976) Aging of bone tissue: mechanical properties. *Journal of Bone and Joint Surgery America*,58,82-6.

[61] An, Y.H, Kang, Q. and Friedman, R.J. (1996) Mechanic al symmetry of rabbit bones studied by bending and indentation testing. *American Journal of Veterinary Research*,57,1786–9.

[62] Ayers RA, Miller, M.R., Simske, S.J. and Norrdin, R.W. (1996) Correlation of flexural structural properties with bone physical properties: a four species survey. *Biomedical Sciences Instrumentation,* 32,251-60.

[63] Xue, W., Liu, X., Zheng, X. and Ding, C. (2005) In vivo evaluation of plasma-sprayed titanium coating after alkali modification. *Biomaterials*,26,3029-37.

[64] Xue, W., Liu, X., Zheng, X. and Ding, C. (2005) In vivo evaluation of plasma-sprayed wollastanite coating, *Biomaterials*, 26,3455-60.

[65] Abrahamsson, I. and Berglundh, T. (2006) Tissue characteristics at microthreaded implants: an experimental study in dogs. *Clinical Implant Dentistry and Related Research,*8,107-13.

[66] Coelho, P.G., Cardaropoli, G., Suzuki, M. and Lemons, J.E. (2009) Histomorphometric evaluation of a nanothickness bioceramic deposition on endosseous implants: a study in dogs. *Clinical Implant Dentistry and Related Research*,11,292-302.

[67] Coelho, P.G., Suzuki, M., Guimaraes, M.V., Marin, C., Granato, R., Gil, J.N. and Miller, R.J. (2010) Early bone healing around different implant bulk designs and surgical techniques: A study in dogs. *Clinical Implant Dentistry and Related Research*,2,202-8.

[68] Granato, R., Marin, C., Gil, J.N., Chuang, S.K., Dodson, T.B., Suzuki, M. and Coelho, P.G. (2009) Thin Bioactive Ceramic-Coated Alumina-Blasted/Acid-Etched Implant Surface Enhances Biomechanical Fixation of Implants: An Experimental Study in Dogs. *Clinical Implant Dentistry and Related Research*, Aug 3.

[69] Huré, G., Aguado, E., Grizon, F., Baslé, M.F. and Chappard, D. (2004) Some biomechanical and histologic characteristics of early-loaded locking pin and expandable implants: a pilot histologic canine study. *Clinical Implant Dentistry and Related Research*,6,33-9.

[70] Persson, L.G., Mouhyi, J., Berglundh, T., Sennerby, L. and Lindhe, J. (2004) Carbon dioxide laser and hydrogen peroxide conditioning in the treatment of periimplantitis: an experimental study in the dog. *Clinical Implant Dentistry and Related Research*,6,230-8.

[71] Rasmusson, L., Kahnberg, K.E. and Tan, A.(2001) Effects of implant design and surface on bone regeneration and implant stability: an experimental study in the dog mandible. *Clinical Implant Dentistry and Related Research*,3,2-8.

[72] Tehemar, S., Hanes, P. and Sharawy, M. (2003) Enhancement of osseointegration of implants placed into extraction sockets of healthy and periodontally diseased teeth by using graft material, an ePTFE membrane, or a combination. *Clinical Implant Dentistry and Related Research,*5,193-211.

[73] Wikesjö, U.M., Sorensen, R.G., Kinoshita, A. and Wozney, JM. (2002) RhBMP-2/alphaBSM induces significant vertical alveolar ridge augmentation and dental implant osseointegration. *Clinical Implant Dentistry and Related Research*,4,174-82.

[74] Wikesjö, U.M., Qahash, M., Thomson, R.C., Cook, A.D., Rohrer, M.D., Wozney, J.M. and Hardwick, W.R. (2003) Space-providing expanded polytetrafluoroethylene devices define alveolar augmentation at dental implants induced by recombinant human bone morphogenetic protein 2 in an absorbable collagen sponge carrier. *Clinical Implant Dentistry and Related Research,5*,112-23.

[75] de Vicente, J.C., Recio, O., Martín-Villa, L., Junquera, L.M., López-Arranz, J.S. (2006) Histomorphometric evaluation of guided bone regeneration around implants with SLA surface: an experimental study in beagle dogs. *International Journal of Oral and Maxillofacial Surgery*,35,1047-53.

[76] De Maeztu, M.A., Braceras, I., Alava, J.I., Sánchez-Garcés, M.A. and Gay-Escoda, C. (2007) Histomorphometric study of ion implantation and diamond-like carbon as dental implant surface treatments in beagle dogs. *International Journal of Oral and Maxillofacial Implants*,22,273-9.

[77] De Pauw, G.A., Dermaut, L.R., Johansson, C.B. and Martens, G.A. (2002) Histomorphometric analysis of heavily loaded and non-loaded implants. *International Journal of Oral and Maxillofacial Implants,*17,405-12.

[78] Dubruille, J.H., Viguier, E., Le Naour, G., Dubruille, M.T., Auriol, M. and Le Charpentier, Y. (1999) Evaluation of combinations of titanium, zirconia, and alumina implants with 2 bone fillers in the dog. *International Journal of Oral and Maxillofacial Implants*,14,271-7.

[79] Kim, S.G., Chung, C.H., Kim, Y.K., Park, J.C. and Lim, S.C. (2002) Use of particulate dentin-plaster of Paris combination with/without platelet-rich plasma in the treatment of bone defects around implants. *International Journal of Oral and Maxillofacial Implants*, 17,86-94.

[80] Kohal, R.J., De LaRosa, M., Patrick, D., Hürzeler, M.B. and Caffesse, R.G. (1999) Clinical and histologic evaluation of submerged and nonsubmerged hydroxyapatite-coated implants: a preliminary study in dogs. *International Journal of Oral and Maxillofacial Implants*,14,824-34.

[81] Nociti, F.H. Jr, Machado, M.A., Stefani, C.M. and Sallum, E.A. (2001) Absorbable versus nonabsorbable membranes and bone grafts in the treatment of ligature-induced peri-implantitis defects in dogs: a histometric investigation. *International Journal of Oral and Maxillofacial Implants,*16,646-52.

[82] Perry, M., Hodges, N., Hallmon, D.W., Rees, T. and Opperman, L.A. (2005) Distraction osteogenesis versus autogenous onlay grafting. Part I: outcome of implant integration. *International Journal of Oral and Maxillofacial Implants*,20,695-702.

[83] Qahash, M., Hardwick, W.R., Rohrer, M.D., Wozney, J.M. and Wikesjö, U.M. (2007) Surface-etching enhances titanium implant osseointegration in newly formed (rhBMP-2-induced) and native bone. *International Journal of Oral and Maxillofacial Implants*,22,472-7.

[84] Rothamel, D., Schwarz, F., Herten, M., Ferrari, D., Mischkowski, R.A., Sager, M. and Becker, J. (2009) Vertical ridge augmentation using xenogenous bone blocks: a histomorphometric study in dogs. *International Journal of Oral and Maxillofacial Implants*,24,243-50.

[85] Todescan, F.F., Pustiglioni, F.E., Imbronito, A.V., Albrektsson, T. and Gioso, M. (2002) Influence of the microgap in the peri-implant hard and soft tissues: a histomorphometric study in dogs. *International Journal of Oral and Maxillofacial Implants*,17,467-72.

[86] You, T.M., Choi, B.H., Zhu, S.J., Jung, J.H., Lee, S.H., Huh, J.Y., Lee, H.J. and Li, J. (2007) Platelet-enriched fibrin glue and platelet-rich plasma in the repair of bone defects adjacent to titanium dental implants. *International Journal of Oral and Maxillofacial Implants*,22,417-22.

[87] Berglundh, T., Abrahamsson, I. and Lindhe, J. (2005) Bone reactions to longstanding functional load at implants: an experimental study in dogs. *Journal of Clinical Periodontology*,32,925-932.

[88] Blanco, J., Nunez, V., Aracil, L., Munoz, F. and Ramos, I. (2008) Ridge alterations following immediate implant placement in the dog: flap versus flapless surgery. *Journal of Clinical Periodontology*,35,640-648.

[89] Kim, S.G., Kim, W.K., Park, J.C., Kim, H.J. (2002) A comparative study of osseointegration of Avana implants in a demineralized freeze-dried bone alone or with platelet-rich plasma. *Journal of Oral and Maxillofacial Surgery*,60,1018-25.

[90] Kohri, M., Cooper, E.P., Ferracane, J.L. and Waite, D.F. (1990) Comparative study of hydroxyapatite and titanium dental implants in dogs. *Journal of Oral and Maxillofacial Surgery,* 48,1265-73.

[91] Lew, D., Marino, A.A., Startzell, J.M. and Keller, J.C. (1994) comparative study of osseointegration of titanium implants in corticocancellous block and corticocancellous chip grafts in canine ilium. *Journal of Oral and Maxillofacial Surgery*,52,952-8.

[92] Suzuki, M., Guimaraes, M.V., Marin, C., Granato, R., Gil, J.N. and Coelho, P.G. (2009) Histomorphometric evaluation of alumina-blasted/acid-etched and thin ion beam-deposited bioceramic surfaces: an experimental study in dogs. *Journal of Oral and Maxillofacial Surgery*,67,602-7.

[93] Schwarz, F., Olivier, W., Herten, M., Sager, M., Chaker, A. and Becker, J. (2007) Influence of implant bed preparation using an Er:YAG laser on the osseointegration of titanium implants: a histomorphometrical study in dogs. *Journal of Oral Rehabilitation*,34,273-81.

[94] Granato, R., Marin, C., Suzuki, M., Gil, J.N., Janal, M.N. and Coelho, P.G. (2009) Biomechanical and histomorphometric evaluation of a thin ion beam bioceramic deposition on plateau root form implants: an experimental study in dogs. *Journal of Biomedical Materials Research B Applied Biomaterials*,90,396-403.

[95] Hetherington, V.J., Lord, C.E. and Brown, S.A. (1995) Mechanical and histological fixation of hydroxylapatite-coated pyrolytic carbon and titanium alloy implants: a report of short-term results. *Journal of Applied Biomaterials*,6,243-8.
[96] Wang, X., Mabrey, J.D. and Agrawal, C.M. (1998) An interspecies comparison of bone fracture properties. *Biomedical Materials Engineering*,8,1-9.
[97] Kuhn, J.L., Goldstein, S.A., Ciarelli, M.J. and Matthews, L.S. (1989) The limitations of canine trabecular bone as a model for human: a biomechanical study. *Journal of Biomechanics*,22,95-107.
[98] Odgaard, A., Hvid, I. and Linde, F. (1989) Compressive axial strain distributions in cancellous bone specimens. *Journal of Biomechanics*,22,829-35.
[99] Kang, Q., An, Y.H. and Friedman, R.J. (1998) The mechanical properties and bone densities of canine cancellous bone. *Journal of Materials Science Materials in Medicine*,33,263-7.
[100] Goldstein, S.A., Wilson, D.L., Sonstegard, D.A. and Matthews, L.S. (1983) The mechanical properties of human tibial trabecular bone as a function of metaphyseal location. *Journal of Biomechanics*,16,965-9.
[101] Ciarelli, M.J. Goldstein, S.A., Kuhn, J.L., Cody, D.D. and Brown, M.B. (1991) Evaluat ion of orthogonal mechanical properties and density of human trabecular bone from the major metaphyseal regions with materials testing and computed tomography. *Journal of Orthopedic Research*,9,674-82.
[102] Aerssens, J., Boonen, S., Lowet, G. and Dequeker, J. (1998) Interspecies differences in bone composition, density, and quality: potential implications for in vivo bone research. *Endocrinology*,139,663-670.
[103] Gong, J.K., Arnold, J.S. and Cohn, S.H. (1964). Composition of trabecular and cortical bone. *Anatomical Records*,149,325-332.
[104] Polig, E. and Jee, W.S. (1989) Bone structural parameters, dosimetry, and relative radiation risk in the beagle skeleton. *Radiation Research*,120,83-101.
[105] Fernandez-Tresguerres-Hernandez-Gil, I., Alobera-Gracia, M.A., del Canto Pingarron, M. and Jerez, L.B. (2006) Physiological bases of bone regeneration I. Histology and physiology of bone tissue. *Medicina Oral Patología Oral Cirugía Bucal*,11,E47-E51.
[106] Kimmel, D.B. and Jee, W.S. (1982) A quantitative histologic study of bone turnover in young adult beagles. *Anatomical Records*,203,31-45.
[107] Magee, F.P., Longo, J.A. and Hedley, A.K. (1989) The effect of age on the interface strength between porous coated implants and bone. *Transactions of Orthopaedic Research Society,*14, 575.
[108] Martens, M., Van Audekercke, R., Delport, P., De Meester, P. and Mulier, J.C. (1983) The mechanical characteristics of cancellous bone at the upper femoral region. *Journal of Biomechanics*,16,971-83.
[109] Linde, F., Hvid, I. and Pongsoipetch, B.(1989) Energy absorptive properties of human trabecular bone specimens during axial compression. *Journal of Orthopedic Research*, 7,432-9.
[110] Vahey, J.W., Lewis, J.L. and Vanderby, Jr. R. (1987) Elastic moduli, yield stress, and ultimate stress of cancellous bone in the canine proximal femur. *Journal of Biomechanics*,20,29-33.
[111] Rubin, C.T., Lanyon, L.E. (1984) Regulation of bone formation by applied dynamic loads. *Journal of Bone and Joint Surgery*,66-A,397-402.

[112] Lanyon, L.E., Hampson, W.G., Goodship, A.E. and Shah, J.S. (1975) Bone deformation recorded in vivo from strain gauges attached to the human tibial shaft. *Acta Orthopedica Scandinavia*,46,256-68.

[113] Burr, D.B., Milgrom, C., Fyhrie, D., Forwood, M., Nyska, M., Finestone, A., Hoshaw, S., Saiag, E. and Simkin, A.l. (1996) In vivo measurement of human tibial strains during vigorous activity. *Bone*,18,405-10.

[114] Aamodt, A., Aamodt, A., Lund-Larsen, J., Eine, J., Andersen, E., Benum, P. and Husby, OS. (1997) In vivo measurements show tensile axial strain in the proximal lateral aspect of the human femur. *Journal of Orthopedic Research*, 15,927-31.

[115] Manley, P.A., Schatzker, J. and Sumner-Smith, G. (1982) Evaluat ion of tension and compression forces in the canine femur in vivo. *Archives of Orthopedic Trauma Surgery*,99,213-6.

[116] Szivek, J.A., Johnson, E.M. and Magee, F.P. (1992) In vivo strain analysis of the greyhound femoral diaphysis. *Journal of Investigative Surgery*,5,91-108.

[117] Rubin, C.T. and Lanyon, L.E. (1982) Limb mechanics as a function of speed and gait: a study of functional strains in the radius and tibia of horse and dog. *Journal of Experimental Biology*,101,187-211.

[118] Frost HM. (1990) Skeletal structural adaptations to mechanical usage (SATMU): 1. Redefining Wolff's law: the bone modelling problem. *Anatomical R*ecords,226,403-413.

[119] Cehreli, M., Sahin, S., Akça, K. (2004) Role of mechanical environment and implant design on bone tissue differentiation: current knowledge and future contexts. *Journal of Dentistry*,32,123-32.

[120] Newman, E., Turner, A.S. and Wark, J.D. (1995) The potential of sheep for the study of osteopenia: current status and comparison with other animal models. *Bone*,16,277S-284S.

[121] Swindle, M.M., Smith, A.C. and Hepburn, B.J. (1988) Swine as models in experimental surgery. *Journal of Investigative Surgery*,1,65-79.

[122] Hickey, J.S., O'Neal, R.B., Scheidt, M.J., Strong, S.L., Turgeon, D. and Van Dyke, T.E. (1991) Microbiologic characterization of ligature-induced peri-implantitis in the microswine model. *Journal of Periodontology*,62,548-53.

[123] Sennerby, L., Odman, J., Lekholm, U. and Thilander, B. (1993) Tissue reactions towards titanium implants inserted in growing jaws. A histological study in the pig. *Clinical Oral Implants Research,*4,65-75.

[124] Bousdras, V.A., Walboomers, F., Jansen, J.A., Cunningham, J.L., Blunn, G., Petrie, A., Jaecques, S., Naert, I.E., Sindet-Pedersen, S. and Goodship, A.E. (2007) Immediate functional loading of single-tooth TiO2 grit-blasted implant restoration. A controlled prospective study in a porcine model. Part II: Histology and histomorphometry. *Clinical Implant Dentistry and Related Research,*9,207-16.

[125] Verdonck, H.W., Meijer, G.J., Laurin, T., Nieman, F.H., Stoll, C., Riediger, D., Stoelinga. P.J. and de Baat, C. (2008) Implant stability during osseointegration in irradiated and non-irradiated minipig alveolar bone: an experimental study. *Clinical Oral Implants Research*,19,201-6.

[126] Gruber, R.M., Ludwig, A., Merten, H.A., Pippig, S., Kramer, F.J. and Schliephake, H. (2009) Sinus floor augmentation with recombinant human growth and differentiation

factor-5 (rhGDF-5): a pilot study in the Goettingen miniature pig comparing autogenous bone and rhGDF-5. *Clinical Oral Implants Research*,20,175-82.

[127] Gahlert, M., Röhling, S., Wieland, M., Sprecher, C.M., Kniha, H. and Milz, S. (2009) Osseointegration of zirconia and titanium dental implants: a histological and histomorphometrical study in the maxilla of pigs. *Clinical Oral Implants Research*,20,1247-53.

[128] Scarano, A., Piattelli, A., Assenza, B., Quaranta, A., Perrotti, V., Piattelli, M. and Iezzi, G. (2010) Porcine bone used in sinus augmentation procedures: a 5-year retrospective clinical evaluation. *Journal of Oral and Maxillofacial Surgery*,68,1869-73.

[129] Rohlmann, A., Zilch, H., Bergmann, G. and Kölbel, R. (1980) Material properties of femoral cancellous bone in axial loading. Part I: time independent properties. *Archives of Orthopedic and Trauma Surgery,*97,95-102.

[130] Mosekilde, L. and Mosekilde, L. (1986) Normal vertebral body size and compressive strength: relations to age and to vertebral and iliac trabecular bone compressive strength. *Bone*,7,207-12.

[131] Kopperdahl, D.L. and Keaveny, T.M. (1998) Yield strain behavior of trabecular bone. *Journal of Biomechanics*,31,601-8.

[132] Ko, C.C., Douglas, W.H. and Cheng, Y.S. (1995) Intrinsic mechanical competence of cortical and trabecular bone measured by nanoindentation and microindentation probes. In: *ASME Bioengineering Conference*, BED Vol.29, ASME Press, New York, 1995.

[133] Mosekilde, L., Kragstrup, J. and Richards, A. (1987) Compressive strength, ash weight, and volume of vertebral trabecular bone in experimental fluorosis in pigs. *Calcified Tissue International,*40,318-22.

[134] Mosekilde, L., Weisbrode, S.E., Safron, J.A., Stills, H.F., Jankowsky, M.L., Ebert, D.C., Danielsen, C.C., Sogaard, C.H., Franks, A.F., Stevens, M.L., Paddock, C.L. and Boyce, R.W. (1993) Calcium-restricted ovariectomized Sinclair S-1 minipigs: an animal model of osteopenia and trabecular plate perforation. *Bone*,14,379-382.

[135] Laiblin, C. and Jaeschke, G. (1979) Klinisch-chemische Untersuchungen des Knochen- und Muskelstoffwechsels unter Belastung bein Göttinger Miniaturschwein - eine experimentelle Studie (Clinical-chemical investigations of the metabolism of bone and muscle under stress in the Göttiningen miniature pig – an experimental study), Berl Münch Tierärztl Wschr 92: 124.

[136] Limeback, H., Schlumbohm, C., Sen, A. and Nikiforuk, G. (1992) The effects of hypocalcemia/hypophosphatemia on porcine bone and dental hard tissues in an inherited form of type 1 pseudo-Vitamin D deficiency rickets. *Journal of Dental Research*,71,346-352.

[137] Robinson, C., Kirkham, J., Weatherell, J.A., Richards, A., Josephsen K. and Fejerskov, O. (1988) Mineral and protein concentrations in enamel of the developing permanent porcine dentition, *Caries Research*,22,321-326.

[138] Robinson, I.B. and Sarnat, B.G. (1955) Growth pattern of the pig mandible; a serial roentgenographic study using metallic implants. *American Journal of Anatomy*,96,37-64.

[139] Weaver, M.E., Sorenson, F.M. and Jump, EB. (1962) The miniature pig as an experimental animal in dental research. *Archives of Oral Biology*,7,17-23.

[140] Haas, R., Mailath, G., Dörtbudak, O. and Watzek, G. (1998) Bovine hydroxyapatite for maxillary sinus augmentation: analysis of interfacial bond strength of dental implants using pull-out tests. *Clinical Oral Implants Research*,9,117-22.

[141] Bosu, W.T., Johansson, E.D. and Gemzell, C. (1973) Peripheral plasma levels of oestrone, oestradiol-17ß and progesterone during ovulatory menstrual cycles in the rhesus monkey with special reference to the onset of menstruation. *Acta Endocrinologica*, 74,732-742.

[142] Bouvier, M and Hylander, W.L. (1981) Effect of bone strain on cortical bone structure in macaques (Macaca mulatta). *Journal of Morphology*, 167,1-12.

[143] Bouvier, M and Hylander, W.L. (1996) The mechanical or metabolic function of secondary osteonal bone in the monkey Macaca fascicularis. *Archives of Oral Biology*, 41,941-950.

[144] Schou, S., Holmstrup, P., Jørgensen, T., Skovgaard, L.T., Stoltze, K., Hjørting-Hansen, E. and Wenzel A. (2003) Implant surface preparation in the surgical treatment of experimental peri-implantitis with autogenous bone graft and ePTFE membrane in cynomolgus monkeys. *Clinical Oral Implants Research*,14,412-22.8.

[145] Carr, A.B., Gerard, D.A. and Larsen, P.E. (2000) Histomorphometric analysis of implant anchorage for 3 types of dental implants following 6 months of healing in baboon jaws. *International Journal of Oral and Maxillofacial Implants*, 15,785-91.

[146] Watzak, G., Zechner, W., Ulm, C., Tangl, S. , Tepper, G. and Watzek, G. (2005) Histologic and histomorphometric analysis of three types of dental implants following 18 months of occlusal loading: a preliminary study in baboons. *Clinical Oral Implants Research*,16,408-16.

In: Biomechanics of Dental Implants
Editor: Murat Çehreli

ISBN 978-1-62100-780-7

Chapter 3

Nanotopography in Dental Implant Surfaces

Gustavo Mendonça, Daniela B. S. Mendonça and Lyndon F. Cooper

1. Introduction

Implant surface character is one implant design factor affecting the rate and extent of osseointegration [1-4]. The process of osseointegration is now well described both histologically and at the cellular level. Beginning in the late 1960's the focused efforts of PI Branemark led to the detailed microscopic characterization of interfacial bone formation at machined titanium endosseous implants [5,6]. These concepts of osseointegration focused the profession on a proscribed surgical technique and the biocompatible nature of a defined titanium implant surface. Bone formation at the endosseous implant surface was considered a positive outcome that was contrasted to fibrous encapsulation, a negative and undesired result [7]. The main clinical advantage of osseointegration was the predictable clinical result that occurred when an osseous interface was reproducibly formed and maintained at the titanium surface of load bearing dental implants [8].

Osseointegration emerged as widely accepted in clinical dentistry, and is the basis for dental implant success. The low rate of implant failure in dense bone of the parasymphyseal mandible [9-12] has not been fully recapitulated by subsequent data from studies involving more challenging clinical situations [13,14]. Anecdotal reports of difficulty in achieving high rates of implant success in select patient populations (e.g. smokers, diabetics, osteoporosis) were supported by initial reports [15-18]. The cause of these failures, while not precisely determined, was largely attributed to a failure in bone formation in support of osseointegration. Challenging osseointegration with new protocols such as immediate placement and immediate loading may require further control of bone formation and osseointegration [13]. While it is presently acknowledged that implant, anatomic, biologic, systemic or functional, as well as clinician-related factors, are important determinants of endosseous implants success, a major interest in implant design factors is evident and clinical efforts to improve implant success have been focused on increasing the amount of bone that forms at the endosseous implant surface.

Osseointegration is dependent upon the adhesion of a fibrin blood clot and the population of the implant surface by blood-derived cells and mesenchymal stem cells orchestrated in a manner that result in osteoid formation and its subsequent mineralization [19-21]. A seamless progression of changing cell populations and elaboration and modification of the tissue / implant interface eventually results in bone forming in direct contact with the implant surface. Precisely how much of the implant surface directly contacts bone, how rapidly this bone accrual occurs, and the mechanical nature of the bone / implant connection is influenced by the nature of the implant surface itself [22].

The implant surface is implicated in this complex process of osseointegration in a number of different ways. Early investigations revealed the biocompatible nature of the cpTitanium implant [23], and revealed some pragmatic advantages for cpTitanium over other suitable materials [24]. Molecular investigations have contributed to defining cellular responses to titanium as "compatible" and advantageous. For example, Suska and colleagues [25] showed relatively low inflammatory signaling within cells in tissues adjacent to cpTitanium implants and suggested that this is a part of the osseointegration process. During the first 10 – 20 years of applied endosseous implant experience, the concept that cpTitanium implant biocompatibility supported clinical osseointegration success dominated clinical thinking.

Today, a growing aspect of endosseous implant surface research is focused on further enhancing the activity of bone forming cells at the tissue implant interface, using a variety of different approaches. Clearly, cpTitanium surfaces can be modified to direct specific cellular responses such as osteogenesis. More specifically, cpTitanium implant surfaces can be made to direct the osteoinduction of adherent progenitor cells. While one approach is the immobilization of bioactive peptides or growth factors, such as RGDs and BMPs [26-28], other approaches have embraced the use of nanoscale surface engineering to induce intrinsic osteoinductive signaling of the surface adherent cells. The purpose of this review is to explore how nanotechnology applications to the cpTitanium implant surface may provide new opportunities to create endosseous implant surfaces with greater specific control of adherent cell and adjacent tissue fate.

2. Significance of Implant Surface Topography

Early studies clearly demonstrated that the enhancement of implant surface topography improves osseointegration as exemplified by Buser and colleagues [29]. The observation that a micron-scale rough surface prepared by grit blasting and subsequent acid etching was capable of rapid and increased bone accrual reiterated an earlier report that a TiO_2- grit blasted surface also supported more rapid and increased bone accrual at cpTitanium implants [30]. These observations suggested that cpTitanium was not only "bioinert" or "biocompatible", but could influence bone regenerative capacity at the endosseous implant / bone interface.

Three different lines of thinking have evolved to better interpret or explain how surface topography at the micron-scale can increase bone to implant contact. One is the biomechanical theory of Hannson and Norton [31], the second is the concept of contact osteogenesis [32], and the third is a surface signaling hypothesis supported by many cell culture investigations [1,33,34].

A biomechanical theory has been proposed to explain implant surface effects on osseointegration. Hannson [31] described by a hypothetical construct considering the geometry of the surface features. Animal studies revealed that this theory was correct; an ideal implant surface should be densely covered with pits of approximately 1.5μm depth and 3-5μm diameter. Several studies on implant topography effects on bone to implant contact [35,36] support this concept. The mechanical interlocking of bone improves the performance of endosseous implants. Increased bone to implant contact is related to improved physical interaction of micron level rough implants with bone [37,38]. What has not been fully elucidated is how mechanical signaling in the unmineralized tissue of forming bone and adjacent connective tissue is affected by the implant surface. The bonding of bone to the implant surface is not implicated as a mechanism of enhancing the early physical associations of the implant with bone.

Surface topography may act by stabilization of the fibrin clot during initial healing of bone around endosseous implants [39]. The physical interlocking of fibrin fibers with the surface features provide an avenue for bone forming cells to the implant/bone interface. Topology enhancement may stabilize the forming tissue scaffolds for conduction of cells toward and onto the implant surface (contact guidance) [40,41].

Importantly, micronscale surface features can affect adherent cell [42-45]. Many investigations have revealed that surface adhesion controls adherent cell function. Increased surface topography effectively enhances ECM synthesis contributing to improved osseointegration response [45-58]. Signaling from the surface involves ECM proteins-receptors (integrin) interactions [42]. Cellular differentiation is influenced by changes in cell shape [59]. Evidence that integrins are involved in cellular responses to implant surfaces has been obtained using MG63 cell culture studies [60], and knocking down integrin α2 reduced osteoblast cells maturation [61]. The Rho family of GTPases (Rho, Rac, and Cdc42 GTPases) has also been implicated in the process of transferring the signaling to the cell [62-64].

Micron level topography effects on increased bone to implant contact are observed in vivo [29,65], and in human clinical histology [66]. Micron-scale topographic modification of the cpTitanium surface is prevalent in clinical dentistry [35,36]. Some clinical data demonstrates that micron level surface topography results in greater accrual of bone at the implant surface [67,68]. These surfaces have been generally interpreted to be biocompatible devices with beneficial influence on bone formation.

2.1. The Technologies of Nanotopography

Nanotechnology is defined as the creation of functional materials, devices and systems through control of matter on the nanometer scale (1–100nm), and exploitation of novel phenomena and properties (physical, chemical, biological) at that length scale (National Aeronautics and Space Administration). Nanotechnology involves one-dimensional concept (nanodots, nanowires) or the self-assembly of more complex structures (nanotubes). Implant surfaces typically display the two-dimensional distribution of adhered or formed nanoparticle or nanostructures. These nanofeatures are arranged in an organized (isotropic) or unorganized (anisotropic) manner, depending on the method of manufacture (lithography optical methods vs. coatings and chemicals). The embellishment of the surface with nanometer scale features leads to alteration in the physicochemical behavior (e.g. bone bonding) or biochemical events (e.g. altered protein adsorption, cell adhesion with changes in cell behavior).

Nanoscale manipulation of surfaces causes chemical changes of the bulk surface material. The mechanical and topographic surface features are inter-related [35]. It can be extremely difficult to isolate chemistry or charge effects induced by the nanotopography [69]. Nonetheless, the superimposition of nanofeatures on bulk surface topography does alter cell and tissue responses [70].

Several approaches are currently prevalent in the experimental application of nanoscale features to endosseous implants (Table 1). One physical approach involves the compaction of nanoparticles of TiO_2 versus micron level particles to yield surfaces with nanoscale grain boundaries [55]. An advantage of this method is that it preserves the chemistry of the surface among different topographies. The molecular process of self-assembled monolayers (SAMs) is a second approach. SAMs are formed by the spontaneous surface chemisorptions and/or self associating positioning of molecules onto surfaces [71]. The exposed functional end group could be an osteoinductive or cell adhesive molecule. An example of this is the use of cell adhesive peptide domains (RGD domains) appended to SAMs composed of polyethylene Glycol (PEG) and applied to the Titanium implant surfaces [28].

Table 1. Methods for Creating Nanotopography on cpTitanium Implants

Methods		Characteristics
Chemical methods	Acid etching	Combined with other methods (sandblasting and/or Pedoxidation) can impart nanofeatures to the surface and remove contaminants.
	Peroxidation	Produces a titania gel layer. Both chemical and topography changes are imparted.
	Alkali treatment (NaOH)	Produces a sodium titanate gel layer allowing hydroxyapatite deposition. Both chemical and topography changes are imparted.
	Anodization	Can impart nanofeatures to the surface creating a new oxide layer (based on the material used).
Physical approaches	Compaction of nanoparticles	Conserves the chemistry of the surface among different topographies. Not readily applied over implant surfaces
	Ion Beam Depostion	Can impart nanofeatures to the surface based on the material used.
Self assembly of Monolayers		The exposed functional end group could be a molecule with different functions (an osteoinductive or cell adhesive molecule).
Nanoparticle Deposition	Sol-Gel (colloidal particle adorption)	Can modify the thickness and chemistry of the surface. Creates a thin-film of controlled chemical characteristics. Atomic-scale interactions display strong physical interactions.
	Discrete crystalline deposition	Superimposes a nanoscale surface topographical complexity on the surface.

Chemical treatment creates nanoscale topography and is a third method that can be used. Peroxidation (H_2O_2) or acid oxidation, such as hydrofluoric acid (HF) have been used to create cpTitanium nanotopography [2,72-76]. The use of H_2O_2 with acid etching creates nanostructures of amorphous titanium oxide on the implant surface [77] (Figure 1). Treatment with HF also creates discrete nanostructures on TiO_2 grit blasted surfaces [78]. Several cell culture studies [43,70,79], preclinical investigations [47,80], and clinical studies [81] support the observation that HF treatment of TiO_2 grit blasted titanium implants is associated with rapid bone accrual at the implant surface. NaOH treatment catalyzes the production of titanium nanostructures outward from the titanium surface [82]. Treatment with a NaOH solution produces a sodium titanate gel layer on the Ti surface while H_2O_2 produces a titania gel layer.

Deposition of a coating is another approach to create nanofeatures on a titanium dental implant [83] (Figure 2). Sol-gel transformation techniques deposit nanometer scale calcium phosphate accretions to the implant surface [84,85]. Alumina, titania, zirconia and other materials can be applied as well [69,86,87]. The atomic-scale interactions assure strong physical association of nanoscale surface features [83,88-90]. The quantum interaction of high electron density at the atomic level can enforce high bond strength between the substrate and nanoscale coating. Examples of this have been reported for the calcium phosphate (CaP) nanofeatures using different methods for coating of Ti alloy implant surfaces [91]. The risk of nanoscale coating and detachment and related toxicity was addressed by Gutwein and Webster [48] who showed that nanoparticles of titania and alumina had less negative impact in cell viability and proliferation compared to micron-particles.

A fifth approach to creating nanoscale topography on Titanium is the use of optical methods (typically lithography) reliant on wavelength specific dimensions to achieve the appropriate nanoscale modification [82]. The present use of lasers to promote micron level groove on an implant surface can produce micron level modification of the implant surface [92] associated with altered interfacial responses. All are relevant to the endosseous dental implant surface and experimental examples of each can be identified (below).

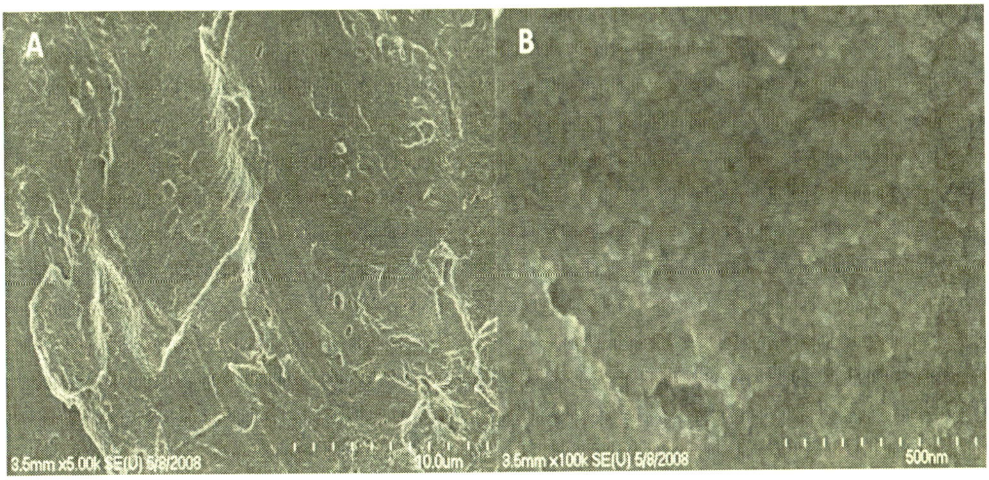

Figure 1. Surface prepared by grit blasting with 100μm Al_2O_3 particles followed by H_2SO_4/H_2O_2 treatment. (A) Demonstrates the micron-scale characteristics of the surface created by grit blasting. (B) Shows the nanotopography imparted to the surface due to the H_2SO_4/H_2O_2 treatment.

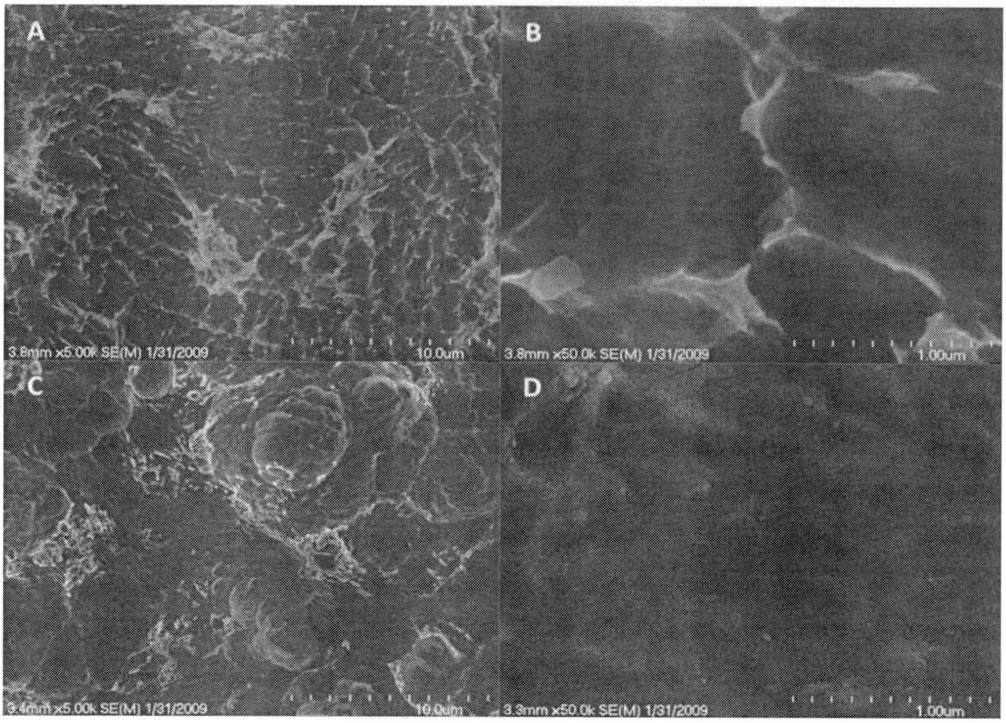

Figure 2. Atomic layer deposition (ALD) on acid etched cpTitanium. (A) and (B) 5,000X and 50,000x images of acid etched cpTitanium. (C) and (D) 5,000X and 50,000x images of acid etched cpTitanium with Al_2O_3 coating done by ALD.

2.3. The Actions of Nanotopography

Nanotopography-specific effects on adherent cell behavior (Table 2) have been demonstrated using a wide range of cell types including epithelial cells, fibroblasts, myocytes and osteoblasts [93-97]. Sereval investigators have focused in studying the influence of nanotechnology on osteoblast-related behavior *in vitro* and *in vivo*. These findings have also been summarized in Tables 3 and 4. Nanotopography has direct and indirect effect on cell behavior. It mimics the cell environment [98] favoring protein adsorption [99] and; therefore modulates cell/surface interactions and cell fate [100] (Figures 3 and 4).

Table 2. Biologic influences of Nanotopography

Biologic effects	Reference
Protein adsorption	[99]
Cell adhesion and spreading	[101]
Selectivy of Adhesion	[93]
Cell signaling pathways	[102]
Cell motility and proliferation	[103]
Osteoinduction and differentiation	[70]
Bone formation	[69]

Table 3. Reported osteogenic responses to Nanotopography – *In vitro*

Cell response	Reference
Change in Signaling	[104,105]
Changes in cell cytoskeleton	[106-109]
Decreased Apoptosis	[48]
Increased Adhesion	[50,54,93,110-123]
Ca deposition	[50,53,57,117,118,120-122,124,125]
Increased Proliferation	[122,125-128]
Increased Differentiation	[53,57,69,70,76,87,119-125,128-133]
Increased Osteoblast Specificity	[51,94,128,134,135]
Increased tensile test resistance	[136,137]

Table 4. Reported osteogenic responses to Nanotopography – *In vivo*

size / nanofeatures	Tissue response	material / fabrication	Animal model	ref.
AAT texture showed micropores and an overlapped nanometric net of filaments	Increased Bone-to-implant contact	cpTi / alkali etching process with CaP solution (biomimetic)	Sheep	[127]
Not shown	Increased Bone-to-implant contact	cpTi / HA - Ion Beam Assisted Deposition (IBAD)	Rabbit	[138]
~100nm features on Ti	Increased Bone-to-implant contact	cpTi / TiO_2 Blasting / HF treatment	Dog	[80]
~100nm features on Ti	Increased Differentiation	cpTi / TiO_2 Blasting / HF treatment	Ratus novergicus	[70]
Not shown	Increased osseoactivity	cpTi / HA - Ion Beam Assisted Deposition (IBAD)	Dog	[139]
Discrete deposition of HA nanoparticles (20–40 nm) on Ti substrate	Increased Push-out test resistance	cpTi / dual acid etch / coated with CaP by DCD	Ratus novergicus	[140]
Not shown	Increased removal torque – Bone-to-implant contact – Bone volume	cpTi / Sandblast / HA - Ion Beam Assisted Deposition (IBAD)	Rabbit	[141]
20–100 nm range of the features (HA)	Increased tensile test resistance	cpTi and Ti6Al4V / acid etch / coated with CaP by DCD	Ratus novergicus	[91]
24 to 30 nm features	No difference comparing Nano-Titania to Nano-HA	Sol-gel coating of cpTi with Titania or HA	Rabbit	[142]
Deposition of HA nanoparticles	Increased Bone-to-implant contact	Sol-gel coating of cpTi with HA	Rabbit	[143]
20–100 nm range of the features (HA)	Increased tensile test resistance	cpTi and Ti6Al4V / acid etch / coated with CaP by DCD	Ratus novergicus	[136]
20 – 50 nm surface features	Increased removal torque – Bone-to-implant contact	Sol-gel coating of cpTi with Alumina	Ratus novergicus	[69]
20–100 nm range of the features (HA)	Increased tensile test resistance	cpTi and Ti6Al4V / acid etch / coated with CaP by DCD	Ratus novergicus	[137]

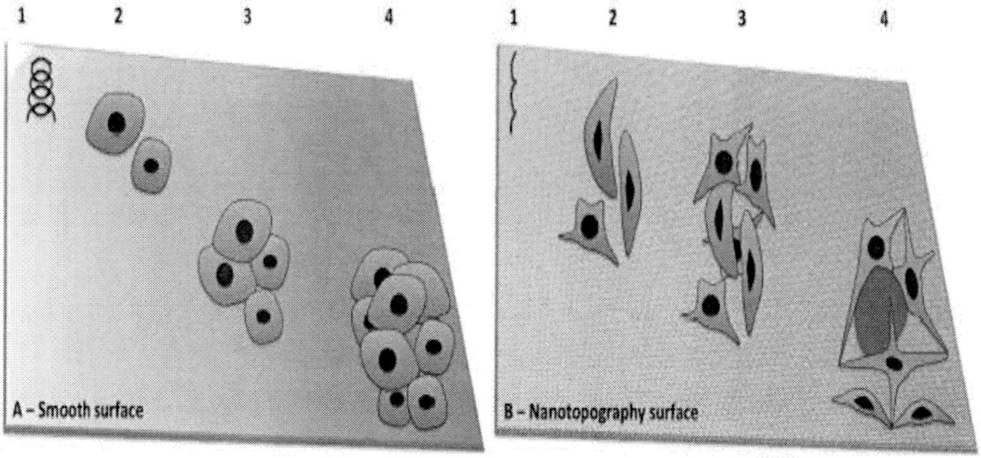

Figure 3. Depiction of effects of nanotopography observed at the implant surface. Protein adsorption is modulated by nanotopography altering cell adhesion, specificity and signaling. For osteoblast, several investigators have shown nanotopography enhances cell differentiation. 1- Protein adsorption, 2- Cell adhesion, 3- Cell proliferation, 4- Cell differentiation/mineralization.

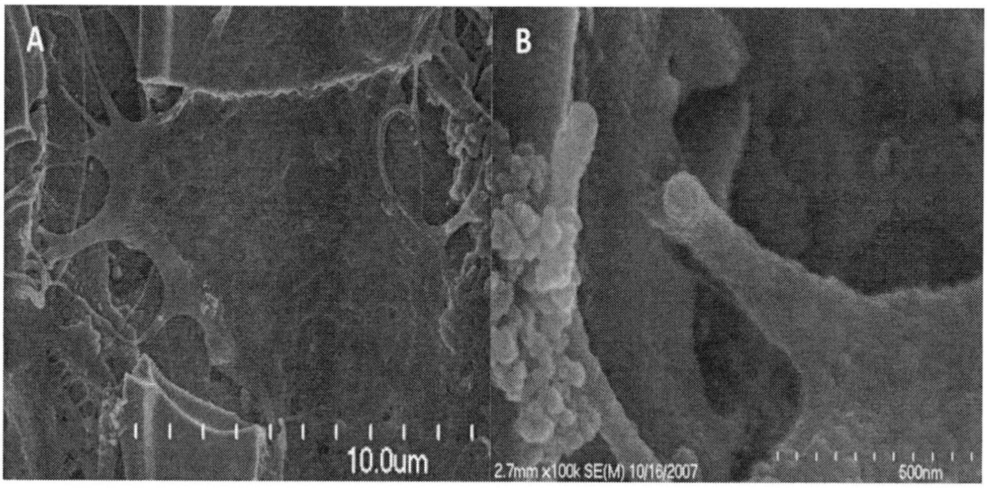

Figure 4. Cell-Nanotopography interactions. Affinity of cells for nanoscale features. TiO_2 produced by sol-gel coating created nanofeatures on the surface. These features are interactive points for lamellipodia and supporting cell spreading. A = 5,000x image of adherent cell. B = 50,000x magnification of the cell with nanofeatures.

3. Biomimetics

Nanotopography can contribute to the mimicry of natural cellular environments which promote rapid bone accrual [144]. The structure of the epithelial basement membrane contains pores approximating 70–100nm [145]. Cell adhesion to these membranes is an example of nanoscale interaction. The surface roughness of bone is approximately 32nm making it within the nanoscale range [98,104,119,146]. These further exemplify an anisotropic arrangement of nanofeatures. Intentionally placing molecular structures at such resolution on an endosseous implant may be achieved with anisotropic arrangements and is expected to increase osteoconduction towards the implant surface [136].

3.1. Protein Adsorption

Surface roughness at the nanoscale is an important determinant of protein interactions that ultimately direct cell activity in control of tissue formation at implant surfaces [147]. The adsorption of proteins depend significantly on surface nanostructure and that the relevant morphological parameter regulating the protein adsorption process is the size of the nanometric pores as they generate the conditions for protein nucleation inside the pores [99]. In a study using SAMs biofunctionalized with RGD, Cavalcanti-Adam and colleagues [101] found that nanofeature spacing modulates focal adhesion (FA) formation. Webster and colleagues [54,93] observed an increased vitronectin adsorption on nanostructured surfaces when compared to conventional surfaces. Self-assembled monolayers (SAM) control protein adsorption; hydrophobic groups are more likely to adsorb albumin, which is not replaced by ECM proteins. This disrupts cell adhesion. Hydrophilic surfaces allowed for the interchange of adsorbed albumin by ECM proteins [148,149]. In a similar manner, hydrophilic surfaces engendered fibronectin adsorption and greater osteoblast cell adherence [71].

3.2. Cell Adhesion and Spreading

Cells respond differently to the scale of roughness, and surface nanofeatures affect both cell adhesion and cell motility. Underlying substratum topography influences cell behavior by both direct and indirect interactions [150]. Indirect interactions are enacted through proteins adsorption phenomena described above. Changes in initial protein/surface interaction control osteoblast adhesion [114]. Cell binding to protein domains of adhesive ECM proteins involves receptors termed integrin receptors that transmit signals through collection of proteins on the cytoplasmic face of the contact, termed focal contacts [151,152]. Direct interactions involving the integrin receptors with the surface may also transmit signals to control adhesion, spreading and motility. Teixeira and colleagues [97], showed the cellular responses to nanoscale and submicron topographic cues are context dependent. These and other studies demonstrate that integrin-mediated adhesion is sensitive to nanoscale features. Osteoprogenitor cells adhesion were enhanced on poly-L-lactide (PLLA) and polystyrene (PS) surface with nano-scale and micro-scale roughness compared to smooth surfaces [153]. Dalby and co-workers [129] investigated primary human osteoblast cell behavior on nanopitted surfaces. They found that an ordered arrays of nanopits reduced cell spreading and randomization of the pits led to more cell spreading. Existing data suggest that nanotopography may work at a linear scale that facilitates the mechano-transduction signaling mechanisms of the adherent osteoblast.

3.3. Selectivity of Adhesion

Investigators have demonstrated the relative diminution of fibroblast compared to osteoblast adhesion when nano- and micron- structured surfaces were evaluated [50,93,135]. For example, on nano-sized materials, the affinity ratio between osteoblasts and fibroblasts was 3 to 1. In the conventional materials the ratio was 1 to 1 [93]. Similar results with other cell types such as smooth muscle cells and chondrocytes have also been reported [134]. The implications for tissue responses at bone and mucosal surfaces of the dental implant/abutment are intriguing. Bacterial adhesion and proliferation is also diminished on nanophase materials [96]. Decreased bacterial colonization on nanostructured TiO_2 and ZnO is observed even though these surfaces promote osteoblast adhesion and differentiation. Nanotopography also influenced the colonization pattern of a surface by two different types of bacteria. The surface allowed for the colonization of *S aureus* while prevented *P. aeruginosa* to adhere to the surface [154]. These observations imply that further development of the implant and the abutment surface can be explored in terms of biofilm accumulation and controlling peri-implantitis.

3.4. Cell Signaling Pathways

The cytoskeleton plays an important role in sensing and transmitting the signal from implant surface to the cell nucleus (mechano-transduction). In fact nanotopography alone is able to induce osteoblast differentiation [100]. Integrins are clearly involved in this process [97]. The ERK/MAPK signaling pathway is affected by human osteoblasts and MSCs interactions with nanoscale surfaces [102,155]. Other studies of stem-cell differentiation on nanoscale surface suggest that cell shape alterations involve small GTPase and RhoA [156-158]. RhoA/ROCK play a significant role in regulating cytoskeletal dynamics, mediating actin cytoskeletal tension and stress fiber formation by activating myosin light chain kinase [156,157,159]. This activation of RhoA/ROCK and the cytoskeletal isometric tension that may accompany their activation are important factors in mesenchymal stem cell fate [157,158,160]. Little is known about the exact mechanisms controlling cells response to nanoscale surfaces.

3.5. Cell Motility and Proliferation

The motility of fibroblasts and MSCs is altered by subtle changes in nanostructured surfaces [161,162]. Nanoscale topography affects the mechanical properties of the adherent cell. This may involve remodeling of the cytoskeleton or more complex biophysical changes in the cell membrane. Fujita and colleagues [103] demonstrated the role of filopodia in probing the cell environment on surfaces with nanotopography. The direction of nanogrooves influenced the stability of focal adhesion and cell orientation. Webster and colleagues [93] also observed increased osteoblast proliferation on the nanoscale (alumina, titania and hydroxyapatite) materials tested. Another study also demonstrated a reduced proliferation of fibroblasts compared to endothelial cells on nanopatterned surfaces [163]. Nanoscale surface-to-cell signaling may alter one or more pathways that control proliferation. One example is the cross talk between integrin-signaling and the predominant MAP kinase pathways affecting cell proliferation [164].

3.6. Osteoinduction and Differentiation

Nanotopography can dramatically influence cellular differentiation. The gene expression pattern indicative of differentiation by osteoblasts is modulated on nanostructured surfaces. Immunolabeled osteopontin and BSP were found in higher concentration in nanostructured surfaces [130]. Isa and co-workers [43] compared adherent palatal mesenchyme cell differentiation when cultured on a hydrophilic micron-scale topography cpTi surface or a nanoscale cpTi surface. Both surfaces supported osteoblastic differentiation, however, Runx2 expression (the key transcription factor controlling osteoblast differentiation) was increased on the nanoscale surface only. Others in vitro and in vivo studies have also demonstrated the upregulation in Runx2 expression and ALP activity [69,70,128]. Also, many other genes were upregulated in nanostructured surfaces as a response to Runx2 levels, such as, BSP, OPN, OCN [69,76,87,133].

3.7. Bone Formation

Early indications of nanoscale topography advantages were reported by Webster [52]. They revealed that alkaline phosphatase synthesis and calcium mineral content increased in cell layers formed on nano-sized materials after 21 and 28 days. Increased bone formation was measured for nanoscale rough implant surfaces in animal models [69,136,140,165]. Other studies also found early bone formation and increased torque removal when implant surfaces were added with different nanotopographies [69,142,143,165-167].

4. The Clinical Application of Nanotopography

From the many different methods discussed here (see table 1) to impart nanoscale features to the implant surface, several nanoscale modifications have reached the clinical environment. Nanotopography affect specific cells behavior and overall enhances osteoblastic differentiation to promote osseointegration (See Tables 3 and 4). Initial clinical stability is a prerequisite that requires other overall implant design considerations. The investigations of Meirelles and co-workers [143,165,166] suggest that nanometer-scale topography alone is not sufficient to assure robust osseointegration. It is possible that micron level roughness is of additional value to the process of osseointegration. The theoretical consideration of how forming tissues interlock with micron-level topographic elements [32], and how mechanical stimulation of forming tissues is imparted by such topographic elements [31] represent ideas that may not be fully displaced by the introduction of nanotopographic modification to the endosseous implant surface (Figure 5). Despite all the progress made in implant surface technology, one other application nanotopography remains largely unexplored. How nanotechnology and nanotopography can be used to enhance the tissue-abutment interface, maintain soft tissue stability and also prevent bacterial colonization.

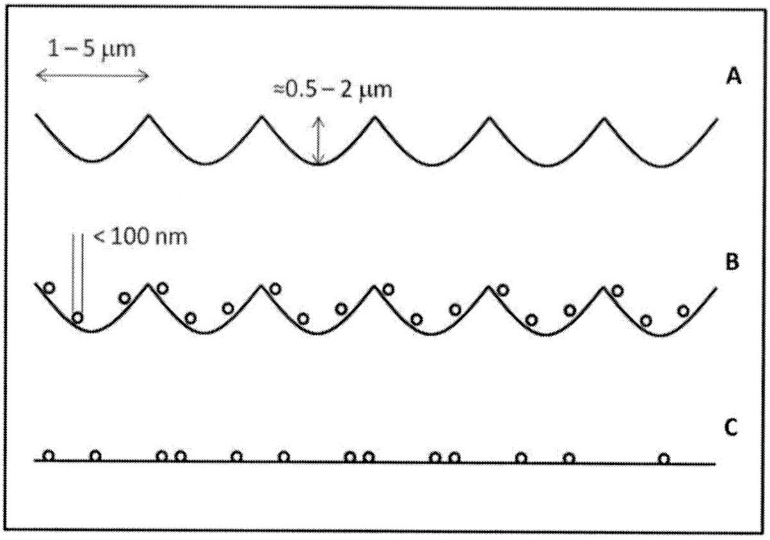

Figure 5. Nanotopography involves materials that have a size range between 1-100nm (10^{-9}m) at any significant dimension. Depiction of micron and nanoscale added to the surface. (A) Micron-scale only present at the surface. (B) Nanoscale features added to a surface with microtopography. (C) Nanoscale features on a smooth surface.

Presently, several nanotopography surface modifications are used to enhance osseointegration response with clinical dental implants [168]. The OsseoSpeed surface (Astra Tech AB, Molndal, Sweden) possesses nanostructured features created by TiO_2 blasting followed by a proprietary hydrofluoric acid treatment [70,79]. Across a micron rough titanium surface, 50–100nm surface accretions of Titanium oxide are observed by SEM analysis. Greater osteoblastic gene expression (Runx2, Osterix, Alkaline Phosphatase and Bone Sialoprotein) was measured in cells adherent to the nanoscale HF treated surface compared to the micron scale surface [70]. This nanotopography is associated with the elevated levels of gene expression that indicate rapid osteoblastic differentiation. *In vivo* studies using this modified titanium surface is associated with increased bone formation and torque removal values [47]. Histomorphometric evaluations demonstrated higher bone-to-implant contact for the nanoscale OsseoSpeed implants compared to the micronscale TiOBlast implants (Astra Tech AB, Molndal, Sweden) at 1 month (35%±14 vs 26%±8) and 3 months (39%±11 vs 31%±6) after placement in rabbit tibiae. Berglundh and colleagues [80] showed in the canine gap model of osseointegration that new bone formed in the voids within the first 2 weeks of healing was greater for HF-modified (OsseoSpeed) implants than for TiOblast implants. Most recent investigations show that this nanotopography promotes high levels of IGF-2 and BMP2 and BMP6 expression by adherent human mesenchymal stem cells for prolonged periods of time in culture [133].

In a clinical study, 634 patients received 1860 OsseoSpeed™ implants [81]. The initial report indicated 4% surfaces had signs of inflammation (BOP) with plaque present on 12% of sites. 21 patients have lost a total of 25 implants (15 in maxilla and 10 in mandible) for a CISR of 98.7% from placement. Evaluation of this effectiveness trial performed in more than

100 practices is ongoing. High success in challenging situations such as immediate placement and loading was also reported [169,170]. Another study also demonstrated an increased osteoblast-related gene expression for Osseospeed when compared to TiOBlast after 7 days of implant placement [171].

Another nanoscale surface implant presently available in the clinical marketplace involves a CaP nanoparticle modification of a minimally rough titanium alloy implant (NanoTite, 3i Implant Innovations, Palm Beach Gardens, FL). The surface involves particulate sol-gel deposition method using discrete crystalline deposition (DCD) of calcium phosphate (CaP) with 50–100 nm. Mendes and co-workers [91] measured bone ingrowth for implants modified using this technology in a rat tibia model using a well defined bone chamber model. The extend of bone ingrowth was 26.95% and 29.73% for cpTi and Ti alloy modified surfaces compared to the 12.01% cpTi and 16.97% Ti alloy chambers. Mendes and colleagues [91] further showed bone bonding behavior; DCD, surfaces had statistically greater tensile detachment force (e.g.; 11.30 N nanoscale DCD vs. 1.90 N control). The histological evaluation of clinical implants revealed bone to implant contact of 19%±14.2% and 32.2%±18.5% for the Osseotite control and the Nanotite experimental implants, respectively [172]. Clinically, Goene and co-workers [173] observed greater bone formation at 4 and 8 weeks and concluded that the addition of a nanometer-scale calcium phosphate treatment to a dual acid-etched implant surface appeared to increase the extent of bone development after 4 and 8 weeks of healing. The authors suggest that this rapid accrual of bone at the implant expedites the implant healing period and supports early loading protocols. During a one-year prospective clinical trial this nanoscale surface presented with a 99.2% cumulative survival rate in immediate loading situations [174]. Another surface using a similar method of CaP deposition has also be used and *in vitro* and *in vivo* data demonstrates an increased osteoblast differentiation as well [175,176].

Ion Bean Assisted Deposition (IBAD) has also been used to create a commercially available dental implant surface [139]. This technique creates a thin-film over the implant surface by deposition of the chemical element of interest. In one available study, the bone formation (measured by tetracycline labeling quantification) was higher in the experimental group than in the control group (sand-blasted/acid-etched) after two (13.56% vs 24.04%) and four weeks (14.22% vs 27.39%) [139]. Another *in vivo* study also presented an increased torque removal and BIC for this surface treatment [177]. An example of this type of surface modification is presented on the Nanotite surface of Bicon Implants (Nanotite, Bicon Inc. Boston, MA). These very different chemical and physical approaches all impart nanoscale features to existing endosseous cpTitanium implant surfaces.

These initial reports with high survival rates for nanoscale topography implants along with other *in vitro* and *in vivo* studies provide information into advantages for dental implant therapy. The high survival in effectiveness trials involving the HF modified TiO_2 grit blasted surface implant and in challenging clinical examinations may reflect greater control of initial bone formation due to the rapid differentiation of osteoblastic cells observed in laboratory studies [178]. The potential impact of bone bonding measured in preclinical studies requires further studies and long-term clinical evaluations are also urged; however, for each example of current nanoscale implant surfaces of available implants, cell culture, histological, and clinical data suggests that nanoscale surfaces offer incremental advantages to clinical problems where rapid bone accrual at the implant surface provide solutions. The main application of these "more bioactive" implant surfaces (modified with nanotechnology approaches) would be the challenging situations with deficiency in bone height and quality.

CONCLUSION

It seems clear so far that the development of an implant/bone interface may be influenced by both nanoscale and micron scale parameters of topography. The role of surface parameters (both chemistry and topography) requires consideration of molecular interactions (ionic and biologic) with the surface, cell adhesion and local biomechanical features of the established interface. Proposed changes include enhanced wettability, altered protein adsorption, and therefore, altered cell adhesion and potential mineralization phenomenon. These changes in wettability and altered protein adsorption are likely to involve both integrin and non-integrin receptors in affecting cell behavior. Several methods have been used to modify titanium substrates with nanoscale features. Cell culture studies reveal that there is a range of nanotopography that promotes improved differentiation of osteoblasts and osteoinduction. Additionally, nanoscale alterations may promote bone-bonding behavior at the implant/bone interface; enhancing interfacial bone formation measured as bone-to-implant contact. In the present, a grit-blasted / hydrofluoric acid modified titanium endosseous implant with nanoscale features and three grit-blasted / acid etched / calcium phosphate nanofeature-modified titanium implants are available for clinical use. The risks and benefits of manipulating biomaterial interfaces at the nanoscale will be defined in the future and long-term clinical evaluation of such endosseous devices is necessary. Yet, how nanotopography enhances adherent cell behavior on titanium endosseous implants remains to be fully clarified.

REFERENCES

[1] Cooper, L.F. (1998) Biologic determinants of bone formation for osseointegration: clues for future clinical improvements. *Journal of Prosthetic Dentistry*,80,439-449.
[2] Nanci, A., Wuest, J.D., Peru, L., Brunet, P., Sharma, V., Zalzal, S. and McKee, M.D. (1998) Chemical modification of titanium surfaces for covalent attachment of biological molecules. *Journal of Biomedical Materials Research*,40,324-335.
[3] Boyan, B.D., Schwartz, Z. and Hambleton, J.C. (1993) Response of bone and cartilage cells to biomaterials in vivo and in vitro. *Journal of Oral Implantology*,19,116-122.
[4] Schwartz, Z., Swain, L.D., Marshall, T., Sela, J., Gross, U., Amir, D., Muller-Mai, C. and Boyan, B.D. (1992) Modulation of matrix vesicle enzyme activity and phosphatidylserine content by ceramic implant materials during endosteal bone healing. *Calcified Tissue International*,51,429-437.
[5] Branemark, P.I., Adell, R., Breine, U., Hansson, B.O., Lindstrom, J. and Ohlsson, A. (1969) Intra-osseous anchorage of dental prostheses. I. Experimental studies. *Scandinavian Journal of Plastic and Reconstructive Surgery*,3,81-100.

[6] Linder, L., Albrektsson, T., Branemark, P.I., Hansson, H.A., Ivarsson, B., Jonsson, U. and Lundstrom, I. (1983) Electron microscopic analysis of the bone-titanium interface. *Acta Orthopedica Scandinavia*,54,45-52.

[7] Albrektsson, T. and Sennerby, L. (1990) Direct bone anchorage of oral implants: clinical and experimental considerations of the concept of osseointegration. *International Journal of Prosthodontics*,3,30-41.

[8] Adell, R., Eriksson, B., Lekholm, U., Branemark, P.I. and Jemt, T. (1990) Long-term follow-up study of osseointegrated implants in the treatment of totally edentulous jaws. *International Journal of Oral and Maxillofacial Implants*,5,347-359.

[9] Adell, R., Lekholm, U., Rockler, B., Branemark, P.I. (1981) A 15-year study of osseointegrated implants in the treatment of the edentulous jaw. *International Journal of Oral Surgery*,10,387-416.

[10] Albrektsson, T., Dahl, E., Enbom, L., Engevall, S., Engquist, B., Eriksson, A.R., Feldmann, G., Freiberg, N., Glantz, P.O. and Kjellman, O. (1988) Osseointegrated oral implants. A Swedish multicenter study of 8139 consecutively inserted Nobelpharma implants. *Journal of Periodontology*,59,287-296.

[11] Goodacre, C.J., Kan, J.Y. and Rungcharassaeng, K. (1999) Clinical complications of osseointegrated implants. *Journal of Prosthetic Dentistry*,81,537-552.

[12] Zarb, G.A. and Schmitt, A. (1990) The longitudinal clinical effectiveness of osseointegrated dental implants: the Toronto study. Part III: Problems and complications encountered. *Journal of Prosthetic Dentistry*,64,185-194.

[13] Morton, D., Jaffin, R. and Weber, H. (2004) Immediate restoration and loading of dental implants: clinical considerations and protocols. *International Journal of Oral and Maxillofacial Implants*,19 Suppl,103-108.

[14] Tolstunov, L. (2006) Dental implant success-failure analysis: a concept of implant vulnerability. *Implant Dentistry*,15,341-346.

[15] Jaffin, R.A. and Berman, C.L. (1991) The excessive loss of Branemark fixtures in type IV bone: a 5-year analysis. *Journal of Periodontology*,62,2-4.

[16] Bain, C.A. (1996) Smoking and implant failure--benefits of a smoking cessation protocol. *International Journal of Oral and Maxillofacial Implants*,11,756-759.

[17] Fiorellini, J.P., Chen, P.K., Nevins, M. and Nevins, M.L. (2000) A retrospective study of dental implants in diabetic patients. *International Journal of Periodontics and Restorative Dentistry*,20,366-373.

[18] Grant, B.T., Amenedo, C., Freeman, K., Kraut, R.A. (2008) Outcomes of placing dental implants in patients taking oral bisphosphonates: a review of 115 cases. *Journal of Oral and Maxillofacial Surgery*,66,223-230.

[19] Masuda, T., Salvi, G.E., Offenbacher, S., Felton, D.A. and Cooper, L.F. (1997) Cell and matrix reactions at titanium implants in surgically prepared rat tibiae. *International Journal of Oral and Maxillofacial Implants*,12,472-485.

[20] Meyer, U., Joos, U., Mythili, J., Stamm, T., Hohoff, A., Fillies, T., Stratmann, U. and Wiesmann. HP. (2004) Ultrastructural characterization of the implant/bone interface of immediately loaded dental implants. *Biomaterials*,25,1959-1967.

[21] Berglundh, T., Abrahamsson, I., Lang, N.P. and Lindhe, J. (2003) De novo alveolar bone formation adjacent to endosseous implants. *Clinical Oral Implants Research*,14,251-262.

[22] Le Guehennec, L., Soueidan, A., Layrolle, P. and Amouriq, Y. (2007) Surface treatments of titanium dental implants for rapid osseointegration. *Dental Materials*,23,844-854.

[23] Kasemo, B. (1983) Biocompatibility of titanium implants: surface science aspects. *Journal of Prosthetic Dentistry*,49,832-837.

[24] Johansson, C.B. and Albrektsson, T. (1991) A removal torque and histomorphometric study of commercially pure niobium and titanium implants in rabbit bone. *Clinical Oral Implants Research*,2,24-29.

[25] Suska, F., Gretzer, C., Esposito, M., Emanuelsson, L., Wennerberg, A., Tengvall, P. and Thomsen, P. (2005) In vivo cytokine secretion and NF-kappaB activation around titanium and copper implants. *Biomaterials*,26,519-527.

[26] Becker, J., Kirsch, A., Schwarz, F., Chatzinikolaidou, M., Rothamel, D., Lekovic, V., Laub, M. and Jennissen, H.P. (2006) Bone apposition to titanium implants biocoated with recombinant human bone morphogenetic protein-2 (rhBMP-2). A pilot study in dogs. *Clinical Oral Investigations*,10,217-224.

[27] Schliephake, H., Aref, A., Scharnweber, D., Bierbaum, S., Roessler, S. and Sewing, A. (2005) Effect of immobilized bone morphogenic protein 2 coating of titanium implants on peri-implant bone formation. *Clinical Oral Implants Research*,16,563-569.

[28] Germanier, Y., Tosatti, S., Broggini, N., Textor, M. and Buser, D. (2006) Enhanced bone apposition around biofunctionalized sandblasted and acid-etched titanium implant surfaces. A histomorphometric study in miniature pigs. *Clinical Oral Implants Research*,17,251-257.

[29] Buser, D., Schenk, R.K., Steinemann, S., Fiorellini, J.P., Fox, C.H. and Stich, H. (1991) Influence of surface characteristics on bone integration of titanium implants. A histomorphometric study in miniature pigs. *Journal of Biomedical Materials Research*,25,889-902.

[30] Gotfredsen, K., Hjorting-Hansen, E. and Budtz-Jorgensen, E. (1990) Clinical and radiographic evaluation of submerged and nonsubmerged implants in monkeys. *International Journal of Prosthodontics*,3,463-469.

[31] Hansson, S. and Norton, M. (1999) The relation between surface roughness and interfacial shear strength for bone-anchored implants. A mathematical model. *Journal of Biomechanics*,32,829-836.

[32] Davies, J.E. (2003) Understanding peri-implant endosseous healing. *Journal of Dental Education*,67,932-949.

[33] Puleo, D.A. and Nanci, A. (1999) Understanding and controlling the bone-implant interface. *Biomaterials*,20,2311-2321.

[34] Schwartz, Z., Lohmann, C.H., Oefinger, J., Bonewald, L.F., Dean, D.D. and Boyan, B.D. (1999) Implant surface characteristics modulate differentiation behavior of cells in the osteoblastic lineage. *Advances in Dental Research*,13,38-48.

[35] Albrektsson, T. and Wennerberg, A. (2004) Oral implant surfaces: Part 1--review focusing on topographic and chemical properties of different surfaces and in vivo responses to them. *International Journal of Prosthodontics*,2004;17:536-543.

[36] Albrektsson, T. and Wennerberg, A. (2004) Oral implant surfaces: Part 2--review focusing on clinical knowledge of different surfaces. *International Journal of Prosthodontics*,17,544-564.

[37] Wong, M., Eulenberger, J., Schenk, R. and Hunziker, E. (1995) Effect of surface topology on the osseointegration of implant materials in trabecular bone. *Journal of Biomedical Materials Research*,29,1567-1575.

[38] Wennerberg, A., Ektessabi, A., Albrektsson, T., Johansson, C. and Andersson, B. (1997) A 1-year follow-up of implants of differing surface roughness placed in rabbit bone. *International Journal of Oral and Maxillofacial Implants*,12,486-494.

[39] Park, J.Y., Gemmell, C.H. and Davies, J.E. (2001) Platelet interactions with titanium: modulation of platelet activity by surface topography. *Biomaterials*,22,2671-2682.

[40] Ricci, J.L., Grew, J.C. and Alexander, H. (2008) Connective-tissue responses to defined biomaterial surfaces. I. Growth of rat fibroblast and bone marrow cell colonies on microgrooved substrates. *Journal of Biomedical Materials Research*A,85,313-325.

[41] Lavenus, S., Louarn, G. and Layrolle, P. (2010) Nanotechnology and dental implants. *International Journal of Biomaterials*, 2010:915327.

[42] Schneider, G.B., Perinpanayagam, H., Clegg, M., Zaharias, R., Seabold, D., Keller, J. and Stanford, C. (2003) Implant surface roughness affects osteoblast gene expression. *Journal of Dental Research*,82,372-376.

[43] Isa, Z.M., Schneider, G.B., Zaharias, R., Seabold, D. and Stanford, C.M. (2006) Effects of fluoride-modified titanium surfaces on osteoblast proliferation and gene expression. *International Journal of Oral and Maxillofacial Implants*,21,203-211.

[44] Ogawa, T. and Nishimura, I. (2003) Different bone integration profiles of turned and acid-etched implants associated with modulated expression of extracellular matrix genes. *International Journal of Oral and Maxillofacial Implants*,18,200-210.

[45] Ogawa, T. and Nishimura, I. (2006) Genes differentially expressed in titanium implant healing. *Journal of Dental Research*,85,566-570.

[46] Buser, D., Broggini, N., Wieland, M., Schenk, R.K., Denzer, A.J., Cochran, D.L., Hoffmann, B., Lussi, A. and Steinemann, S.G. (2004)Enhanced bone apposition to a chemically modified SLA titanium surface. *Journal of Dental Research*,83,529-533.

[47] Ellingsen, J.E., Johansson, C.B., Wennerberg, A. and Holmen, A. (2004) Improved retention and bone-tolmplant contact with fluoride-modified titanium implants. *International Journal of Oral and Maxillofacial Implants*,19,659-666.

[48] Gutwein, L.G. and Webster, T.J. (2004) Increased viable osteoblast density in the presence of nanophase compared to conventional alumina and titania particles. *Biomaterials*,25,4175-4183.

[49] Oh, S., Finones, R.R., Daraio, C., Chen, L. and Jin, S. (2005) Growth of nano-scale hydroxyapatite using chemically treated titanium oxide nanotubes. *Biomaterials*,26,4938-4943.

[50] Price, R.L., Gutwein, L.G., Kaledin, L., Tepper, F. and Webster. T.J. (2003) Osteoblast function on nanophase alumina materials: Influence of chemistry, phase, and topography. *Journal of Biomedical Materials Research* A ,67,1284-1293.

[51] Price, R.L., Haberstroh, K.M. and Webster, T.J. (2003) Enhanced functions of osteoblasts on nanostructured surfaces of carbon and alumina. *Medical and Biological Engineering and Computing*,41,372-375.

[52] Webster, T.J., Siegel, R.W. and Bizios, R. (1999) Osteoblast adhesion on nanophase ceramics. *Biomaterials*,20,1221-1227.

[53] Webster, T.J., Ergun, C., Doremus, R.H., Siegel, R.W. and Bizios, R. (2000) Enhanced functions of osteoblasts on nanophase ceramics. *Biomaterials*,21,1803-1810.

[54] Webster, T.J., Schadler, L.S., Siegel, R.W. and Bizios, R. (2001) Mechanisms of enhanced osteoblast adhesion on nanophase alumina involve vitronectin. *Tissue Engineering*, 7,291-301.

[55] Webster, T.J. and Ejiofor, J.U. (2004) Increased osteoblast adhesion on nanophase metals: Ti, Ti6Al4V, and CoCrMo. *Biomaterials*,25,4731-4739.

[56] Schwartz, Z., Nasazky, E. and Boyan, B.D. (2005) Surface microtopography regulates osteointegration: the role of implant surface microtopography in osteointegration. *Alpha Omegan*,98,9-19.

[57] Webster, T.J., Hellenmeyer, E.L. and Price, R.L. (2005) Increased osteoblast functions on theta + delta nanofiber alumina. *Biomaterials*,26,953-960.

[58] Zhao, G., Schwartz, Z., Wieland, M., Rupp, F., Geis-Gerstorfer, J., Cochran, D.L. and Boyan, B.D. (2005) High surface energy enhances cell response to titanium substrate microstructure. *Journal of Biomedical Materials Research* A,74,49-58.

[59] Dike, L.E., Chen, C.S., Mrksich, M., Tien, J., Whitesides, G.M. and Ingber, D.E. (1999) Geometric control of switching between growth, apoptosis, and differentiation during angiogenesis using micropatterned substrates. *In Vitro Cellular and Developmental Biology- Animal*, 35,441-448.

[60] Wang, L., Zhao, G., Olivares-Navarrete, R., Bell, B.F., Wieland, M., Cochran, D.L., Schwartz, Z. and Boyan, B.D. (2006) Integrin beta1 silencing in osteoblasts alters substrate-dependent responses to 1,25-dihydroxy vitamin D3. *Biomaterials*,27,3716-3725.

[61] Olivares-Navarrete, R., Raz, P., Zhao, G., Chen, J., Wieland, M., Cochran, D.L., Chaudhri, R.A., Ornoy, A., Boyan, B.D. and Schwartz, Z. (2008) Integrin alpha2beta1 plays a critical role in osteoblast response to micron-scale surface structure and surface energy of titanium substrates. *Proceedings of National Academy of Science U S A*,105,15767-15772.

[62] Jaffe, A.B. and Hall, A. (2005) Rho GTPases: biochemistry and biology. *Annual Review of Cellular and Developmental Biology*,21,247-269.

[63] Gerecht, S., Bettinger, C.J., Zhang, Z., Borenstein, J.T., Vunjak-Novakovic, G. and Langer, R. (2007) The effect of actin disrupting agents on contact guidance of human embryonic stem cells. *Biomaterials*,28,4068-4077.

[64] Dalby, M.J., Hart, A. and Yarwood, S.J. (2008) The effect of the RACK1 signalling protein on the regulation of cell adhesion and cell contact guidance on nanometric grooves. *Biomaterials*,29,282-289.

[65] Garcia, A.J. and Reyes, C.D. (2005) Bio-adhesive surfaces to promote osteoblast differentiation and bone formation. *Journal of Dental Research*,84,407-413.

[66] Trisi, P., Lazzara, R., Rebaudi, A., Rao, W., Testori, T. and Porter, S.S. (2003) Bone-implant contact on machined and dual acid-etched surfaces after 2 months of healing in the human maxilla. *Journal of Periodontology*,74,945-956.

[67] Cochran, D.L. (1999) A comparison of endosseous dental implant surfaces. *Journal of Periodontology*,70,1523-1539.

[68] Shalabi, M.M., Gortemaker, A., Van't Hof, M.A., Jansen, J.A. and Creugers, N.H.J. (2006) Implant surface roughness and bone healing: a systematic review. *Journal of Dental Research*,85,496-500.

[69] Mendonca, G., Mendonca, D.B., Simoes, L.G., Araujo, A.L., Leite, E.R., Duarte, W.R., Cooper, L.F. and Aragao, F.J. (2009) Nanostructured alumina-coated implant surface:

effect on osteoblast-related gene expression and bone-to-implant contact in vivo. *International Journal of Oral and Maxillofacial Implants*,24,205-215.

[70] Guo, J., Padilla, R.J., Ambrose, W., De Kok, I.J. and Cooper, L.F. (2007) The effect of hydrofluoric acid treatment of TiO2 grit blasted titanium implants on adherent osteoblast gene expression in vitro and in vivo. *Biomaterials*,28,5418-5425.

[71] Scotchford, C.A., Gilmore, C.P., Cooper, E., Leggett, G.J. and Downes, S. (2002) Protein adsorption and human osteoblast-like cell attachment and growth on alkylthiol on gold self-assembled monolayers. *Journal of Biomedical Materials Research*,59,84-99.

[72] Wang, X.X., Hayakawa, S., Tsuru, K. and Osaka, A. (2001) A comparative study of in vitro apatite deposition on heat-, H(2)O(2)-, and NaOH-treated titanium surfaces. *Journal of Biomedical Materials Research*,54,172-178.

[73] Uchida, M., Kim, H., Miyaji, F., Kokubo, T. and Nakamura, T. (2002) Apatite formation on zirconium metal treated with aqueous NaOH. *Biomaterials*,23:313-317.

[74] Guo, J., Padilla, R.J., Ambrose, W., De Kok, I.J. and Cooper, L.F. (2007) The effect of hydrofluoric acid treatment of TiO2 grit blasted titanium implants on adherent osteoblast gene expression in vitro and in vivo. *Biomaterials*,28,5418-5425.

[75] Cooper, L.F., Zhou, Y., Takebe, J., Guo, J., Abron, A., Holmen, A. and Ellingsen, J.E. (2006) Fluoride modification effects on osteoblast behavior and bone formation at TiO2 grit-blasted c.p. titanium endosseous implants. *Biomaterials*,27,926-936.

[76] Mendonca, G., Mendonca, D.B., Aragao, F.J. and Cooper, L.F. (2010) The combination of micron and nanotopography by H(2)SO(4)/H(2)O(2) treatment and its effects on osteoblast-specific gene expression of hMSCs. *Journal of Biomedical Materials Research A*,94,169-179.

[77] Wang, X.X., Hayakawa, S., Tsuru, K. and Osaka, A. (2002) Bioactive titania-gel layers formed by chemical treatment of Ti substrate with a H2O2/HCl solution. *Biomaterials*,23,1353-1357.

[78] Ellingsen, J.E., Thomsen, P. and Lyngstadaas, S.P. (2006) Advances in dental implant materials and tissue regeneration. *Periodontology2000*,41,136-156.

[79] Cooper, L.F., Zhou, Y., Takebe, J., Guo, J., Abron, A., Holmen, A. and Ellingsen, J.E. (2006) Fluoride modification effects on osteoblast behavior and bone formation at TiO2 grit-blasted c.p. titanium endosseous implants. *Biomaterials*,27,926-936.

[80] Berglundh, T., Abrahamsson, I., Albouy, J. and Lindhe, J. (2007) Bone healing at implants with a fluoride-modified surface: an experimental study in dogs. *Clinical Oral Implants Research*,18,147-152.

[81] Stanford, C.M., Johnson, G.K., Fakhry, A., Gratton, D., Mellonig, J.T. and Wanger, W. (2006) Outcomes of a fluoride modified implant one year after loading in the posterior-maxilla when placed with the osteotome surgical technique. *Applied Osseointegration Resarch*,5,50-55.

[82] Zhou, J., Chang, C., Zhang, R. and Zhang, L. (2007) Hydrogels prepared from unsubstituted cellulose in NaOH/urea aqueous solution. *Macromolecular Bioscience*,7,804-809.

[83] Ben-Nissan, B. and Choi, A.H. (2006) Sol-gel production of bioactive nanocoatings for medical applications. Part 1: an introduction. *Nanomedicine (Lond)*,1,311-319.

[84] Liu, D.M., Troczynski, T. and Tseng, WJ. (2001) Water-based sol-gel synthesis of hydroxyapatite: process development. *Biomaterials*,22,1721-1730.

[85] Kim, H., Koh, Y., Li, L., Lee, S. and Kim, H. (2004) Hydroxyapatite coating on titanium substrate with titania buffer layer processed by sol-gel method. *Biomaterials*,25,2533-2538.

[86] Lee, S., Kim, H., Lee, E., Li, L. and Kim, H. (2006) Hydroxyapatite-TiO2 hybrid coating on Ti implants. *Journal of Biomaterials Applications*,20,195-208.

[87] Mendonca, G., Mendonca, D.B., Simoes, L.G., Araujo, A.L., Leite, E.R., Duarte, W.R., Aragao, F.J. and Cooper, L.F. (2009) The effects of implant surface nanoscale features on osteoblast-specific gene expression. *Biomaterials*,30,4053-4062.

[88] Piveteau, L.D., Gasser, B. and Schlapbach, L. (2000) Evaluating mechanical adhesion of sol-gel titanium dioxide coatings containing calcium phosphate for metal implant application. *Biomaterials*,21,2193-2201.

[89] Arias, J.L., Mayor, M.B., Pou, J., Leng, Y., Leon, B. and Perez-Amor, M. (2003) Micro- and nano-testing of calcium phosphate coatings produced by pulsed laser deposition. *Biomaterials*,24,3403-3408.

[90] Choi, A.H. and Ben-Nissan, B. (2007) Sol-gel production of bioactive nanocoatings for medical applications. Part II: current research and development. *Nanomedicine (Lond)*,2,51-61.

[91] Mendes, V.C., Moineddin, R. and Davies, J.E. (2007) The effect of discrete calcium phosphate nanocrystals on bone-bonding to titanium surfaces. *Biomaterials*,28,4748-4755.

[92] Ricci, J.L., Charvet, J., Frenkel, S.R., Chang, R., Nadkarni, P., Turner, J. and Alexander, H. (2000) Bone response to laser microtextured surfaces. In: Davies JE, editor. *Bone Engineering*,p. 8-9.

[93] Webster, T.J., Ergun, C., Doremus, R.H., Siegel, R.W. and Bizios, R. (2000) Specific proteins mediate enhanced osteoblast adhesion on nanophase ceramics. *Journal of Biomedical Materials Research*,51,475-483.

[94] Kay, S., Thapa, A., Haberstroh, K.M. and Webster, T.J. (2002) Nanostructured polymer/nanophase ceramic composites enhance osteoblast and chondrocyte adhesion. *Tissue Engineering*,8,753-761.

[95] Teixeira, A.I., Abrams, G.A., Bertics, P.J., Murphy, C.J. and Nealey, P.F. (2003) Epithelial contact guidance on well-defined micro- and nanostructured substrates. *Journal of Cell Science,*116,1881-1892.

[96] Colon, G., Ward, B.C. and Webster, T.J. (2006) Increased osteoblast and decreased Staphylococcus epidermidis functions on nanophase ZnO and TiO2. *Journal of Biomedical Materials Research* A,78,595-604.

[97] Teixeira, A.I., McKie, G.A., Foley, J.D., Bertics, P.J., Nealey, P.F. and Murphy, C.J. (2006) The effect of environmental factors on the response of human corneal epithelial cells to nanoscale substrate topography. *Biomaterials*,27,3945-3954.

[98] Palin, E., Liu, H. and Webster, T.J. (2005) Mimicking the nanofeatures of bone increases bone-forming cell adhesion and proliferation. *Nanotechnology*,16,1828-1835.

[99] Scopelliti, P.E., Borgonovo, A., Indrieri, M., Giorgetti, L., Bongiorno, G., Carbone, R., Podesta, A. and Milani, P. (2010) The effect of surface nanometre-scale morphology on protein adsorption. *PLoS One*,5,e11862.

[100] Dalby, M.J., Andar, A., Nag, A., Affrossman, S., Tare, R., McFarlane, S. and Oreffo, R.O. (2008) Genomic expression of mesenchymal stem cells to altered nanoscale topographies. *Journal of the Royal Society Interface*,5,1055-1065.

[101] Cavalcanti-Adam, E.A., Volberg, T., Micoulet, A., Kessler, H., Geiger, B. and Spatz, J.P. (2007) Cell spreading and focal adhesion dynamics are regulated by spacing of integrin ligands. *Biophysics Journal*,92,2964-2974.

[102] Biggs, M.J., Richards, R.G., Gadegaard, N., McMurray, R.J., Affrossman, S., Wilkinson, C.D., Oreffo, R.O. and Dalby, M.J. (2009) Interactions with nanoscale topography: adhesion quantification and signal transduction in cells of osteogenic and multipotent lineage. *Journal of Biomedical Materials Research A*,91,195-208.

[103] Fujita, S., Ohshima, M. and, Iwata, H. (2009) Time-lapse observation of cell alignment on nanogrooved patterns. *Journal of the Royal Society Interface*,6 Suppl 3,S269-77.

[104] Lim, J.Y., Dreiss, A.D., Zhou, Z., Hansen, J.C., Siedlecki, C.A., Hengstebeck, R.W., Cheng, J., Winograd, N. and Donahue, H.J. (2007) The regulation of integrin-mediated osteoblast focal adhesion and focal adhesion kinase expression by nanoscale topography. *Biomaterials*, 28,1787-1797.

[105] Zreiqat, H., Valenzuela, S.M., Nissan, B.B., Roest, R., Knabe, C., Radlanski, R.J., Renz, H., Evans, P.J. (2005) The effect of surface chemistry modification of titanium alloy on signalling pathways in human osteoblasts. *Biomaterials*,26,7579-7586.

[106] Monsees, T.K., Barth, K., Tippelt, S., Heidel, K., Gorbunov, A., Pompe, W. and Funk, R.H.W. (2005) Effects of different titanium alloys and nanosize surface patterning on adhesion, differentiation, and orientation of osteoblast-like cells. *Cells Tissues Organs (Print)*,180,81-95.

[107] Hart, A., Gadegaard, N., Wilkinson, C.D.W., Oreffo, R.O.C. and Dalby, M.J. (2005) Filapodial Sensing of Nanotopography in Osteoprogenitor Cells. *European Cell Materials*,10,65-65.

[108] Hart, A., Gadegaard, N., Wilkinson, C.D., Oreffo, R.O. and Dalby, M.J. (2007) Osteoprogenitor response to low-adhesion nanotopographies originally fabricated by electron beam lithography. *Journal of Materials Science- Materials in Medicine*,18,1211-1218.

[109] Berry, C.C., Dalby, M.J., Oreffo, R.O.C., McCloy, D. and Affrosman, S. (2006) The interaction of human bone marrow cells with nanotopographical features in three dimensional constructs. *Journal of Biomedical Materials Research A*,79,431-439.

[110] Webster, T.J. and Ejiofor, J.U. (2004) Increased osteoblast adhesion on nanophase metals: Ti, Ti6Al4V, and CoCrMo. *Biomaterials*,25,4731-4739.

[111] Eisenbarth, E., Velten, D. and Breme, J. (2007) Biomimetic implant coatings. *Biomolecular Engineering*,24,27-32.

[112] Webster, T.J., Ergun, C., Doremus, R.H. and Lanford, WA. (2003) Increased osteoblast adhesion on titanium-coated hydroxylapatite that forms CaTiO3. *Journal of Biomedical Materials Research A*,67,975-980.

[113] Ergun, C., Liu, H., Halloran, J.W. and Webster, T.J. (2007) Increased osteoblast adhesion on nanograined hydroxyapatite and tricalcium phosphate containing calcium titanate. *Journal of Biomedical Materials Research A*,80,990-997.

[114] Balasundaram, G., Sato, M. and Webster, T.J. (2006) Using hydroxyapatite nanoparticles and decreased crystallinity to promote osteoblast adhesion similar to functionalizing with RGD. *Biomaterials*,27,2798-2805.

[115] Chun, A.L., Moralez, J.G., Webster, T.J. and Fenniri, H. (2005) Helical rosette nanotubes: a biomimetic coating for orthopedics? *Biomaterials*,26,7304-7309.

[116] Dong, W., Zhang, T., Epstein, J., Cooney, L., Wang, H., Li, Y., Jiang, Y., Cogbill, A., Varadan, V. and Tian, Z.R. (2007) *Multifunctional Nanowire Bioscaffolds on Titanium. Chemistry of Materials*,19,4454-4459.

[117] Advincula, M.C., Rahemtulla, F.G., Advincula, R.C., Ada, E.T., Lemons, J.E. and Bellis, S.L. (2006) Osteoblast adhesion and matrix mineralization on sol-gel-derived titanium oxide. *Biomaterials*,27,2201-2212.

[118] Webster, T.J. and Smith, TA. (2005) Increased osteoblast function on PLGA composites containing nanophase titania. *Journal of Biomedical Materials Research A*,74,677-686.

[119] Lim, J.Y., Hansen, J.C., Siedlecki, C.A., Runt, J. and Donahue, H.J. (2005) Human foetal osteoblastic cell response to polymer-demixed nanotopographic interfaces. *Journal of the Royal Society Interface*,2,97-9108.

[120] Liu, H., Slamovich, E.B. and Webster, T.J. (2006) Increased osteoblast functions among nanophase titania/poly(lactide-co-glycolide) composites of the highest nanometer surface roughness. *Journal of Biomedical Materials Research A*,78,798-807.

[121] Popat, K.C., Leoni, L., Grimes, C.A., Desai, T.A. (2007) Influence of engineered titania nanotubular surfaces on bone cells. *Biomaterials*,28,3188-3197.

[122] Popat, K.C., Chatvanichkul, K., Barnes, G.L., Latempa, T.J., Grimes, C.A. and Desai, TA. (2007) Osteogenic differentiation of marrow stromal cells cultured on nanoporous alumina surfaces. *Journal of Biomedical Materials Research A*,80,955-964.

[123] Park, J.W., Kim, Y.J., Jang, J.H. and Song, H. (2010) Osteoblast response to magnesium ion-incorporated nanoporous titanium oxide surfaces. Clinical Oral Implants Research, 21,1278-87.

[124] de Oliveira, P.T., Zalzal, S.F., Beloti, M.M., Rosa, A.L. and Nanci, A. (2007) Enhancement of in vitro osteogenesis on titanium by chemically produced nanotopography. *Journal of Biomedical Materials Research A*,80,554-564.

[125] Guida, L., Annunziata, M., Rocci, A., Contaldo, M., Rullo, R. and Oliva, A. (2010) Biological response of human bone marrow mesenchymal stem cells to fluoride-modified titanium surfaces. Clin Oral Implants Res, 21,1234-41.

[126] Briggs, E.P., Walpole, A.R., Wilshaw, P.R., Karlsson, M. and Palsgard, E. (2004) Formation of highly adherent nano-porous alumina on Ti-based substrates: a novel bone implant coating. *Journal of Materials Science- Materials in Medicine*,15,1021-1029.

[127] Chiesa, R., Giavaresi, G., Fini, M., Sandrini, E., Giordano, C., Bianchi, A. (2007) Giardino, R. In vitro and in vivo performance of a novel surface treatment to enhance osseointegration of endosseous implants. *Oral Surgery Oral Medicine Oral Pathology Oral Radiology Endodontics*, 2007;103:745-756.

[128] He, F., Zhang, F., Yang, G., Wang, X. and Zhao, S. (2010) Enhanced initial proliferation and differentiation of MC3T3-E1 cells on HF/HNO3 solution treated nanostructural titanium surface. *Oral Surgery Oral Medicine Oral Pathology Oral Radiology Endodontics*, 110,e13-22.

[129] Dalby, M.J., McCloy, D., Robertson, M., Wilkinson, C.D.W. and Oreffo, R.O.C. (2006) Osteoprogenitor response to defined topographies with nanoscale depths. *Biomaterials*,27,1306-1315.

[130] de Oliveira, P.T. and Nanci, A. (2004) Nanotexturing of titanium-based surfaces upregulates expression of bone sialoprotein and osteopontin by cultured osteogenic cells. *Biomaterials*,25,403-413.

[131] Bigi, A., Nicoli-Aldini, N., Bracci, B., Zavan, B., Boanini, E., Sbaiz, F., Panzavolta, S., Zorzato, G., Giardino, R., Facchini, A., Abatangelo, G. and Cortivo, R. (2007) In vitro culture of mesenchymal cells onto nanocrystalline hydroxyapatite-coated Ti13Nb13Zr alloy. *Journal of Biomedical Materials Research A*,82,213-221.
[132] Tan, J. and Saltzman, W.M. (2004) Biomaterials with hierarchically defined micro- and nanoscale structure. *Biomaterials*,25,3593-3601.
[133] Valencia, S., Gretzer, C. and Cooper, L.F. (2009) Surface nanofeature effects on titanium-adherent human mesenchymal stem cells. *International Journal of Oral and Maxillofacial Implants*,24,38-46.
[134] Price, R.L., Ellison, K., Haberstroh, K.M. and Webster, T.J. (2004) Nanometer surface roughness increases select osteoblast adhesion on carbon nanofiber compacts. *Journal of Biomedical Materials Research A*,70,129-138.
[135] McManus, A.J., Doremus, R.H., Siegel, R.W. and Bizios, R. (2005) Evaluation of cytocompatibility and bending modulus of nanoceramic/polymer composites. *Journal of Biomedical Materials Research A*,72,98-9106.
[136] Mendes, V.C., Moineddin, R. and Davies, J.E. (2009) Discrete calcium phosphate nanocrystalline deposition enhances osteoconduction on titanium-based implant surfaces. *Journal of Biomedical Materials Research A*,90,577-585.
[137] Araujo, M.V., Mendes, V.C., Chattopadhyay, P. and Davies, J.E. (2010) Low-temperature particulate calcium phosphates for bone regeneration. *Clinical Oral Implants Research*,21,632-641.
[138] Jung, Y.C., Han, C.H., Lee, I.S. and Kim, H.E. (2001) Effects of ion beam-assisted deposition of hydroxyapatite on the osseointegration of endosseous implants in rabbit tibiae. *International Journal of Oral and Maxillofacial Implants*,16,809-818.
[139] Coelho, P.G. and Suzuki, M. (2005) Evaluation of an ibad thin-film process as an altenative method for surface incorporation of bioceramics on dental implants. A study in dogs. *Journal of Applied Oral Science*,13,87-92.
[140] Nishimura, I., Huang, Y., Butz, F., Ogawa, T., Lin, L. and JakeWang, C. (2007) Discrete deposition of hidroxyapatite nanoparticles on a titanium implant with predisposing substrate microtopography accelerated osseointegration. *Nanotechnology*,18.
[141] Park, Y., Yi, K., Lee, I., Han, C. and Jung, Y. (2005) The effects of ion beam-assisted deposition of hydroxyapatite on the grit-blasted surface of endosseous implants in rabbit tibiae. *International Journal of Oral and Maxillofacial Implants*,20,31-38.
[142] Meirelles, L., Currie, F., Jacobsson, M., Albrektsson, T. and Wennerberg, A. (2008) The effect of chemical and nanotopographical modifications on the early stages of osseointegration. *International Journal of Oral and Maxillofacial Implants*,23,641-647.
[143] Meirelles, L., Arvidsson, A., Andersson, M., Kjellin, P., Albrektsson, T. and Wennerberg, A. (2008) Nano hydroxyapatite structures influence early bone formation. *Journal of Biomedical Materials Research A*,87,299-307.
[144] You, M.H., Kwak, M.K., Kim, D.H., Kim, K., Levchenko, A., Kim, D.Y. and Suh, KY. (2010) Synergistically enhanced osteogenic differentiation of human mesenchymal stem cells by culture on nanostructured surfaces with induction media. *Biomacromolecules*,11,1856-1862.

[145] Brody, S., Anilkumar, T., Liliensiek, S., Last, J.A., Murphy, C.J and Pandit, A. (2006) Characterizing nanoscale topography of the aortic heart valve basement membrane for tissue engineering heart valve scaffold design. *Tissue Engineering*,12,413-421.

[146] Hansen, J.C., Lim, J.Y., Xu, L., Siedlecki, C.A., Mauger, D.T. and Donahue, H.J. (2007)Effect of surface nanoscale topography on elastic modulus of individual osteoblastic cells as determined by atomic force microscopy. *Journal of Biomechanics*,40,2865-2871.

[147] Park, G.E. and Webster, T.J. (2005) A Review of Nanotechnology for the Development of Better Orthopedic Implants. *Journal of Biomedical Nanotechnology*,1,18-29.

[148] Rodrigues, S.N., Goncalves, I.C., Martins, M.C.L., Barbosa, M.A. and Ratner, B.D. (2006) Fibrinogen adsorption, platelet adhesion and activation on mixed hydroxyl-/methyl-terminated self-assembled monolayers. *Biomaterials*,27,5357-5367.

[149] Arima, Y. and Iwata, H. (2007) Effect of wettability and surface functional groups on protein adsorption and cell adhesion using well-defined mixed self-assembled monolayers. *Biomaterials*,28,3074-3082.

[150] Brunette, D.M. (1988) The effects of implant surface topography on the behavior of cells. *International Journal of Oral and Maxillofacial Implants*,3,231-246.

[151] Fath, K.R., Edgell, C.J. and Burridge, K. (1989) The distribution of distinct integrins in focal contacts is determined by the substratum composition. *Journal of Cell Science*,92(Pt 1),67-75.

[152] Tosatti, S., Schwartz, Z., Campbell, C., Cochran, D.L., VandeVondele, S., Hubbell, J.A., Denzer, A., Simpson, J., Wieland, M., Lohmann, C.H., Textor, M. and Boyan, B.D. (2004) RGD-containing peptide GCRGYGRGDSPG reduces enhancement of osteoblast differentiation by poly(L-lysine)-graft-poly(ethylene glycol)-coated titanium surfaces. *Journal of Biomedical Materials Research A*,68,458-472.

[153] Wan, Y., Wang, Y., Liu, Z., Qu, X., Han, B., Bei, J. and Wang, S. (2005) Adhesion and proliferation of OCT-1 osteoblast-like cells on micro- and nano-scale topography structured poly(L-lactide). *Biomaterials*,26,4453-4459.

[154] Fadeeva, E., Truong, V.K., Stiesch, M., Chichkov, B.N., Crawford, R.J., Wang, J. and Ivanova, E.P. (2011) Bacterial Retention on Superhydrophobic Titanium Surfaces Fabricated by Femtosecond Laser Ablation. *Langmuir*, *27*,3012–3019.

[155] Biggs, M.J., Richards, R.G., Gadegaard, N., Wilkinson, C.D., Oreffo, R.O. and Dalby, M.J. (2009) The use of nanoscale topography to modulate the dynamics of adhesion formation in primary osteoblasts and ERK/MAPK signalling in STRO-1+ enriched skeletal stem cells. *Biomaterials*,30,5094-5103.

[156] Hill, C.S., Wynne, J. and Treisman, R. (1995) The Rho family GTPases RhoA, Rac1, and CDC42Hs regulate transcriptional activation by SRF. *Cell*,81,1159-1170.

[157] McBeath, R., Pirone, D.M., Nelson, C.M., Bhadriraju, K. and Chen, C.S. (2004)Cell shape, cytoskeletal tension, and RhoA regulate stem cell lineage commitment. *Developmental Cell*, 6,483-495.

[158] Sarasa-Renedo, A., Tunc-Civelek, V. and Chiquet, M. (2006) Role of RhoA/ROCK-dependent actin contractility in the induction of tenascin-C by cyclic tensile strain. *Experimental Cell Research*,312,1361-1370.

[159] Riddick, N., Ohtani, K. and Surks, H.K. (2008) Targeting by myosin phosphatase-RhoA interacting protein mediates RhoA/ROCK regulation of myosin phosphatase. *Journal of Cell Biochemistry*,103,1158-1170.

[160] Woods, A. and Beier, F. (2006) RhoA/ROCK signaling regulates chondrogenesis in a context dependent manner. *Journal Biological Chemistry*,281,13134-13140.

[161] Alsberg, E., Feinstein, E., Joy, M.P., Prentiss, M. and Ingber, D.E. (2006) Magnetically-guided self-assembly of fibrin matrices with ordered nano-scale structure for tissue engineering. *Tissue Engineering*,12,3247-3256.

[162] Dalby, M.J., Gadegaard, N., Tare, R., Andar, A., Riehle, M.O., Herzyk, P., Wilkinson, C.D.W. and Oreffo, R.O.C. The control of human mesenchymal cell differentiation using nanoscale symmetry and disorder. *Nature Materials*,6,997-991003.

[163] Csaderova, L., Martines, E., Seunarine, K., Gadegaard, N., Wilkinson, C.D. and Riehle, M.O. (2010) A biodegradable and biocompatible regular nanopattern for large-scale selective cell growth. *Small*,6,2755-2761.

[164] Schwartz, M.A. and Ginsberg, M.H. (2002) Networks and crosstalk: integrin signalling spreads. *Nature Cell Biology*,4,E65-8.

[165] Meirelles, L., Arvidsson, A., Albrektsson, T. and Wennerberg, A. (2007) Increased bone formation to unstable nano rough titanium implants. *Clinical Oral Implants Research*,18,326-332.

[166] Meirelles, L., Albrektsson, T., Kjellin, P., Arvidsson, A., Franke-Stenport, V., Andersson, M., Currie, F. and Wennerberg, A. (2008) Bone reaction to nano hydroxyapatite modified titanium implants placed in a gap-healing model. *Journal of Biomedical Materials Research A*,87,624-631.

[167] Meirelles, L., Melin, L., Peltola, T., Kjellin, P., Kangasniemi, I., Currie, F., Andersson, M., Albrektsson, T. and Wennerberg, A. (2008) Effect of hydroxyapatite and titania nanostructures on early in vivo bone response. *Clinical Implant Dentistry and Related Research*,10,245-254.

[168] Svanborg, L.M., Andersson, M. and Wennerberg, A. (2010) Surface characterization of commercial oral implants on the nanometer level. *Journal of Biomedical Materials Research Part B Applied Biomaterials*,92,462-469.

[169] Oxby, G., Lindqvist, J. and Nilsson, P. (2006) Early Loading of Astra Tech OsseoSpeed Implants Placed in Thin Alveolar Ridges and Fresh Extraction Sockets. *Applied Osseointegration Research*,5,68-72.

[170] Cooper, L.F., Raes, F., Reside, G.J., Garriga, J.S., Tarrida, L.G., Wiltfang, J., Kern, M. and de Bruyn, H. (2010) Comparison of radiographic and clinical outcomes following immediate provisionalization of single-tooth dental implants placed in healed alveolar ridges and extraction sockets. *International Journal of Oral and Maxillofacial Implants*,25,1222-1232.

[171] Bryington, M., Nares, S., Mendonca, G. and Cooper, L.F. (2010) Gene Expression of Adherent Cells to Endosseous Implants in Humans. *Journal Dental Research* (Abstract).

[172] Orsini, G., Piattelli, M., Scarano, A., Petrone, G., Kenealy, J., Piattelli, A. and Caputi, S. (2007) Randomized, controlled histologic and histomorphometric evaluation of implants with nanometer-scale calcium phosphate added to the dual acid-etched surface in the human posterior maxilla. *Journal of Periodontology*,78,209-218.

[173] Goene, R.J., Testori, T. and Trisi, P. (2007) Influence of a nanometer-scale surface enhancement on de novo bone formation on titanium implants: a histomorphometric study in human maxillae. *International Journal Periodontics and Restorative Dentistry*,27,211-219.

[174] Ostman, P.O., Wennerberg, A. and Albrektsson, T. (2010) Immediate occlusal loading of NanoTite PREVAIL implants: a prospective 1-year clinical and radiographic study. *Clinical Implant Dentistry and Related Research*,12,39-47.

[175] Bucci-Sabattini, V., Cassinelli, C., Coelho, P.G., Minnici, A., Trani, A. and Dohan Ehrenfest, DM. (2010) Effect of titanium implant surface nanoroughness and calcium phosphate low impregnation on bone cell activity in vitro. *Oral Surgery Oral Medicine Oral Pathology Oral Radiology Endodontics*, 109,217-224.

[176] Coelho, P.G., Granato, R., Marin, C., Bonfante, E.A., Freire, J.N., Janal, M.N., Gil, J.N. and Suzuki, M. (2010) Biomechanical evaluation of endosseous implants at early implantation times: a study in dogs. *Journal Oral and Maxillofacial Surgery*,68,1667-1675.

[177] Granato, R., Marin, C., Suzuki, M., Gil, J.N., Janal, M.N. and Coelho, P.G. (2009) Biomechanical and histomorphometric evaluation of a thin ion beam bioceramic deposition on plateau root form implants: an experimental study in dogs. *Journal of Biomedical Materials Research Part B Applied Biomaterials*,90,396-403.

[178] Collaert, B., Wijnen, L. and De Bruyn, H. (2011) A 2-year prospective study on immediate loading with fluoride-modified implants in the edentulous mandible. *Clinical Oral Implants Research*, doi: 10.1111/j.1600-0501.2010.02077.x.

In: Biomechanics of Dental Implants
Editor: Murat Çehreli
ISBN 978-1-62100-780-7

Chapter 4

Measuring Implant Stability

Murat Çehreli

1. Introduction

Osseointegration has been defined as the presence of structural and functional bone in the vicinity of a load-carrying implant [1]. From a macroscopic and biomechanical point of view, an implant is considered osseointegrated, if there is no progressive relative motion between the implant and surrounding living bone and marrow under functional levels and types of loading for the entire life of the subject and exhibits deformations of the same order of magnitude, as when the same forces are applied directly to the bone tissue alone. This direct, ankylotic structural and functional relationship between bone and the implant without intervening connective or fibrous tissue could be confirmed at the light microscopy and ultrastructural levels [2,3].

Osseointegrated oral implants are widely used for the rehabilitation of the edentulous patient by means of fixed or removable prosthesis and show predictably high success rates in the event certain preconditions are fulfilled. Albrektsson et al. describe a number of crucial factors that contribute to the achievement of osseointegration. Factors such as a suitable host, biocompatible material, careful implant surgery following a specific and strict protocol, and an appropriate healing time are discussed [4]. Primary implant stability is one of these prerequisites for achievement and maintenance of osseointegration. Primary stability depends on the mechanical engagement of an implant with bone of the osteotomy. Following implant surgery, this mechanical stability is gradually replaced by biologic stability. Transition from primary stability, ruled by the implant design, to secondary stability, provided by new bone formation, occurs during interface healing. The former is a vital requirement for successful secondary stability. Hence, secondary implant stability is the result of osseointegration, occurring after formation of new bone in the area adjacent to the implant. There is, presumably, a period of time in which osteoclastic activity decreases the primary mechanical engagement of the implant, but simultaneous new bone generation has not yet occurred to maintain implant stability. This period, between 2-4 weeks of placement, is referred to as "stability dip" (Figure 1), where the implant would be at high risk for excessive intraosseous micromotion and would theoretically be most susceptible to failure [5].

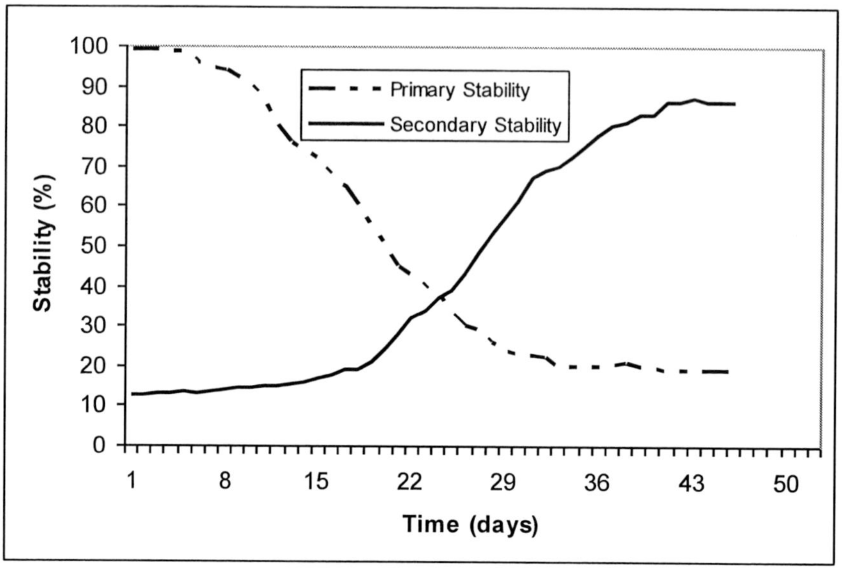

Figure 1. Changeover from primary stability to secondary stability by bone deposition at the interface.

Implant stability, an indirect sign of skeletal tissue integration, is a measure of clinical implant mobility. While optimum implant stability refers to lack of "clinical" mobility, this does not necessarily imply absence of micromotion or displacement in any direction at the micro-scale. Implants, particularly subjected to early and immediate functional loading, and indirect (whole-body) low-amplitude high-frequency biophysical mechanical stimuli might experience "osteogenic" levels of micromovement [6-8], which decrease upon remodeling at the interface. The amount of axial/lateral displacement of a loaded implant can be influenced by several factors such as its design and the density of host bone. At the experimental level, this displacement can be measured directly by displacement sensors or may be predicted means of numeric analysis in a time-dependent manner [9]. The critical threshold of displacement above which fibrous encapsulation and failure of the implant occurs is probably above 150 μm [10], as generally interpreted, although there is not any sound evidence in the context or oral implants.

A freshly-placed and healed implant experiences intraosseous micromovement under physiologic load that is reversible due to elastic properties of the skeletal tissue. As the direction of loads on an implant may be at any random direction (axial, lateral, and rotational), this information obtained may be used to detect or monitor implant stability in a time-dependent manner (Figure 2).

This chapter will focus predominantly on the mechanical evaluation of implant stability. Techniques that have been widely used by several research groups will be included only. Today, standardized periapical radiographs, despite a number of disadvantages such as superimposition of bone surfaces, challenge of taking standardized parallel x-rays, low-resolution to detect bone loss less than 0.1-0.2 mm, could be used to evaluate the stability of an implant non-invasively. In addition, light microscopy and back-scattered scanning electron microscopy assessments of bone-implant samples also provide a great insight into the stability of an implant, but the destructive nature of such analysis do not allow routine clinical use of these techniques.

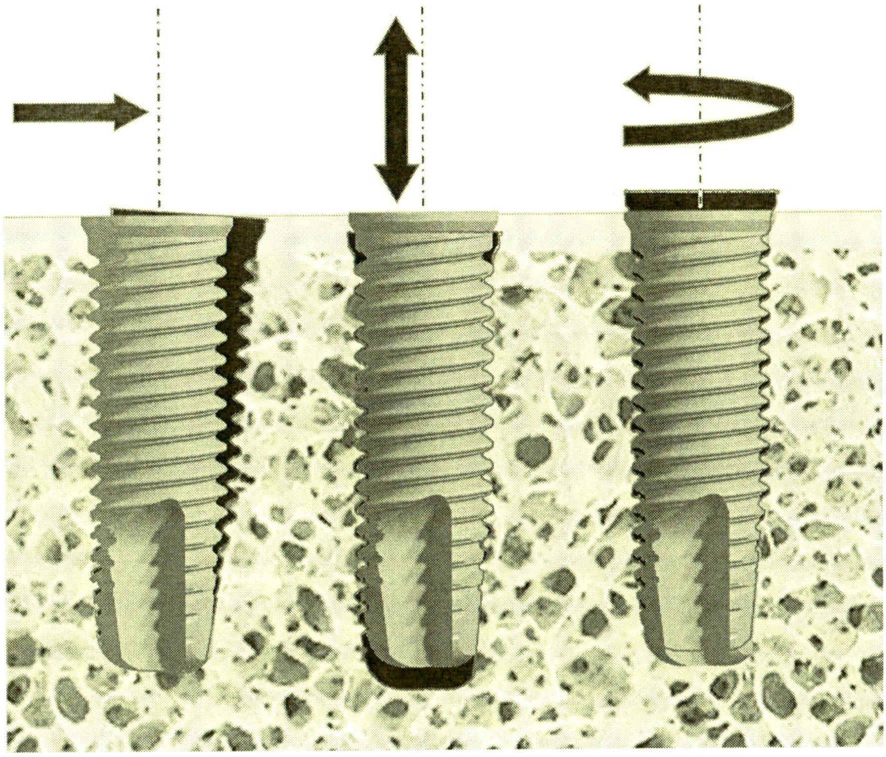

Figure 2. Exaggerated views of possible movement directions of implants subjected to lateral, axial or rotational forces.

Methods for measuring implant stability could be divided into two categories; non-invasive and invasive approaches (Figure 3). Non-invasive methods are presumed not to jeopardize the bone-implant interface and therefore, could be used to monitor implant stability safely in clinical practice to some extent. Invasive methods, however, depend on permanent and catastrophic failure of the bone-implant interface and could be applied only in experimental studies. At present, no gold standard exists for accurate quantification of implant stability at insertion and/or during function.

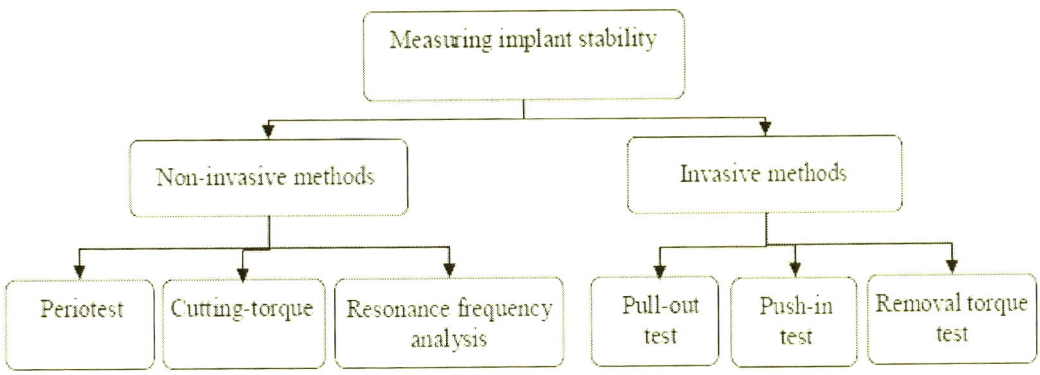

Figure 3. Most frequent non-invasive and invasive methods used to assess implant stability.

2. Non-Invasive Methods to Assess Implant Stability

2.1. Periotest®

Periotest® (Siemens Gulden- Medizintechnik, Bensheim, Germany), was developed by Schulte and co-workers at the University of Tübingen. It is an electronic instrument, originally developed to measure the damping characteristics of periodontal ligament around natural teeth. It has then been used to assess mobility of dental implants [11,12]. Similar to impact impedance methods, it uses an electromagnetically-driven and electronically-controlled rod to impact a natural tooth or implant in much the same way as the manual tapping mobility test. In essence, it measures predominantly the natural frequency and to a lesser extent, the damping characteristics of the tooth or bone-implant interface. The instrument comprises a hand-piece containing a metal rod that is accelerated towards a tooth or an implant. The rod is given a velocity by a propulsion coil that moves at a constant velocity prior to impacting the surface of the test specimen (Figure 4). In clinical practice, it is strictly advised to hold the rod perpendicular to the axis of the tooth or the implant to obtain consistent measurements. The contact duration of the slug on the tooth/implant is measured by an accelerometer. The software in the instrument is designed to relate contact time as a function of tooth mobility. These contact times are in the order of less than a millisecond. The instrument is designed to administer 16 of these impacts in 4 seconds, average the contact times, and calculate the PTV, which is related to T by PTV = 50 T – 21.3, where T is in msec. 6 For a PTV of 0, the contact time T is 0.426 milliseconds, while for a PTV of –4 it is 0.346 milliseconds. The outcome is displayed digitally and audibly as a "descriptive" numeric value (Periotest value-PTV). This value is between -8 (low mobility) to +50 (high mobility) for natural tooth and -8 to +8 for dental implants [13,14].

The bone-implant interface with no intervening soft tissue resembles a serial spring model, which follows Hooke's law and mobility measurements are considered much easier than the natural tooth. The Periotest® instrument has been reported to provide reproducible measurements on natural teeth. However, it may provide unpredictable and inconsistent results on oral implants, mainly due to small variations in the striking point. This inherent limitation has gradually decreased its clinical use in time [15-17]. Moreover, PTVs represent only a narrow range (-8 to +8) over the scale of the instrument and thus, the instrument does not provide high quantitative information regarding implant stability [18]. Considering that the dynamic range is low for implants and most healed implants display a PTV between -2 and -4, the PTV could easily be misinterpreted. Further, the technique does not mirror the mechanical properties of the interface [17] and therefore, its merit on detection of osseointegration is a matter of debate.

2.2. Cutting-Torque or Insertion-Torque Measurements

By measuring the energy (J/mm^3) required for a current-fed electric motor to cut-off a unit of bone volume, the force used to insert a dental implant could be measured. This has been defined as cutting-torque or insertion torque [19,20]. The amount of energy is positively correlated to the density of the bone. The amount of insertion torque relies on the resistance of the surrounding bone tissue to the implant being inserted into the surgical socket. It also relies on the friction between the implant surface with bone.

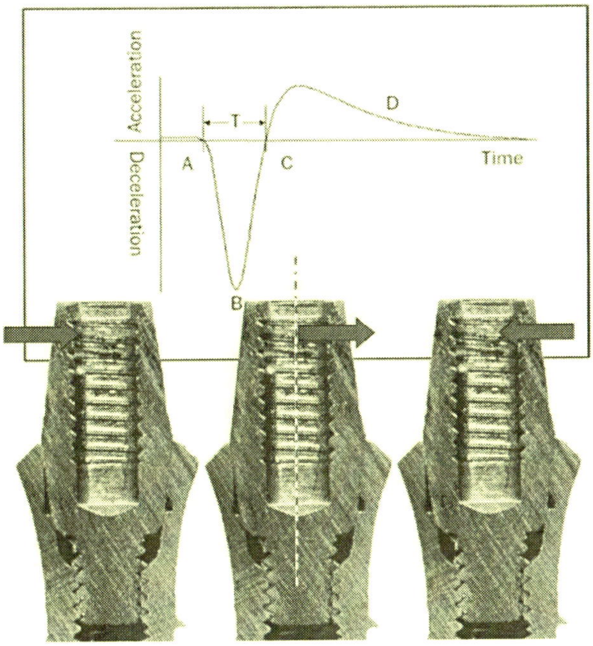

Figure 4. A typical signal from the accelerometer at impact to an implant. The rod first rapidly slows from point A to B). The impact leads to displacement of the implant or in this instance, the abutment only (red line), if the implant is rigidly fixed into a metal block. The displacement occurs toward the direction of force (right) and the maximum displacement is observed at point B. After this point, the abutment moves back to its original position (toward left). During deceleration (region A-B-C), the rod and the test unit are in contact and once the acceleration begins (after point C), the rod and abutment are most likely no longer in contact, where the signal is displays solely the motion of the rod (Adapted from Faulkner et al.[14]).

Therefore, site-specific properties of bone tissue, such as thickness of the cortical bone and density of the bone, the macrodesign of the implant and the difference between the diameters of the implant and the fresh surgical implant socket have decisive effects on the amount of torque output (Figure 5).

Essentially, the technique was developed in conjunction with the use of the Branemark® implant system. The OsseoCare® unit allowed torque-guided incremental insertion of an implant, displayed on the monitor of the device. This implies that one can observe the amount of generated torque at different levels of insertion, although it is only the final torque at complete insertion that matters. The OsseoCare™ unit can apply only a limited amount of torque 20, 30, 40, or 50 N.cm in the surgical mode and 10, 20, 32, and 45 N.cm in the prosthetic mode. The Osseocare™ unit has function measures that can only be used during low-velocity surgery and prosthetic modes. This is an inherent limitation in all commercially-available devices that are used in the clinic. Today, some physiodispensers also incorporate a torque-controller for placement of implants, although the actual amount of torque at complete insertion cannot be measured by these instruments.

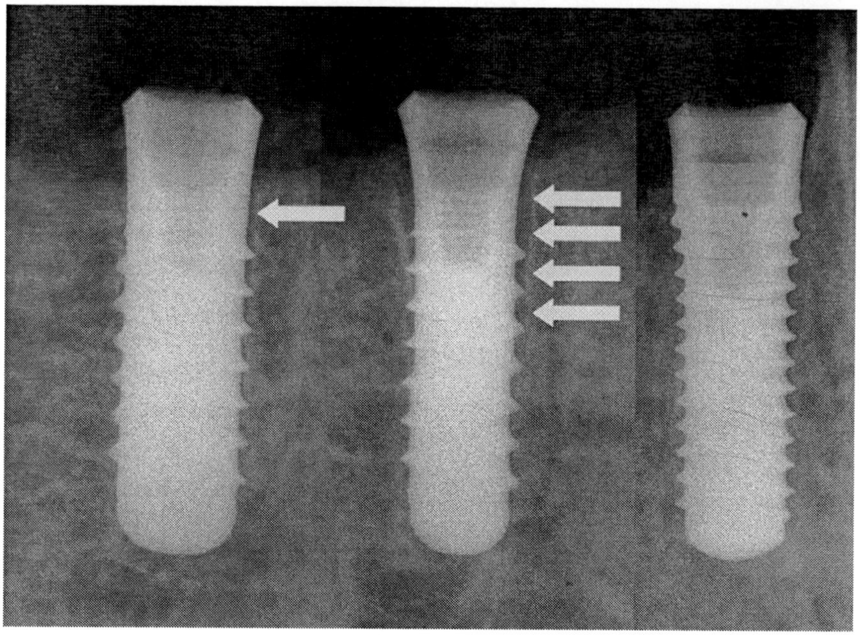

Figure 5. The primary stability of implants is influenced by many parameters such as implant design. When an implant is placed into a fresh extraction socket, the shape of the implant has a ruling effect on the measurements, because there is always a 3-dimensional bone defect, “jumping distance”, around the implants. In this example, the radiographic images of Ø 4.8 mm implant (left), Ø4.1 mm solid screw (middle) and a TE (right) Straumann® implants clearly demonstrate that, an implant resembling the 3-dimensional shape of an extraction socket may potentially have “contact” at the collar region (TE implant), whereas a conventional implant may have little or no contact with bone. In the event an implant has contact with cortical bone, its insertion torque will increase.

Instead of using commercially-available physiodispensers that have inherent limitations, one can easily fabricate a custom-made wrench to measure the insertion torque of an implant with high accuracy. First, the dimensions of the wrench, including the housing for appropriate driver is established, followed by milling out or casting/finishing of the wrench. Two strain-gauges with connecting terminals are bonded in each side of the handle. The strain-gauges are wired into a half bridge configuration for measurements (Figure 6). A calibration experiment is devised to obtain the calibration constant of the wrench and to assess the quality and repeatability of the torque measurements. Given this context, the housing part of the wrench is secured to a clamp, leaving the handle as a cantilever. Different weights, whose corresponding torque values are known, are applied on the handle at a predetermined distance from the strain-gauges. For each weight, strain-gauge measurements are recorded in separate sessions using a data acquisition system with computerized system or a strain indicator. The strain-gauge readings versus elicited torque are obtained and a linear regression to data points is performed to determine the calibration constant, such as 0.126 N.cm/με with $R^2 = 0.99997$. The strain data can be converted to torque units (N.cm) using the general formula:

$$Torque = K \times \varepsilon \qquad \text{(Eq.1)}$$

where K is the calibration constant and ε is the strain-gauge reading. This is also a very important step in commercially-available instruments before carrying out torque experiments.

In none of the published studies involving the use of Osseocare^{TM} unit so far, an attempt to calibrate the device had been undertaken, such as the methods used by O'Sullivan et al. [21,22]. For quantification of final insertion (and highest) torque of an implant, the measurements should be performed when the implant is placed into final position by an approximately half-turn of the torque device to clockwise direction [23-25]. Finally, the gauges and the wires are waterproofed before use in the oral cavity (Figure 7 and 8).

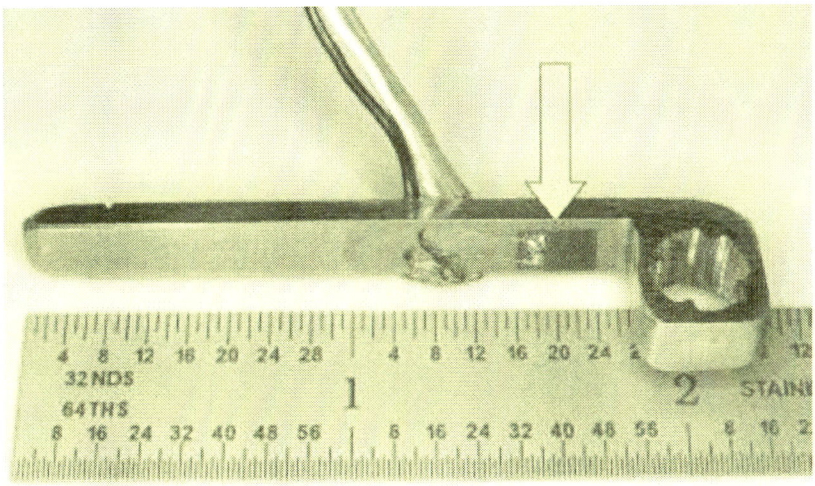

Figure 6. An example of a strain-gauged (yellow arrow) manuel torque wrench, with a housing/aperture for appropriate implant driver.

Today, the cutting-torque measurements are not routinely employed during clinical practice. However, the emerging need to explore the minimum as well as the safest range of insertion torque for immediate loading of implants will probably increase the use of the technique. Beyond this issue, routine measurement of the insertion torque of implants has to be done during surgery, because it might have a decisive role on implant survival. When a screw or press-fit type implant is inserted into bone, which has elastic properties, the impact of the insertion torque might, in some cases, not allow the bone to recover, leading to plastic strain. Such high strain amplitudes might lead to bone breakdown. Consequently, a crater-like bone defect may be observed at early stages of healing. This is where the insertion torque technique might avoid possible trauma. If an implant with high or relatively-high (although there is not such a classification so far) insertion torque is placed, the impact can be removed at early stages of placement by enlarging the diameter of the implant bed to reduce the insertion torque to a biologically acceptable level. Yet, there is not any scientific evidence for a possible torque threshold that would cause bone breakdown around freshly-placed oral implants. This remains as an open field for investigation.

One of the limitations of insertion technique is that it cannot provide any information unless the implant socket is prepared. Moreover, the lower threshold at which an implant may be at highest risk cannot be identified. This objective and quantitative technique can only be applied during placement of an implant. Therefore, the information obtained is restricted to primary stability.

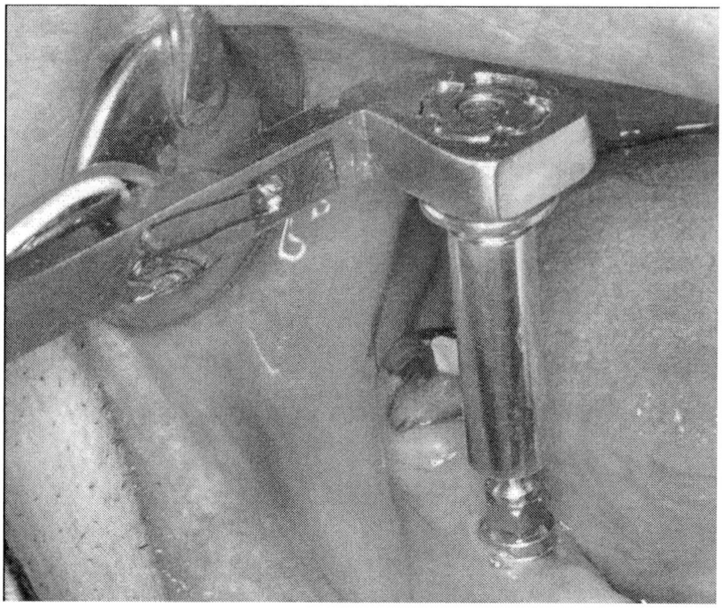

Figure 7. The use of a custom-made torque wrench.

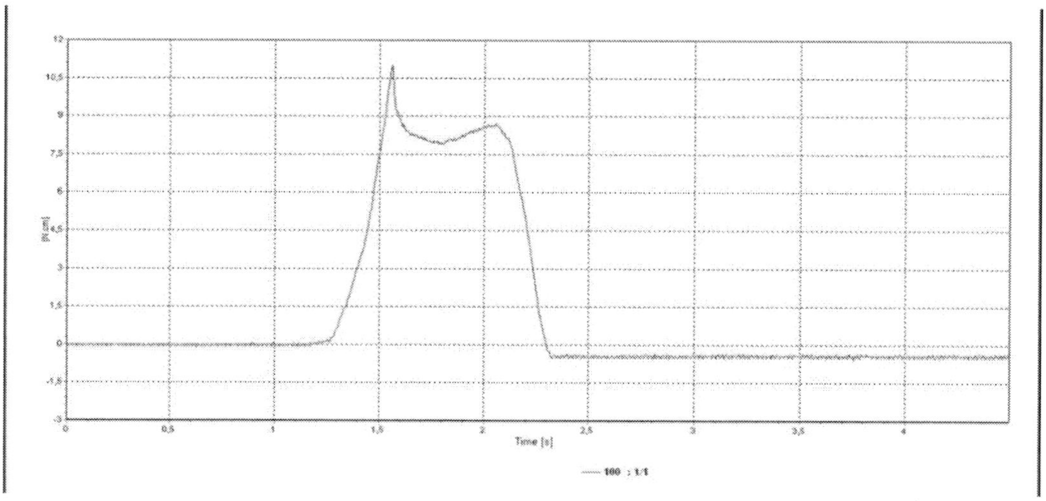

Figure 8. A typical torque versus time graphic of an implant at final insertion into low-density bone. The peak denotes the highest torque applied.

2.3. Resonance Frequency Analysis

Meredith and co-workers [26] introduced the resonance frequency analysis (RFA), a method previously used in construction engineering. It is a non-invasive and non-destructive method for quantitative measurement of implant primary stability and osseointegration by assessing time-dependent changes in implant stability. RFA measurements have been used to document healing changes along the implant-bone interface by measuring the increase/decrease in stiffness of this interface. Moreover, the technique has been used to determine appropriate timing of loading and to identify ‘‘at-risk’’ implants. RFA, based on continual excitation of the implant through dynamic vibration analysis makes use of a transducer connected to an implant, which is excited over a range of sound frequencies with subsequent measurement of vibratory oscillation of the implant. This technique causes the implant to vibrate, while at the same time analyzing implant motion.

The RFA technique and transducers evolved over time. The original electronic technique employed an L-shaped transducer, screwed to an implant body or the abutment that had direct connection between the transducer and the resonance frequency analyzer by wires. The technique relies in application of a very small amount of bending force to the cantilever beam. The transducer was made of stainless steel or commercially pure titanium and included the cantilever beam and two piezoelectric elements. The transducer offset cantilever beam was excited between 5 to 15 kHz sinusoidal frequencies, which caused the beam to displace less than 1 μm, and this response of the cantilever beam was subsequently analyzed by the frequency response analyzer. Because the flexural resonance of the cantilever beam led to change in amplitude in the phase of the received signal, the resonance frequency of the first generation devices was expressed as frequency (in range of 3500-8500 KHz) against amplitude (V). A high value indicates greater stability, whereas a low value indicates instability. One of the major drawback of the first and second generation instruments was that each transducer had its own RF and orientation of the same transducer or use of another transducer led to different measurements. They had to be calibrated by the manufacturer and registration of the implant length was mandatory. It was also not possible to interpret time-dependent RF results. In addition, the system with the response analyzer and a personal computer was considered to be too heavy and not user-friendly. The aim of the third generation RFA was, therefore, to allow reliable chair-side measurements with a user-friendly battery-driven small instrument. Today, the instrument is modified to measure implant stability by telemetry. The instrument makes use of a magnetic peg so called the "Smartpeg" screwed on the implant or the abutment of various systems. The peg is excited and the RF is expressed electromagnetically as Implant Stability Quotient (ISQ units) (the magnetic pulse technique) on a scale from 1 to 100 on the monitor of the instrument (Figure 9). ISQ is a numeric "descriptive" presentation of implant stability determined by the device (Osstell®, Integration Diagnostics, Gothenburg, Sweden), but the values obtained are not directly correlated with any specific physical parameters. A ISQ value higher than 65 indicates a successful implant, whereas <50 might indicate an "at risk" or failing implant. The peg vibrates in two directions, which are approximately perpendicular to each other. This leds to two RFA values, one high and the other low, indicating possible high and low bone support areas for the implant.

RFA is influenced by several factors such as implant design, implant length, implant position, implant surface characteristics, and abutment length. The accuracy of ISQ values may be questioned in some clinical circumstances. It seems that RFA does not mirror the bone-implant contact at deeper parts, but rather at the marginal bone region. The ISQ values may also present false-positive results in different cases, i.e., a regular diameter implant having bone contact at the marginal region may lead to higher ISQ values than a wide implant without any bone contact at the collar region (Figure 5). Despite recent improvements, several investigators also questioned the correlation between RFA with cutting-torque [27-30] removal torque measurements [31,32], bone mineral density and trabecular bone volume and pattern factor [28], and histomorphometry [30,33].

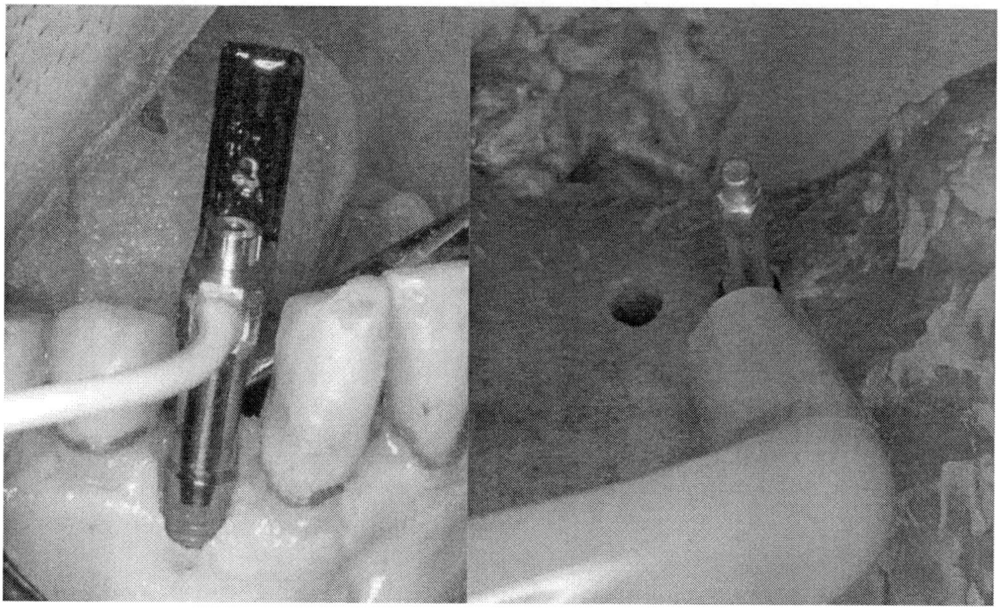

Figure 9. The early L-shaped transducer connected to the instrument with wires (left) and the currently-available wireless transducer allowing telemetric measurements.

2.4. Subjective Assessment

Subjective assessment is another method used to qualitatively assess primary stability of implants. It depends on the tactile sensation of the surgeon, who scores the primary stability, i.e., good, moderate, poor, with regard to the resistance of the implant being inserted. Therefore, it relies on the experience and training of the surgeon on discrimination of bone density. So far, this method has not been used extensively on implants, as it is not an objective method and does not provide a quantitative assessment [34,35]. Moreover, it can only be used during insertion of the implant. Its reliability and correlation with the outcome of non-invasive techniques is still under consideration.

2.5. Percussion Test

The percussion test, based on vibrational-acoustic science and impact-response theory, is probably the simplest way to subjectively assess the stability and osseointegration of an implant. The quality of the sound heard upon an impact on an implant is used to judge the biologic state of the bone-implant interface. A clear ringing sound refers to successful osseointegration, whereas a dull sound may indicate no osseointegration. Application of such an impact force to an implant at early stages of healing may jeopardize the process of osseointegration. Owing to its subjective nature lack of objective quantitative assessment, the reliability of this merely used technique is questionable.

2.6. Correlations between Current Methods

In an attempt to explore possible correlations between current methods used to assess implant stability, a meta-analysis had been performed on studies between 1998-2008 using inclusion/exclusion criteria. The details of the analysis are explained elsewhere [36]. In brief, 47 articles fulfilled the criteria and the studies were categorized as human cadaver studies (n=11), clinical studies (n=15), animal studies (n=15), and in vitro studies (n=5). Among the 47 articles, p values for comparative evaluation of insertion torque and RFA (or ISQ values) were provided in 6 studies [28-30,32,34,37]. The p values of these studies were combined using Fisher's method to obtain χ^2 and one p value. The correlation between insertion torque and RFA was significant (χ^2=30.64 $>\chi^2_{2K}=\chi^2_{12}$=21.026, p=0.0022<0.05). Among cadaver studies, two studies [28,32] provided p values and suggested that the correlation between insertion torque and RFA is insignificant (χ^2=6.73 $<\chi^2_{2K}=\chi^2_4$=9.487, p=0.15>0.05). Among the clinical studies, p values for comparative evaluation of insertion torque and RFA were provided in two studies [34,37]. The results of these studies showed that the correlation between insertion torque and RFA is significant (χ^2=15.86 $>\chi^2_{2K}=\chi^2_4$=9.487, p=0.0032<0.05). Among animal studies, p values for comparative evaluation of insertion torque and RFA was provided in two studies [29,34]. The results of these studies showed that the correlation between insertion torque and RFA is insignificant (χ^2=8.05$<\chi^2_{2K}=\chi^2_4$=9.487, p=0.09>0.05). One study [32] provided p value for comparison of insertion torque and removal torque value and found a significant correlation (p=0.001<0.05). The same study [32] provided p value for comparison of RFA and reverse torque test values and did not find a significant correlation (p=0.319>0.05; r=0.405). One study [28] provided p-value for comparison of Periotest® and insertion torque and Periotest® and RFA. A significant correlation was found between Periotest® and insertion torque (p=0.015<0.05), whereas the correlation between Periotest® and RFA was insignificant (p=0.28>0.05).

As fixed effect model: r=0.577 (standard error: 0.03 p=0.000), random effect model: r=0.554 (standard error: 0.109 p=0.000), and Q value=94.57 (p=0.000<0.05) were found, the studies had heterogeneity. As the random effect model led to r=0.554, insertion torque and RFA showed a direct relationship of 55.4%, which was statistically significant (p=0.000<0.05). In the event the cadaver studies were excluded from this analysis, and the outcome re-evaluated, the studies had heterogeneity, as fixed effect model: r=0.629 (standard error: 0.032 p=0.000), random effect model: r=0.637 (standard error: 0.099 p=0.000), and Q value=48.44 (p=0.000<0.05). The random effect model had r=0.637 and showed that the clinical studies detected a direct relationship of 63.7% between insertion torque and RFA (p=0.000<0.05). In the event the analysis was undertaken for the cadaver studies, homogeneity was detected, as fixed effect model: r=0.726 (standard error: 0.029 p=0.000), random effect model: r=0.726 (standard error: 0.029 p=0.000) and Q value=0.457 (p=0.499<0.05). The fixed effect model yielded r=0.726 and therefore, showed a direct relationship of 72.6% between insertion torque and RFA (p=0.000<0.05). Among animal studies, one study [30] provided a r= -0.149.

For comparison of insertion torque and reverse torque test values, r values have been provided in two studies [31,32]. The studies had homogeneity, as fixed effect model: r=0.876 (standard error: 0.114 p=0.000), random effect model: r =0.876 (standard error: 0.114

$p=0.000$), and Q value=0.117 ($p=0.732>0.05$). These studies revealed a direct relationship of 87.6% between insertion torque and reverse torque test values, which was statistically significant ($p=0.000<0.05$). Based upon a meta-analysis of dental implant literature that revealed 47 studies that met strict inclusion criteria, a statistically significant correlation between insertion torque and RFA was found.

3. Invasive Methods to Assess Implant Stability

3.1. Pull-Out and Push-out Test

The structural durability of the bone-implant interface is a prerequisite for long-term uneventful functioning. Because bone tissue has lower modulus of elasticity than titanium alloy as well as zirconium implants, it is more prone to failure. A typical failure due to difference in elastic moduli includes microfractures in the trabeculae or high strain amplitudes leading to microfractures in the cortical bone. The holding power or anchorage of an implant relies on the mechanical properties of the implant itself and the shear strength of the bone tissue. In this regard, push-out and pull-out tests are frequently used to test the mechanical competence of primary stability or biological integration of endosseous implants or screws. These methods are suitable to describe the interfacial bonding between bioactive materials and the host bone related to the bone-implant biochemical interactions, besides the mechanical interlocking related to surface roughness and topologic characteristics. Because these tests cannot reveal the biomechanical characterization of endosseous implants alone, they are usually employed as a part of the experiment.

The tests are undertaken either immediately after sacrifice of the animals or in fresh-frozen samples retrieved from those animals. The latter approach refers to storing of the bone-implant samples at -21°C to preserve the mechanical properties of bone tissue. The specimens are then thawed at room temperature for 24h before mechanical tests are carried out [38,39]. In the event the tissues are kept frozen, the mechanical properties are preserved for a limited period of time, but freezing unembalmed tissue only suspends the process of decay. Therefore, a series of experiments cannot be performed on the same tissue after freezing/thawing [40]. In addition, unembalmed bodies have considerable disadvantages such as rapid deterioration and possible unacceptable health hazards. Use of formalin-fixed bone specimens result in false data regarding the mechanical properties of the bone-implant interface [39]. Formalin is an extremely reactive electrophilic chemical that reacts with different cross-linking functional groups in tissue proteins, nucleic acid, and polysaccharides, which creates stable and irreversible methylene crosslinks [41]. There is strong evidence that formalin significantly alters the mechanical/biomechanical behavior of the ligaments, spine, and bone of animals [42-45].

Implant/screw pull-out or push-out test refers to quantification of the force required to pull-out or push-out, respectively, an implant inserted in bone. It is used to invasively/destructively test the biomechanics of the bone-implant interface. This is influenced by several factors including the local material properties of the bone, cortical bone thickness, unicortical versus bicortical anchorage, bone density, and time-dependent mechanical properties of the differentiating tissue at the interface. In addition, implant macrodesign, diameter and length, thread depth, surface texture, chemical composition, the insertion angle of the implant, and pull-out or push-out speed influence the maximum pull-out and push-out force. As bone consists of a complex arrangement of trabeculae and anisotropic property, the angulation and location of the test implant/screw are important parameters that should be considered in this analysis. Specimen preparation as well as carrying out these tests is relatively simple. A universal testing machine operated under displacement control is used to apply axial load with usually 0.5-2 mm/sec crosshead speed. A hook up system and a relatively simple support jig are used for the pull-out and push-out tests, respectively [46] (Figure 10). The pull-out and push-out tests create an increasing shear stress at the interface until failure, which indicates the maximal interface loading capacity (Figure 11).

Frequently, the end point of these tests is the ultimate shear strength of the bone-implant interface, σu, calculated by the general formula [42];

$$\sigma u = F\text{max}/A \qquad \text{(Eq.2)}$$

where Fmax is the maximum push-out or pull-out force and A is the nominal interface area.

For a cylindrical implant,

$$A = \pi DL \qquad \text{(Eq.3)}$$

where D is the outer diameter of the cylindrical implant and L is the length of the implant in contact with bone. The results are usually reported in terms of force per unit area (i.e., the maximum shear stress). As a simplified approach to compare two implants with the same test conditions, Fmax can be used alone or the yield strength, where the load-deformation curve becomes nonlinear.

The ultimate shear strength of the interface, quantities called the "apparent shear stiffness," the "apparent elastic modulus," or the "interface stiffness" have been calculated as either [42];

$$\Delta F/d \qquad \text{(Eq.4)}$$

or

$$(\Delta F/A)/d \qquad \text{(Eq.5)}$$

where d is the displacement. Berzins and Summer [46] point out that these calculations are based on the assumption that the interface is infinitely thin, and that shear depends upon an angular deformation and the interface must therefore have a finite thickness. Accordingly, Berzins and Summer [46] and Branemark [47] pointed out that assessing only ultimate stress is insufficient for describing the mechanical properties of the bone-implant interface. They developed formulas to assess the elasticity of the peri-implant bone by calculating the interface shear modulus from pull-out or push-out measurements.

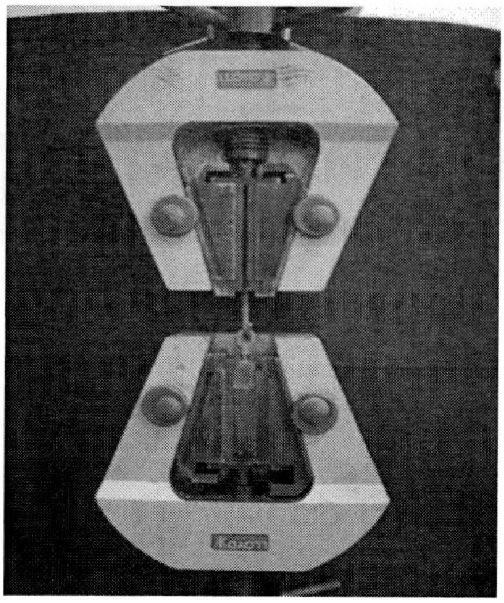

Figure 10. An example for a pull-out test setup. In this experiment the maximum pull-out force of a luting agent is tested. Note the hookup system to apply axial force.

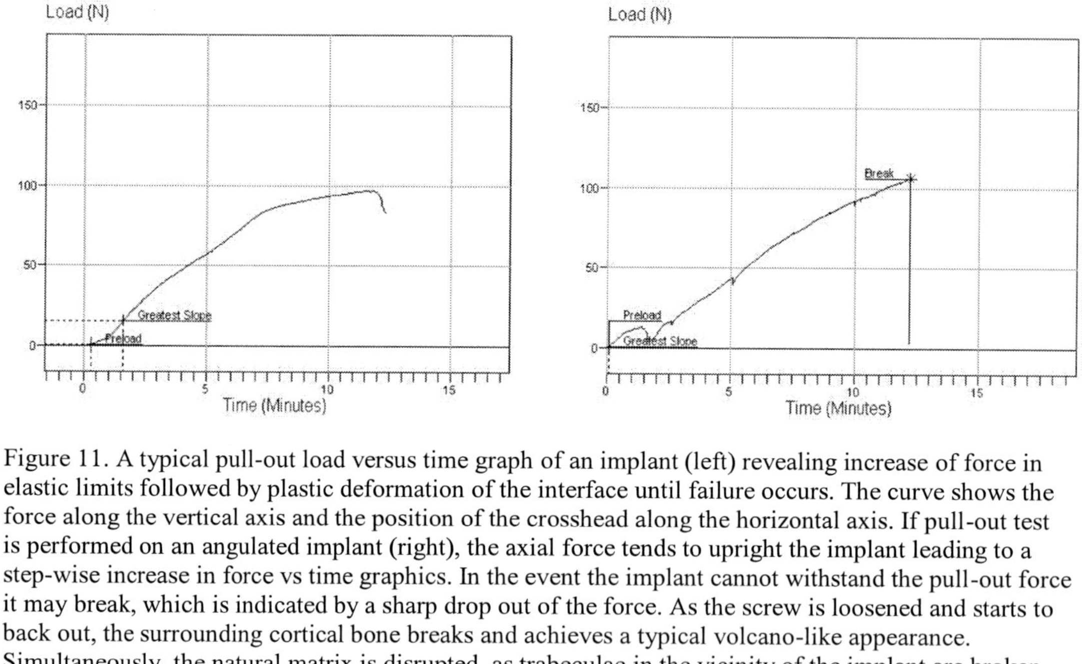

Figure 11. A typical pull-out load versus time graph of an implant (left) revealing increase of force in elastic limits followed by plastic deformation of the interface until failure occurs. The curve shows the force along the vertical axis and the position of the crosshead along the horizontal axis. If pull-out test is performed on an angulated implant (right), the axial force tends to upright the implant leading to a step-wise increase in force vs time graphics. In the event the implant cannot withstand the pull-out force it may break, which is indicated by a sharp drop out of the force. As the screw is loosened and starts to back out, the surrounding cortical bone breaks and achieves a typical volcano-like appearance. Simultaneously, the natural matrix is disrupted, as trabeculae in the vicinity of the implant are broken. Eventually, the removed implant yields a large void in bone at the termination of the test.

According to Berzins and Summer [46,48],

$$G = F((\ln R_2 - \ln R_1)/2\pi dL) \quad \text{(Eq.6)}$$

where G is the interface shear modulus, F is the applied force, R_1 is the radius of the implant, R_2 is the radius of the support hole, and d and L are as defined above. Energy absorption at the interface to failure can be measured as the area under the force-displacement curve.

Branemark [47] stated that for pull-out tests, any approximate theory for estimation of shear stresses will probably be less realistic than the simple torsion case, even though pull-out test set-up is frequently used. Therefore, the aim of the aforementioned formula was not to estimate true shear stress. Branemark stated that the formula should be interpreted as a parameter used in that specific experiment to compare the mechanical quality of bone surrounding the implant, i.e., after different healing times. This estimate of bone quality in the mechanical sense is performed by normalizing pull-out load against some measure of bone volume around the implant. This approach is very similar to estimating average shear stress and can be readily computed for an assumed cylinder with a diameter equal to the outer diameter of the threads and a length equal to the axial length of the implant in bone. The effective length is calculated as measured cortical thickness multiplied with times the relative amount of bone present in the area between the threads. An average shear stress in pull-out, τ_p, can be estimated by the formula,

$$\tau_p = F/2\pi r_o h(\%\mathrm{ba}) \quad \text{(Eq.7)}$$

where F= maximal pull-out load, r_o= outer fixture radius, h= thickness, and %ba= percentage of bone in threads. Branemark also stated that a shear modulus in pull-out, G_p can be estimated assuming that shear occurs in the region between the outer edge of the threads and the hollow hemisphere that supports the bone.

$$G_p = \tau_p/\varepsilon/r_h - r_o \quad \text{(Eq.8)}$$

where ε= extension or deformation at maximal pull-out load, r_h= radius of the hole in the support (Figure 12).

Shear modulus can also be calculated from pull-out or push-out tests in a direct manner. The bone-implant interface stiffness can be calculated as the quotient of the measured force and the deformation of the bone-implant unit, shear modulus as the quotient of shear stress and interface strain [46,49]. Shear modulus can be viewed as the slope of a tangent at the approximately linear part of the shear stress/strain curve. In order not to determine the shear modulus from an unloading curve and to take into account the viscoelastic properties of the bone-implant interface, pre-conditioning of the peri-implant bone is suggested by cyclic loading [50].

3.2. Push-in Test

Similar to previous mechanical invasive tests, implant push-in test assesses the characteristics of osseointegration by the breakpoint load at the implant-tissue interface and - to some extent-may differentiate between the effects of different implant surface topographies. These mechanical tests can give an indication of resistance to applied shearing loads (inward or outward), but such analytical assessments cannot provide information about the type and quality of osseointegration. In the push-in test, the direction of the load is the just the opposite in comparison with the pull-out test. Unlike the pull-out or push-out tests, this test has not been used extensively to assess the bone-implant interface [51,52].

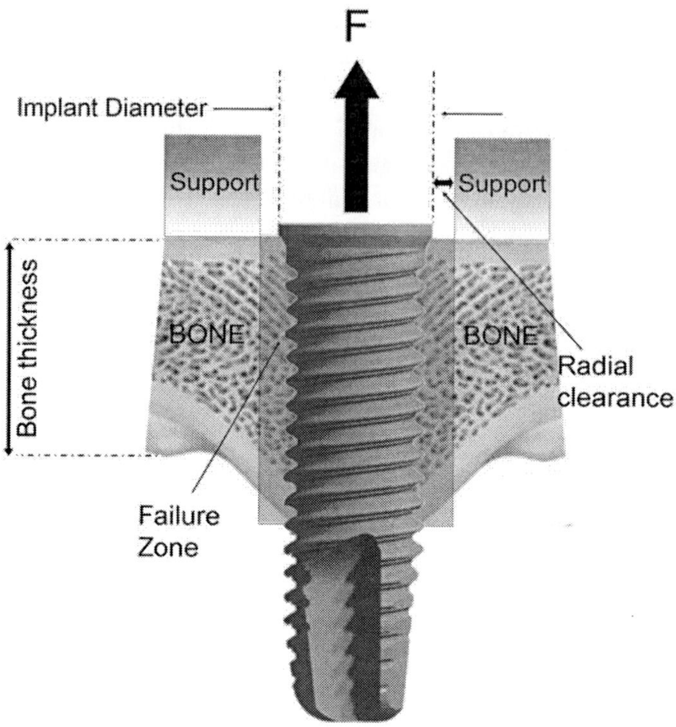

Figure 12. Schematic illustration of the pull-out test set-up of Branemark [43]. A pull-out load, F, is applied to the implant. The radial clearance between the implant and the support is a critical parameter in such experiments. The bone in the failure zone is subjected to shear forces, and tensile and compressive stresses might also occur (Adapted from Branemark [47]).

3.3. Removal Torque Test

Correct torsional testing of metallic medical screws/implants are described in ASTM standard F543-00 and ISO6475, where insertion or removal torque is measured. The torsion test determines mechanical properties such as torsional yield strength, maximum (insertion or removal) torque, and breaking angle. Removal torque test is used to probe the interface mechanics in animal studies, and rarely in humans [53]. The removal torque test has the theoretical advantage of a uniform stress field at the bone-implant interface, although the experimental setup is more elaborate than other tests cited above. Probably it was first Johansson and Albrektsson [54], who first described the technique. Removal torque test provdes quantitative information on the degree of interface healing and stiffness. As torque is applied, the interface experiences sliding, and this generates shear stresses at the interface. An average shear stress can be calculated by dividing the force applied to the interface by the interface area. This relies on the fact that an implant cannot have 100% contact with bone tissue. An estimated interfacial shear stress, τ_t, can be calculated by the following formulas [47]:

$$\tau_t = M/r/A_T(\%bc) \qquad \text{(Eq.9)}$$

where M= applied torque, r=mean thread radius, A_T= total implant arca, and %bc= percentage bone contact with A_T. Branemark [47] states that, the thickness of the interface must be known. Suppose the interface thickness is, i, then;

$$G_t = \tau_b i / 2\pi r(\varphi_b/360) \quad \text{(Eq. 10)}$$

where τ_b=shear stress in rotation at breakpoint, φ_b= angular deformation at breakpoint (the point of change in the slope from linear increase to a distinct decrease of the angle in the torque vs angle graph). Removal torque test has been extensively employed to assess stiffness of the interface at different healing times, using different implant designs and surfaces. As this test might result in failure of the interface, its clinical use to assess healing is not suitable. Removal torque testing might not be the most suitable evaluation of implant osseointegration, because the underlying biomechaincal phenomena, i.e., shear stress condition is very complex at the interface. Moreover, push-out tests seem to show a stronger correlation with implant surface roughness than torque tests and could potentially mirror the bone-implant response [55]. Nevertheless, push-out tests require the use of cylindrical implants to estimate interfacial shear stresses. As currently-available root-form implant systems include a degree of taper toward the apical part, torque testing needs to be performed and the results should be supported with histologic/histomorphometric findings [56].

REFERENCES

[1] Brånemark, P.I., Hansson, B.O., Adell, R., Breine, U., Lindström, J., Hallén, O. and Ohman, A.(1977) Osseointegrated implants in the treatment of the edentulous jaw. Experience from a 10-year period. *Scandinavian Journal of Plastic and Reconstructive Surgery*,16,1-132.

[2] Steflik, D.E., McKinney, R.V. Jr, Koth, D.L. and Singh, B.B. (1984) The biomaterial-tissue interface: a morphological study utilizing conventional and alternative ultrastructural modalities. *Scanning Electron Microscopy*,(Pt 2), 547-55.

[3] Meyer, U., Joos, U., Mythili, J., Stamm, T., Hohoff, A., Fillies, T., Stratmann, U. and Wiesmann, H.P. (2004) Ultrastructural characterization of the implant/bone interface of immediately loaded dental implants. *Biomaterials*,25,1959-67.

[4] Albrektsson, T., Brånemark, P.-I., Hasson, H.A., Lindstrom, J. (1981) Osseointegrated titanium implants. Requirements for ensuring a longlasting direct bone to implant anchorage in man. *Acta Orthopedica Scandinavia*,52,155-170.

[5] Raghavendra, S., Wood, M.C. and Taylor, T.D. (2005) Early wound healing around endosseous implants: a review of the literature. *International Journal of Oral and Maxillofacial Implants*,20,425-31.

[6] Cehreli, M., Sahin, S. and Akça, K. (2004) Role of mechanical environment and implant design on bone tissue differentiation: current knowledge and future contexts. *Journal of Dentistry*,32,:123-32.

[7] Akca, K., Sarac, E., Baysal, U., Fanuscu, M., Chang, T.L. and Cehreli, M. (2007) Micro-morphologic changes around biophysically-stimulated titanium implants in ovariectomized rats. *Head and Face Medicine*,16;3:28.

[8] Ogawa, T., Zhang, X., Naert, I., Vermaelen, P., Deroose, C.M., Sasaki, K. and Duyck, J. (2010) The effect of whole-body vibration on peri-implant bone healing in rats. *Clinical Oral Implants Research*, Oct 6. doi: 10.1111/j.1600-0501.2010.02020.x.

[9] Eser, A., Tonuk, E., Akca, K. and Cehreli, M.C. (2010) Predicting time-dependent remodeling of bone around immediately-loaded dental implants with different designs. *Medical Engineering and Physics*,32,22-31.

[10] Szmukler-Moncler, S., Salama, H., Reingewirtz, Y. and Dubruille, J.H. (1998) Timing of loading and effect of micromotion on bone-dental implant interface: review of experimental literature. *Journal of Biomedical Materials Research*,43,192-203.

[11] d'Hoedt, B., Lukas, D., Mühlbradt, L., Scholz, F., Schulte, W., Quante, F. and Topkaya, A. (1985) Periotest methods--development and clinical trial. *Deutsche Zahnärztliche Zeitschrift*, 40,113-25.

[12] Schulte,W., Lukas, D. (1993) Periotest to monitor osseointegration and to check the occlusion in oral implantology. *Journal of Oral Implantology*,19,23-32.

[13] Lukas, D. and Schulte, W. (1990) Periotest-A dynamic procedure for the diagnosis of the human perioduntium. *Clinical Physics and Physiological Measurement*,11,65-75.

[14] Faulkner, G.M., Giannitsios, D., Lipsett, A.W. andWolfaardt, J.F. (2001) The use and abuse of the Periotest for 2-piece implant/abutment systems. *International Journal of Oral and Maxillofacial Implants*,16,486-494.

[15] Elias, J., Brunski, J. and Scarton, H. (1996) A dynamic modal testing technique for noninvasive assessment of bone-dental implant interfaces. *International Journal of Oral and Maxillofacial implants*,11,728-734.

[16] Ramp, L., Reddy, M. and Jeffcoat, R. (2000) Assessment of osseointegration by nonlinear dynamic response. *International Journal of Oral and Maxillofacial implants,*15,197-208.

[17] Cehreli, M.C., Akca, K., Iplikcioglu, H. and Sahin, S. (2004) Dynamic fatigue resistance of implant-abutment junction in an internally-notched morse-taper oral implant: influence of abutment design. *Clinical Oral Implants Research*,15,459-65.

[18] Olive, J. and Aparcio, C. (1990) The Periotest method as a measure of osseointegrated oral implant stability *International Journal of Oral and Maxillofacial Implants*,5:390-400.

[19] Johansson, P. and Strid, K.G. (1994) Assessment of bone quality from placement resistance during implant surgery. *International Journal of Oral and Maxillofacial implants,*9,279-288.

[20] Friberg, B., Sennerby, L., Roos, J., Johansson, P. Strid, C.G. and Lekholm, U. (1995) Evaluation of bone density using cutting resistance measurements and microradiography: an in vitro study in pig ribs. *Clinical Oral Implants Research*,6,164-71.

[21] O'Sullivan, D., Sennerby, L., Jagger, D. and Meredith, N. (2004) A comparison of two methods of enhancing implant primary stability. *Clinical Implant Dentistry and Related Research*,6,48-57.

[22] O'Sullivan, D., Sennerby, L. and Meredith, N. (2004) Influence of implant taper on the primary and secondary stability of osseointegrated titanium implants. *Clinical Oral Implants Research*,15,474-480.

[23] Cehreli, M.C., Akca, K. and Tonuk, E. (2004) Accuracy of a manual torque application device for morse-taper implants: A technical note. *International Journal of Oral and Maxillofacial Implants*,2004;19:743-8.

[24] Cehreli, M.C., Akkocaoglu, M., Comert, A., Tekdemir, I. and Akca, K. (2005) Human ex vivo bone tissue strains around natural teeth versus immediate oral implants. *Clinical Oral Implants Research*,16,540-548.

[25] Cehreli, M.C., Comert, A., Akkocaoglu, M., Tekdemir, I. and Akca, K. (2006) Towards the limit of quantifying low-amplitude strains on bone and in coagulum around immediately-loaded implants in extraction sockets. *Medical and Biological Engineering and Computing*,44,86-94.

[26] Meredith, N., Rasmusson, L., Sennerby, L. and Alleyne, D. (1996) Mapping implant stability by resonance frequency analysis. *Medical Science Research*,24,191-193.

[27] Andreaza da Cunha, H., Francischone, E.C. and Filho, H.N. (2004) A comparison between cutting torque and resonance frequency in the assessment of primary stability and final torque capacity of standard and TiUnite single-tooth implants under immediate loading. *International Journal of Oral and Maxillofacial Implants*,19,578-585.

[28] Nkenke, E., Hahn, M., Weinzierl, K., Radespiel-Troger, M., Neukam, F.W. and Engelke, K. (2003) Dental implant stability and histomorphometry: a correlation study in human cadavers. *Clinical Oral Implants Research*,14,601-609.

[29] Nkenke, E., Lehner, B., Fenner, M., Roman, F.S., Thams, U., Neukam, F.W. and Radespiel-Tröger, M. (2005) Immediate versus delayed loading of dental implants in the maxillae of minipigs: Follow-up of implant stability and implant failures. *International Journal of Oral and Maxillofacial Implants*,20,39-47.

[30] Schliephake, H., Sewing, A and Aref, A. (2006) Resonance frequency measurements of implant stability in the dog mandible: experimental comparison with histomorphometric data. *International Journal of Oral and Maxillofacial Surgery*,35,941-946.

[31] Akkocaoglu, M., Uysal, S., Tekdemir, I., Akca, K., Cehreli, M.C. (2005) Implant design and intraosseous stability of immediately placed implants: a human cadaver study. *Clinical Oral Implants Research*,16,202-209.

[32] Akca, K., Akkocaoglu, M., Comert, A., Tekdemir, I. and Cehreli, M.C. (2007) Bone strains around immediately loaded implants supporting mandibular overdentures in human cadavers. *International Journal of Oral and Maxillofacial Implants*,22,101-9.

[33] Ito, Y., Sato, D., Yoneda, S., Ito, D., Kondo, H. and Kasugai, S. (2008) Relevance of resonance frequency analysis to evaluate dental implant stability: simulation and histomorphometrical animal experiments. *Clinical Oral Implants Research*,19,9–14.

[34] Alsaadi, G., Quirynen, M., Michiels, K., Jacobs, R and van Steenberghe, D. (2007) A biomechanical assessment of the relation between the oral implant stability at insertion and subjective bone quality assessment. *Journal of Clinical Periodontology*,34,359-366.

[35] Degidi, M., Daprile, G. and Piattelli, A. (2010). Determination of primary stability: a comparison of the surgeon's perception and objective measurements. *International Journal of Oral and Maxillofacial Implants*,25,558-61.

[36] Cehreli, M.C., Karasoy, D., Akca, K. and Eckert, S.E. (2009) Meta-analysis of methods used to assess implant stability. *International Journal of Oral and Maxillofacial Implants*,24,1015-32.

[37] Friberg, B, Sennerby, L., Meredith, N. and Lekholm, U. (1999) A comparison between cutting torque and resonance frequency measurements of maxillary implants. A 20-month clinical study. *International Journal of Oral and Maxillofacial Surgery*,28,297-303.

[38] Cowin SC. (Ed) Bone Mechanics Handbook, CRC Press, 2001;20:9-11.

[39] Cömert, A., Kökat, A.M., Akkocaoğlu, M., Tekdemir, I., Akça, K. and Cehreli, MC. (2009) Fresh-frozen vs. embalmed bone: is it possible to use formalin-fixed human bone for biomechanical experiments on implants? *Clinical Oral Implants Research*,20,521-5.

[40] Anderson, S.D. (2006) Practical light embalming technique for use in the surgical fresh tissue dissection laboratory. *Clinical Anatomy*,19,8-11.

[41] Fox, C.H., Johnson, F.B., Whiting, J. and Roller, P.P.(1985) Formaldehyde fixation. *Journal of Histochemistry and Cytochemistry*,33,845-853.

[42] Viidik, A. and Lewin, T. (1966) Changes in tensile strength characteristics and histology of rabbit ligaments induced by different modes of postmortal storage. *Acta Orthopedica Scandinavia*,37,141–155.

[43] Edmondston, S.J., Singer, K.P., Day, R.E., Breidahl, P.D. and Price, R.I. (1994) Formalin fixation effects on vertebral bone density and failure mechanics:an in vitro study of human and sheep vertebrae. *Clinical Biomechanics,* 9,175-179.

[44] Currey, J.D., Brear, K., Zioupos, P. and Reilly, G.C.(1995) Effect of formaldehyde fixation on some mechanical properties of bovine bone. *Biomaterials*,16,1267-1271.

[45] Wilke, H.-J., Krischak, S. and Claes, L.E. (1996) Formalin fixation strongly influences biomechanical properties of the spine. *Journal of Biomechanics*,29,1629-1631.

[46] Berzins, A., Sumner, D.R. (2000) Implant pushout and pullout tests. In: An, Y., Draughn, R.A. (Eds.), Mechanical Testing of Bone and the Bone-Implant-Interface. CRC Press, Boca Raton. 463-476.

[47] Branemark, R. (1996) A biomechanical study of osseointegration. In vivo measurements in rat, rabbit, dog and man. Thesis, Göteborg University:78-80.

[48] Berzins, A., Shah, B., Weinans, H., and Sumner, D.R. (1997) Nondestructive measurements of implant–bone interface shear modulus and effects of implant geometry in pullout tests. *Journal of Biomedical Materials Research*,34,337.

[49] Keller, T.S., Liebschner, M.A. (2000) Tensile and compression testing of bone. In: An, Y., Draughn, R.A. (Eds.), Mechanical Testing of Bone and the Bone-Implant-Interface. CRC Press, Boca Raton. 175–205.

[50] Linde, F., Hvid, I., 2000. Nondestructive mechanical testing of cancellous bone. In: An, Y., Draughn, R.A. (Eds.), Mechanical Testing of Bone and the Bone-Implant-Interface. CRC Press, Boca Raton, pp. 151–157.

[51] Evans, M., Spencer, M., Wang, Q., White, S.H. and Cunningham, J.L. (1990) Design and testing of external fixator bone screws. *Journal of Biomedical Engineering*,12,457-462.

[52] Brosh, T., Persovski, Z. and Binderman, I. (1995) Mechanical properties of bone implant interface: an in vitro comparison of the parameters at placement and at 3 months. *International Journal of Oral and Maxillofacial Implants*,10,729-735.

[53] Cehreli, M., Akkocaoglu, M. and Akca, K. (2006) Numerical simulation of in vivo intraosseous torsional failure of a hollow-screw oral implant. *Head and Face Medicine*,4,2:36.

[54] Johansson, C. and Albrektsson, T. (1987) Integration of screw implants in the rabbit: a 1-year follow-up of removal torque of titanium implants. *International Journal of Oral and Maxillofacial Implants,*2,69-75.

[55] Thompson, J.I., Gregson, P.J., Revell, P.A. (1999) Analysis of push-out test data based on interfacial fracture energy. *Journal of Material Science: Materials in Medicine*,10,863-868.

[56] Shalabi, M.M., Gortemaker, A., Van't Hof, M.A., Jansen, J.A. and Creugers, N.H. (2006) Implant surface roughness and bone healing: a systematic review. *Journal of Dental Research*,85,496-500.

In: Biomechanics of Dental Implants
Editor: Murat Çehreli

ISBN 978-1-62100-780-7

Chapter 5

Animal Experimental Findings on the Effect of Mechanical Load on Peri-Implant Tissue Differentiation and Adaptation

Katleen Vandamme, Ignace Naert and Joke Duyck

1. Introduction

The use of endosseous oral implants, fixed into bone by means of direct bone contact, became a predictable treatment modality during the eighties. Since then, the goal of achieving an optimal bone-implant interface or speeding up the biological healing response has been approached by optimization of implant surface micro-roughness, chemical composition, surface energy and wettability and by application of mechanical stimuli during the healing stage. It has been shown that the immediate implant loading protocol might achieve similar clinical success rates as those noted in the traditional delayed protocols [1-3]. Nevertheless, in order to achieve this predicatively high success rates, a careful selection of cases for the immediate loading treatment plan is required, with optimization of the surgery-, host-, implant- and occlusion-related factors [1,4].

Loading during healing does not necessarily lead to fibrous encapsulation of implants. Some degree of micromotion is tolerated [5], but a certain threshold should not be surpassed [6-8]. Inversely, a certain amount of loading may even enhance osseointegration. Experimental and clinical evidence of the anabolic effect of mechanical loading on fracture healing might assist in this research track [9-13]. Since trabeculae are damaged both in fracture healing and peri-implant healing, fracture healing can be used as a model for peri-implant tissue regeneration. Excessive movement is in general counterproductive for fracture healing. However, increased interfragmentary movements are capable of enhancing fracture healing. The loading parameters possessing an osteogenic effect on fracture healing in specific bone conditions are only partly explored [12,13]. Compared with the complicated mechanical environment of a fracture callus, the model of implant osseointegration allows characterization and quantification of bone formation in a location-specific highly reproducible manner and is therefore a valuable model for a better understanding of the mechanobiology governing other clinical contexts, such as fracture healing.

In order to explore the stimulating effect of mechanical loading on the bone regeneration at the tissue-implant interface, rigorous well-controlled experimental testing is required. The limitation of bone *in vitro* culture is the lack of controlled physiological loading. No *in vitro* cell culture system is able to produce loading that simulates the *in vivo* situation. Considering also the fact that animal data have a much higher clinical relevance compared to *in vitro* data [14], the use of animal models is an essential step in the testing of loading protocols prior to implementation into the clinic. Furthermore, the use of mathematical models (*in silico*) in biology and medicine has increased enormously the last decade. Mathematical models can propose and test possible regeneration mechanisms, but corroboration of these models by comparison with experimental data still is necessary to determine their predictive power. The selected *in vivo* models should adopt calibrated well-controlled mechanical stimuli enabling the identification of the impact of the specific parameters, i.e. the nature, magnitude, rate, frequency and duration of the stimulus, on mechanical mediated bone tissue formation and adaptation.

In this chapter, the animal experimental findings on the mechanobiological rules governing implant osseointegration during immediate, early and delayed implant loading, conducted by the BIOMAT Research Cluster of the Department of Prosthetic Dentistry of the Katholieke Universiteit Leuven (Belgium) will be reviewed.

2. Mechanobiology at the Tissue-Implant Interface

2.1. Definition

Based on the ideas of Roux [15], who suggested that the regulation of the biological processes depend on signals to cells generated by mechanical loading, the interaction of biology and mechanics is nowadays referred to as mechanobiology. The central question in mechanobiology is how load-bearing tissues like bone are produced, maintained and adapted by cells as an active response to biophysical stimuli in their environment.

When assessing the biological relevance of stress and strain, “Wolff’s law” [16] – the arrangement of the bone tissues is such that bones are organized optimally to resist to loads imposed by functional activities – emerges. This law, which never has been proven, has been recasted in various ways using modern engineering terms. For example, Frost [17] presumes that bone has a control system that acts via modelling and remodelling to maintain constancy of the mechanical environment of cells when external loading conditions change. The strain magnitude was implemented as the main determinant of bone formation, stating that 1500-2500 $\mu\varepsilon$ is the range of “minimum effective strain for mechanically controlled bone remodelling” (mechanostat principle). A key scientific challenge in the field of implant dentistry is to establish whether and how the mechanical morphogenesis and homeostasis of bone works in the peri-implant environment. Confounding experimental variables such as implant size, shape, material, topography, etc. and other mechanical parameters than the strain magnitude should be considered.

2.2. Theoretical Mechanobiology

2.2.1. Mechanobiological Models of Tissue Differentiation

Using simple experimental models of a fracture callus, Pauwels [18] proposed a quantitative hypothesis for the influence of mechanical factors on tissue differentiation. Based on the observations that shear forces result in a change of shape of tissue, while hydrostatic pressure results in a change of volume, Pauwels stated that shear forces stimulate mesenchymal cell differentiation into fibroblasts while hydrostatic compression stimulates differentiation into chondrocytes. Shear deformation may influence tissue differentiation by deforming cells, thus altering gene expression and synthetic activity. Volumetric strains may affect cell differentiation by inhibiting capillary blood flow, resulting in a decreased tissue oxygen tension. No specific mechanical stimulus for the formation of bone was identified. Based on this model, Carter et al. [19] and Giori et al. [20] further developed the concepts relating tissue differentiation to mechanical loading. They specifically theorized the importance of cyclic tissue loading. With good vascular supply, low levels of cyclic hydrostatic stress and tensile strain permit the direct formation of bone. However, even in a low stress environment, bone cannot be formed without a sufficient blood supply. Instead, cartilage is formed.

Other mechanoregulatory models were proposed and were tested for experimental situations such as fracture healing [21-24] or distraction osteogenesis [25-27]. None of these models however takes the biological aspects of bone regeneration into account. Bailon-Plaza and van der Meulen [28] proposed a bioregulatory model of bone regeneration. This model assumes an ideal mechanical environment and describes fracture healing as a process regulated solely by osteogenic and chondrogenic growth factors. Some of the most recent models tried to combine both mechanical and biological factors [29]. More sophisticated finite element models related to peri-prosthetic interfaces and incorporating biological parameters like growth factors are under development [30].

2.2.2. Mechanobiological Model for Peri-Implant Tissue Differentiation

The theories of Carter et al. [19] and Giori et al. [20] described above may explain why poorly anchored implants subjected to cyclic loads develop a soft tissue interface. It is possible that the relative motion at the interface produce large distortional changes in the mesenchymal cells, leading to a differentiation of fibroblasts instead of osteoblasts. The fact that the tissues surrounding oral implants do not differentiate into chondroid tissue can be explained by the adequately oxygenated interfacial tissues.

Wiskott and Belser [31] proposed, after reviewing experimental set-ups on the osteogenic response in relation to the stress magnitude and time dependency, a theoretical relationship between bone formation and the magnitude of interfacial micro-movement (Figure 1). Without interfacial motion, osteoblasts will form bone. When subjected to minute shear stresses, mineralized material deposition will increase. Beyond an optimum value, bone formation will decline and ultimately cease completely. However, neither the actual magnitude of micro-motion nor other mechanical parameters, implant-related characteristics or timing of the onset of loading after installation were considered.

2.3. Experimental Mechanobiology

2.3.1. Mechanical Parameters Controlling Bone Biology

The parameters involved in mechanically mediated bone adaptation are well understood. Results from a number of animal experiments have suggested that the magnitude [32,33], rate [34], distribution [35-40], frequency [41-45] and duration [46-47] of the dynamic strain stimulus [48] are among the important determinants in bone adaptation.

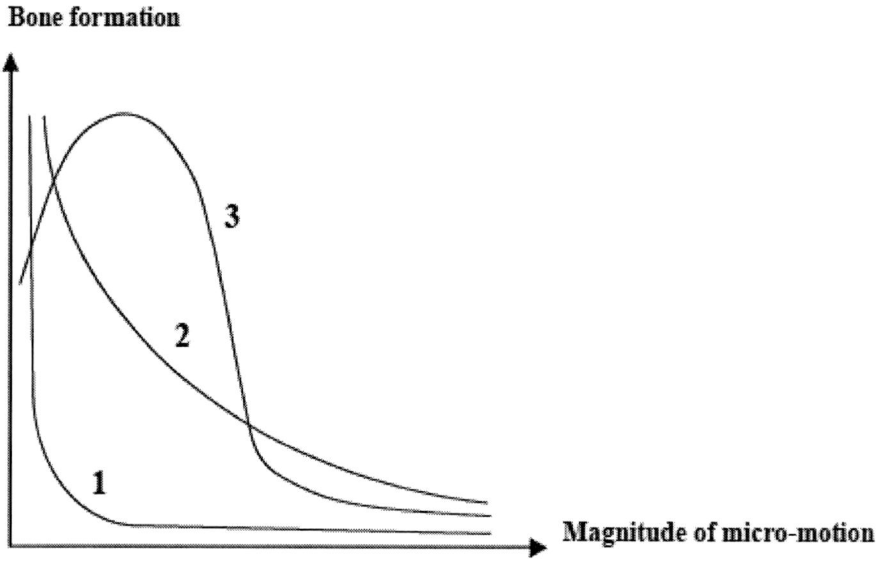

Figure 1. Theoretical relationships between bone formation and magnitude of interfacial movement. (1) Complete immobilization, which is unrealistic. Pattern (2) is possible but there are indications that some degree of micro-movement is beneficial to osseointegration. Pattern (3) (Pilliar et al. [5] would match the proposed concept of stress that induces strain (i.e. deformation and micro-movement) which in turn induces bone formation. (from Wiskott and Belser [31], with permission).

However, which of these biomechanical parameters specifically trigger the initial healing process is hardly defined.

2.3.1. Mechanical Parameters Controlling Bone Biology at the Tissue-Implant Interface

The findings from the basic mechanobiologic concepts suggest that bone formation at the implant surface is permitted and perhaps promoted in areas exposed to low to moderate tensile strain. The effect of micro-motion – a relative displacement between an implant and surrounding tissues – on the bone-dental implant interface has been discussed by Szmukler-Moncler et al. [8]. The authors gathered data from animal studies on the histological status of implants loaded during the healing period. Reports were summarized indicating that early loading induces fibrous tissue encapsulation as well as reports claiming just the opposite. Experimental evidence was found that the extent of micro-motion during the initial healing phase interferes with the concomitant bone reaction. Attempts have been made to define the threshold of micro-movement during the healing phase at the interface [5,7]. The literature suggests that for implants with a bio-inert surface, e.g. turned titanium, the critical threshold for successful implant integration lies somewhere between 50 and 150 μm. However, not only the magnitude of micro-motion but rather the interaction of different mechanical parameters (magnitude, frequency, rate, etc), together with the timing of loading relative to the time of implantation, the implant design and surface properties, and the properties of the host tissues will determine how the peri-implant tissues will respond to the mechanical stimuli. Only an efficient force transfer between the implant and the surrounding tissues will result in a good biomechanical coupling and positively affect bone formation and adaptation, ultimately leading to successful implant integration.

In order to investigate the impact of these variables, different animal models developed by the BIOMAT Research Cluster of the Department of Prosthetic Dentistry (Katholieke Universiteit Leuven) are presented here.

3. Mechanobiology at the Tissue-Implant Interface during Immediate Implant Loading

In clinical situations, immediate implant loading has been defined as "a situation where the suprastructure is attached to the implants and placed in occlusion within 72 h after implant installation" [49]. For the conducted animal studies described below, immediate implant loading was defined as a well-controlled loading of the implant by a custom-designed external loading device starting at the day itself or 1 day after implant installation.

3.1. Animal Model 1: The Rabbit Bone Chamber Model

3.1.1. Animal Model and Sample Processing

The rabbit bone chamber methodology has been developed for investigation of the sensitivity of peri-implant tissue differentiation to well-controlled immediate implant loading. The model is described in detail in Duyck et al. [50]. The bone chamber primarily consists of dual-structure perforated hollow cylinders with a centrally positioned implant (Figure 2). In a first stage, the outer cylinder – called the outer bone chamber – is installed in the proximo-medial tibia of New Zealand White rabbits and is allowed to osseointegrate. The bone chamber has an outer diameter of 1 cm and on average 9 mm – of a total height of 12 mm – is inserted into the bone. During this healing period, the outer bone chamber is filled with a solid Teflon cylinder to prevent tissue invasion through the perforations. After 3 months of healing, this Teflon cylinder is removed and replaced by an inner cylinder – called the inner bone chamber. Through the perforations, bone grows into the bone chamber. The implant, located in the centre of the bone chamber, is displaced in a well-controlled manner by means of a loading device. During implant loading, the implant slides in a Teflon bearing. At the end of the experiment, the implant and its surrounding tissues are harvested by removing the inner form the outer bone chamber. The outer bone chamber is re-used for further experimentation with a new inner bone chamber. Owing to its double structure, repeated sampling of the regenerated tissue in the bone chamber is possible. This allows conducting several experiments within the same animal. The custom-designed loading device can easily be screw-mounted onto the bone chamber and be connected to the implant. The loading device

(Figure 2) consists of a piezo translator (preloaded closed loop LVPZT translator, P-841.60, ALT, Best, The Netherlands), which can induce a displacement of up to 90 μm and a load cell (XFTC 100-M5M-1000N, FGP Sensors, Les Clayes sous Bois, France) with a capacity of 1000 N in tension and 100 N in compression. The displacement is controlled by a strain gauge onto the piezo translator itself. A closed loop control assures the required displacement. The load can be applied with a frequency of up to 50 Hz. A controlled axial load or displacement applied onto the implant induces shear stresses and strains in the tissue surrounding the implant. Variations in load parameters such as displacement of the implant, forces onto the implant, frequency of loading, and amount and distribution of load cycles are possible. Both input (defined load parameters) and output (applied forces on and displacement of the implant) are controlled by software written in TestPoint (Norton, MA, USA).

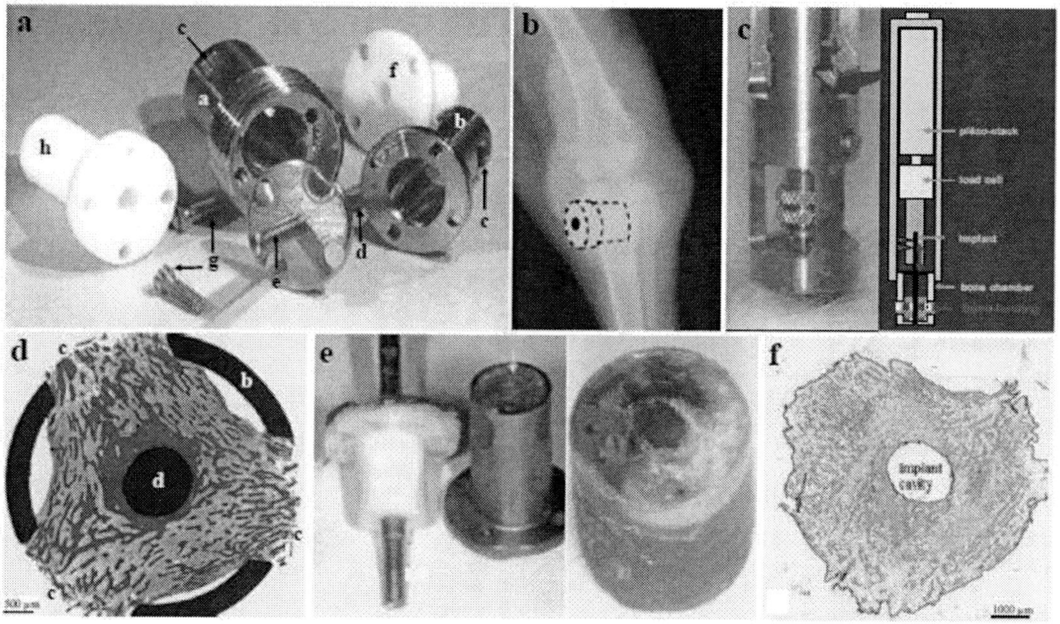

Figure 2. a. Experimental bone chamber model (Duyck et al. [50]). (a) outer bone chamber, (b) inner bone chamber, (c) perforations, (d) implant, (e) point of implant fixation, (f) Teflon bearing, (g) screws (to hold the inner into the outer bone chamber), (h) Teflon cylinder (space maintainer during integration of the outer bone chamber); b. Radiological view of a bone chamber, installed in the proximal tibia of a New Zealand white rabbit; c. Loading device (ESAT-MICAS, K.U.Leuven) screw-retained onto the outer bone chamber, in situ on the medial side of the proximal tibia (left); Schematic drawing of the bone chamber and loading device components (right); d. Histological section of an MMA-embedded inner bone chamber with loaded implant harvested 9 weeks after installation (Vandamme et al. [51]), (b) inner bone chamber, (c) perforations, (d) implant. Stevenel's blue and Von Gieson's picrofuchsin stain. e. Inner bone chamber content after removal of metal parts and embedded in paraffin (Duyck et al. [52]). f. Haematoxylin and eosin-stained section cut perpendicular to the implant axis, showing the central cavity where the implant used to be. The tissue area represents all available tissue in the bone chamber, outlined in blue (Duyck et al. [52]).

In order to obtain immediate implant loading conditions, loading of the implants was initiated the same day as the inner bone chamber installation. Evaluation of the tissue growth into the bone chamber lumen and the bone formation around and in contact with the implant was performed through histological and histomorphometrical analyses on methylmetacrylate (MMA)- or paraffin-embedded samples. The specimens allocated for MMA embedding were fixed in a $CaCO_3$-buffered formalin solution and dehydrated in an ascending series of ethanol concentrations over 18 days. Embedding was performed by infiltration of a benzoylperoxide (0.018 %)-methylmetacrylate solution over 7 days. Sectioning transversely perpendicular to the implant and grinding were performed using a diamond saw. Five sections at the level of the perforations for each harvested bone chamber were reduced to a final thickness of 20 to 30 µm by micro-grinding and polishing using a cutting-grinding device. The sections were stained with a combination of Stevenel's blue and Von Gieson's picrofuchsin (Figure 2d). The specimens allocated for paraffin embedding were fixed in 2% paraformaldehyde after sacrifice of the animals. These tissue blocks were decalcified in 0.5M EDTA (pH 7.4)/phosphate buffered saline (PBS) at 4°C before dehydration and embedded in paraffin. The implants were gently removed from the paraffin-embedded tissue blocks and the tissue plug was removed from the inner bone chamber after removing the inner bone chamber bottom. The effect of this implant removal was evaluated in a previous study [53] by scanning electron microscopy. Only very few small sponge-like structures without cells were observed, probably consisting of fibrous tissue or cell remnants. The remaining surface of the implants was comparable with the surface of a sterile implant. The samples were re-embedded immediately after implant removal to refill the central cavity and cut to 4µm-thick sections perpendicular to the implant axis. Three sets of 2 sections at different levels along the implant were taken per sample and stained with haematoxylin and eosin (Figure 2f).

Histological examinations were performed under a light microscope (Leitz Laborlux S, Wetzlar, Germany) at a magnification of 40x, 100x and 400x. The assessments of the histomorphometrical proportions were performed by means of a high-sensitivity colour video camera (JVC TK-1280E, JVC, Ibaraki-ken, Japan; AxioCam MRc5, Zeiss, Göttingen, Germany) mounted on the light microscope and by means of a colour image analyzing histometry software package (Image Pro Plus®, Media Cybernetics, Silver Spring, MD, USA). Besides the well-known parameter bone-to-implant contact (BIC; %), the parameter 'bone area fraction' (BAF; %) was recorded and defined as the percentage of a defined area occupied by bone tissue. BAF was recorded for the entire bone chamber area. On the paraffin-embedded sections (study described in paragraph 3.1.2.3), BIC could not be quantified due to the implant removal. The BAF in the 100 µm zone around the implant was therefore used as a measure of peri-implant bone formation. In the numerical study reported in paragraph 3.1.3.2., BAF was also additionally calculated for a 200 µm circumferential zone around the implant Finally, the ratio of mineralized trabeculae in all these zones was calculated (MB, 'mineralized bone'; %).

3.1.2. Experimental Studies

Several bone chamber studies (each of them consisting of several experiments) were performed, exploring the effect of immediate loading, of the implant design and surface properties in immediate loading conditions and of the magnitude of micro-motion on the peri-implant tissue response. The investigated variable(s) in each study are highlighted in bold in the column headings.

3.1.2.1. Effect of Immediate Implant Loading on Peri-Implant Tissue Differentiation

A first study was conducted to explore whether the mechanobiological rule of stimulation of bone formation by moderate mechanical loading also holds true for the regenerating tissue surrounding implants. Immediate loading experiments were designed for a cylindrical unthreaded turned c.p. titanium implant (Ra-value: 0.45 μm) [54]. The implant was loaded twice a week by a displacement of maximum 50 μm with a load frequency of 1 Hz for up to 800 load cycles, for 12 consecutive weeks (Table 1). The tissue response around immediately loaded *versus* unloaded implants was compared. It was hypothesized that this controlled immediate loading of the implant would enhance the peri-implant bone formation.

Table 1. Loading parameters for 2 experiments with an unthreaded turned implant, installed in the tibia of mature New Zealand white rabbits (n=14). The varying parameter is highlighted in bold

	Duration of the experiment	# loading sessions/wk	**Displacement (μm)**	Frequency (Hz)	# cycles/ loading session
Exp1:12w-L	12 wk	2	30-50	1	400-800
Exp2:12w-U	12 wk	/	/	/	/

12w-L, immediate loading for 12 weeks; 12w-U, unloading for 12 weeks.

The histomorphometrical results revealed that when immediate loading was applied to the implant, significantly more bone was formed in the bone chamber lumen (BAF and MB) and at the implant surface (BIC). This study validated the bone chamber model as being sensitive and suitable for assessment of the impact of mechanical stimuli on peri-implant tissue differentiation and bone healing. Controlled immediate implant loading with moderate stimulation conditions accelerated the tissue mineralization in the vicinity and at the surface of the implant, thereby confirming the study hypothesis.

3.1.2.2. Effect of Implant Design on Peri-Implant Tissue Differentiation

The role of the implant geometry, a determining factor for the anchorage and organisation of the tissues at the implant surface, in the stimulation of bone formation in immediate implant loading was investigated in another study with the bone chamber model [55]. Immediate loading experiments were designed for an unthreaded *versus* threaded turned c.p. titanium implant (Ra-value: 0.70 μm). The implant was loaded three times a week by a displacement of 30 μm with a load frequency of 1 Hz for 400 load cycles. The tissue response around immediately loaded *versus* unloaded implants with the 2 different implant designs for a time interval of 9 weeks was compared (Table 2). It was hypothesized that the enhancement of peri-implant tissue mineralization through mechanical stimuli is dependent on the implant design.

Analogous to the results of previous study, it was observed that more bone was formed in the bone chamber lumen (BAF and MB) and at the implant interface (BIC) when loading the implant. Moreover, more BIC was observed for an immediately loaded threaded implant compared to an unthreaded one. The threaded implant design promoted osseointegration by providing a favorable local mechanical environment for bone formation at the implant surface compared with the unthreaded implant design. The study hypothesis was affirmed. In terms of clinic considerations, threaded implants do not only favor primary stability, but also improve the so-called secondary stability achieved by physiological loading compared to the unthreaded implant design.

Table 2. Loading parameters for 3 experiments with an unthreaded *versus* threaded implant, installed in the tibia of mature New Zealand white rabbits (n=10). The varying parameters are highlighted in bold

	Implant design	Duration of the experiment	# loading sessions/wk	**Displacement (μm)**	Frequency (Hz)	# cycles/ loading session
Exp1: nTL	Unthreaded	9 wk	3	30	1	400
Exp2: TL	Threaded	9 wk	3	30	1	400
Exp3: TU	Threaded	9 wk	/	/	/	/

nTL, unthreaded immediately loaded implant; TL, threaded immediately loaded implant; TU, threaded unloaded implant.

3.1.2.3. Effect of Implant Surface Roughness on Peri-Implant Tissue Differentiation

In the studies dealt with before, the implant surface roughness was kept low in order to investigate the pure effect of mechanical loading, without taking the risk of the tissue response to mechanical loading being overruled by implant surface modifications. Two series of bone chamber experiments were undertaken to investigate the effect of the implant surface roughness on the peri-implant tissue differentiation in immediate loading.

For the first series of experiments, threaded turned (Ra-value: 0.70 μm) *versus* threaded roughened (Friadent® plus, Mannheim, Germany; Ra-value of 2.75 μm) c.p. titanium implant were used. The tissue response around immediately loaded *versus* unloaded implants with the 2 different surface properties for a time interval of 9 weeks was compared (Table 3) [56]. The hypothesis to be tested was that roughened implants enhance the peri-implant bone tissue formation compared to implants with turned surfaces for both unloaded and immediately loaded conditions.

For these experiments, it was observed that the surface topography had no impact on the bone tissue response inside the bone chamber area (BAF and MB). At the implant interface however, a roughened surface seemed to generate *per se* (without loading) tissue interlocking and favored the establishment of osseointegration (presence/absence of BIC). However, the bone-forming cells' response to an altered surface micro-topography was mechanically state dependent: well-controlled immediate implant loading stimulated the establishment of osseointegration for turned as well as for roughened threaded implants, thereby attenuating the influence of the surface micro-topography. The study hypothesis could only partially be confirmed.

The second series of experiments [52] used an identical study protocol as the one of previous series of experiments, except for the duration of (un)loading (6 instead of 9 wks) (Table 4), and the sample processing (paraffin- instead of PMMA-embedded sections). The same hypothesis as in the previous series of experiments was formulated.

Table 3. Loading parameters for 4 experiments with a threaded turned *versus* a threaded roughened implant, installed in tibia of mature New Zealand white rabbits (n=10). The varying parameters are highlighted in bold

	Implant roughness	Duration of the experiment	# loading sessions/wk	**Displacement (µm)**	Frequency (Hz)	# cycles/ loading session
Exp1: nRU	0.70 µm	9 wk	/	0	/	/
Exp2: nRL	0.70 µm	9 wk	3	30	1	400
Exp3: RU	2.75 µm	9 wk	/	0	/	/
Exp4: RL	2.75 µm	9 wk	3	30	1	400

nRU, turned unloaded implant; nRL, turned immediately loaded implant; RU; roughened unloaded implant; RL, roughened immediately loaded implant.

Table 4. Loading parameters for 4 experiments with a threaded turned *versus* threaded roughened implant, installed in tibia of mature New Zealand white rabbits (n=6). The varying parameters are highlighted in bold

	Implant roughness	Duration of the experiment	# loading sessions/ wk	**Displacement (µm)**	Frequency (Hz)	# cycles/ loading session
Exp1: nRU	0.45 µm	6 wk	/	0	/	/
Exp2: nRL	0.45 µm	6 wk	3	30	1	400
Exp3: RU	2.75 µm	6 wk	/	0	/	/
Exp4: RL	2.75 µm	6 wk	3	30	1	400

nRU, turned unloaded implant; nRL, turned immediately loaded implant; RU; roughened unloaded implant; RL, roughened immediately loaded implant.

The results of the histomorphometric analyses revealed that BAF was not affected by the experimental conditions. A higher ratio of mineralised trabeculae (MB) in the bone chamber lumen in response to loading was observed, but not in response to the surface roughness. In the 100 µm wide peri-implant zone however, the surface roughness seemed to promote the formation and mineralisation of the bone trabeculae around loaded implants. The latter result is in contradiction with the results of the 1st series of experiments, where a masking of the effect of the implant surface roughness by the load-related bone reaction was suggested. Both studies differed not only in analysis approach (PMMA-embedded sections enabling the measurement of BIC (1st series of experiments) *versus* paraffin-embedded sections excluding the recording of BIC, with MB in a 100 µm zone as alternative (2nd series of experiments)), but also in the duration of loading (9 *versus* 6 weeks respectively). Based on the consistent findings in the performed bone chamber studies of the bone-stimulating effect of loading, implant loading in the 2nd series of experiments, with an additional 3 weeks would probably have resulted in more bone surrounding and in contact with the implant, both for the turned and the roughened implant. More bone in contact with the implant would result in a similar biomechanical coupling for both implant designs and would blur the effect of the implant surface roughness. As long as a good mechanical coupling between a turned implant and its surrounding tissues is not established, the stimulation of initial peri-implant bone formation is restricted for the immediately loaded threaded turned implant compared to the roughened one.

3.1.2.4. Effect of Magnitude of Loading on Peri-Implant Tissue Differentiation

Determination of the threshold of micro-motion for successful peri-implant tissue regeneration was aimed for in another 2 series of bone chamber experiments. The 1st series of experiments used an unthreaded turned titanium implant, the 2nd one a threaded roughened titanium implant. Owing to the different implant designs and surface properties, the force transfer from the loaded implant to the surrounding tissues is certainly different and different values for the micro-motion upper threshold promoting implant osseointegration are expected.

For the first series of experiments [57], the tissue response around unloaded *versus* immediately loaded unthreaded turned implants (Ra-value: 0.45 μm) with increasing loading amplitude up to 90 μm for a time interval of 6 weeks was considered (Table 5). For the second series of experiments [51], the tissue response around unloaded *versus* immediately loaded threaded roughened implants (Friadent® plus, Mannheim, Germany; Ra-value of 2.75 μm) with increasing loading amplitude up to 90 μm for a time interval of 9 weeks was compared (Table 6).

Conflicting results were obtained for the amount of bone formed in the bone chamber lumen (BAF) for the untreaded turned implant: equal volumes of bone were formed for the unloaded and the 90 μm-loaded implants, both significantly different with the lower BAF values when applying 30 μm implant micromotion.

Table 5. Loading parameters for 4 experiments with an unthreaded turned titanium implant, installed in tibia of mature New Zealand white rabbits (n=10)

	Duration of the experiment	# loading sessions/wk	Displacement (μm)	Frequency (Hz)	# cycles/ loading session
Exp1: 0 μm	6 wk	/	0	/	/
Exp2: 30 μm	6 wk	2	30	1	800
Exp3: 60 μm	6 wk	2	60	1	800
Exp4: 90 μm	6 wk	2	90	1	800

0 μm, unloaded implant; 30 μm, immediately loaded implant with a displacement of 30 μm; 60 μm, immediately loaded implant with a displacement of 60 μm; 90 μm, immediately loaded implant with a displacement of 90 μm.

Table 6. Loading parameters for 3 experiments with a threaded roughened titanium implant, installed in tibia of mature New Zealand white rabbits (n=10)

	Duration of the experiment	# loading sessions/wk	Displacement (μm)	Frequency (Hz)	# cycles/ loading session
Exp1: 0 μm	9 wk	/	0	/	/
Exp2: 30 μm	9 wk	3	30	1	400
Exp3: 90 μm	9 wk	3	90	1	400

0 μm, unloaded implant; 30 μm, immediately loaded implant with a displacement of 30 μm; 90 μm, immediately loaded implant with a displacement of 90 μm.

The number of mineralized bone trabeculae (MB) increased with increasing micromotion with significantly higher values for the 60 μm- and 90 μm-conditions compared to the unloaded situations. On the other hand, micromotion of an unthreaded turned implant exerted a negative influence on the establishment of BIC. It is suggested that the impairment of osseointegration in immediate loading was caused by an unfavourable biomechanical coupling between the unthreaded turned implant and its surrounding tissue.

For the threaded roughened implant however, straightforward results in favour of enhancement of tissue mineralization in the vicinity and at the surface of the implant through immediate loading have been observed. An increasing bone inductive capacity of immediate loading was observed with increasing micromotion up to 90 μm.

3.1.3. Numerical Simulation Studies

Current research in the field of (oral) implant osseointegration is predominantly of an experimental nature. Owing to its arduous character, mathematical modelling of biological processes as a parallel approach has been introduced in several biomedical domains (e.g., cancer research, drug development). Mathematical models allow investigators not only to optimize the experimental set-up, but also to test new hypotheses and to design optimal treatment strategies, which can subsequently be verified experimentally.

3.1.3.1. Numerical Model of the Bone Chamber

For the computational modelling of the bone chamber set-up, an axisymmetric (longitudinal section through the perforation) (for the 2 performed numerical studies described below) and a 3D Finite Element (FE) model (only for the study described in paragraph 3.1.3.3) of a threaded and unthreaded implant and the tissue inside the bone chamber were created. An example of the 3D finite element model and the applied boundary conditions are shown (Figure 3a). Implant loading took place in an axial displacement controlled manner and was simulated by the imposition of displacement boundary conditions. The tissue was modelled to be connected to the implant and the bone surrounding the chamber, but not to the inner bone chamber walls and the apex of the implant. At the perforation sites, free fluid flow (zero pore pressure) was allowed.

All tissues were modelled as biphasic materials. More information on the FE model can be found in Geris et al. [58,59].

3.1.3.2. Numerical Simulation of the Effect of Implant Design on Peri-Implant Tissue Differentiation

The research hypothesis underlying this study [58] was that existing mechanoregulatory models [21-23] are able to predict peri-implant tissue differentiation in the bone chamber for unthreaded and threaded implants. The goal of the study therefore was to corroborate each of the previously mentioned mechanoregulatory models by applying them to the bone chamber set-up and comparing the simulation results both qualitatively and quantitatively with the experimental observations.

The experimental conditions modeled correspond to those reported in section 3.1.2.2. The simulation results were processed in analogy with the histology to facilitate a quantitative comparison between simulations and experiments. For this study, the tissue response (BAF; MB) has been additionally quantified in a 200 μm wide zone around the implant. As reported previously, no histological differences were found between the two implant geometries for the global amount of bone formation in the entire chamber (BAF). However, a significantly larger amount of BIC was observed for the threaded implant compared to the unthreaded one. In the simulations, a larger amount of bone was also predicted to be in contact with the threaded implant. However, other experimental observations could not be predicted. The simulation results showed a distribution of cartilage, fibrous tissue and (im)mature bone, depending on the mechanoregulatory model that was applied. In reality, no cartilage was observed. Adaptations to the differentiation models did not lead to a better correlation between experimentally observed and numerically predicted tissue distribution patterns. The hypothesis that the existing mechanoregulatory models were able to predict the patterns of tissue formation in the *in vivo* bone chamber could not be fully sustained.

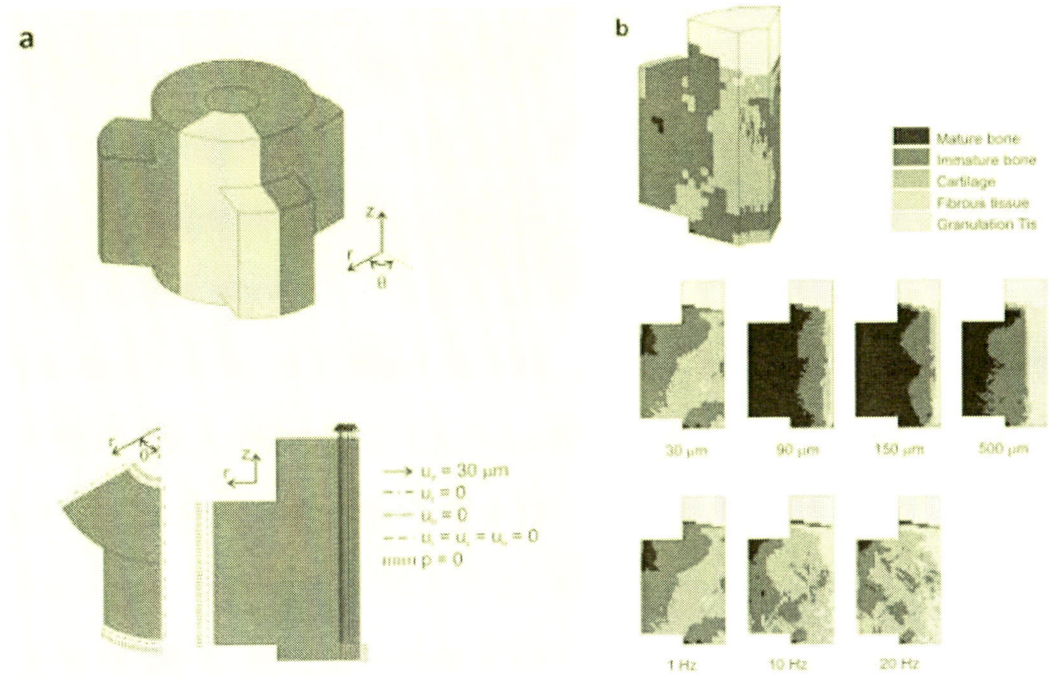

Figure 3. a. 3D FE model of the tissue inside the bone chamber (above); the boundary conditions for the FE analyses (u: displacement, p: pressure) (below); b. Tissue distribution throughout the bone chamber as predicted by the tissue differentiation simulation for the 3D model after 9 weeks of loading of an unthreaded turned implant (30 μm, 1 Hz, 400 cycles, 3x/wk); Simulation outcomes for different implant displacement magnitudes (middle) and frequencies (below) using the 2D axisymmetric model.

3.1.3.3. Numerical Simulation of the Effect of the Amplitude and Frequency of Loading on Peri-Implant Tissue Differentiation

This study aimed to numerically investigate the effects of the implant displacement magnitude and frequency on peri-implant tissue differentiation in the bone chamber model using the mechano-regulatory model of Prendergast et al. (1997) (Geris et al. 2009). The experimental conditions modeled correspond to those reported in sections 3.1.2.2, where an unthreaded implant displacement of 30 μm, 1 Hz, 400 cycles was applied 3x/wk for 9 wks, and to other conditions (3.1.2.4), where implant displacements of 30, 60 and 90 μm, 1 Hz, 800 cycles were applied 2x/wk for 6 wks to an unthreaded implant. For comparison with the histological findings, the value for BAF in the entire bone chamber was calculated from the simulation results. In addition to these experimental conditions tested in the bone chamber, loading conditions with higher loading frequencies (up to 20 Hz) and implant displacements (up to 500 μm) were modeled. Corresponding to literature data, implant displacement magnitudes larger than 90 μm predicted the formation of fibrous tissue encapsulation of the implant (Figure 3b). In contradiction to findings in orthopedic implant osseointegration [43,60], implant displacement frequencies higher than 1 Hz did not favor the formation of peri-implant bone in the chamber. The predictions for these extreme loading conditions need to be confirmed by additional bone chamber experiments.

3.2. Animal Model 2: The Rabbit Cortical Tibia Model

3.2.1. Animal Model and Sample Processing

The second animal model for investigation of the peri-implant tissue response to immediate loading is based on the guinea pig model with percutaneous tibial implants as described by Jansen and de Groot [61]. Percutaneous implants (Ti6Al4V TiO_2-blasted; Ra-value: 1.74 μm) with dimensions of 10.5 x 1.8 mm were used and placed bi-cortically in the diaphyses of the rabbit tibiae [62]. In order to obtain a good primary stability of the implants, a cavity with a diameter slightly smaller than the implant's diameter was drilled. Implants were inserted by manual torque. The implants made direct contact with the host bone of the upper and lower cortices, while the mid part of the implant faced the bone marrow cavity. The endosseous implant part was threaded, while the percutaneous part was turned and designed to fit a mechanical stimulation lever.

Mechanical stimulation of the implants started 1 day after implant installation. Unloaded implants served as control. A sinusoidally varying bending moment was applied with a force-controlled electro-mechanical shaker (Model 4810, Brüel and Kjaer, Naerum, Denmark). The loading device was screw-retained onto the implant with the horizontal lever part parallel to the long axis of the tibia. In this way, the loading direction corresponded with the long axis of the implants. At a distance of 20mm distally from the attachment point implant-loading device, the electro-mechanical shaker applied the sinusoidally varying force through a piezo load cell model PCB 208B03 (PCB Piezotronics, Depew, NY, USA). Force was transferred from the load cell to the lever through an aluminium-threaded pin. A preload of approximately 1N ensured continuous contact with the lever. Signals from the load cell were amplified by a PCB 480D06 amplifier (PCB Piezotronics) and captured. Control software, including a force-controlled feedback loop, data acquisition and visualization was implemented as a TestPoint application. Loaded implants were stimulated at 2.2 N with a frequency of 3 Hz for 1800 cycles 5 days/week. The exact strain values at the peri-implant cortical bone of the loaded samples, i.e. 333 μstrain (calculated from cadaver strain gauge measurements - see paragraph 4), were proven to be anabolic to the peri-implant tissues.

The implant and the peri-implant tissue were harvested 3, 7, 14, 28 and 42 days after implant installation. Immediately after animal sacrifice, tibial specimens were isolated and fixed in 2% paraformaldehyde. The bone segments were decalcified in 0.5M EDTA (pH 7.4)/ PBS at 4°C and paraffin-embedded, whereupon the implants were gently unscrewed. Next, the samples were re-embedded and 4 μm sections were made (HM360; Microm, Walldorf, Germany) parallel to the long axis of the tibial bone and the implant. Sections were stained with hematoxyline and eosine for characterization of the healing and adaptive processes at the cellular level. Histomorphometry was conducted on three sections per tissue sample using an image analyzing system (KS400 V3.00; Kontron Electronik, Munich, Germany). The following measurements were performed (Figure 4): (i) the region of osteocytic cell damage (visualized as a loss of nuclear staining in the haematoxylin-eosin-stained sections); (ii) the area occupied by basic multicellular units (BMU, organized structures of multiple osteoclasts and osteoblasts together with blood vessels); (iii) bone neo-formation in the cortex; (iv) endosteal bone neo-formation; (v) bone density of the endosteal bone neo-formation; and (vi) the thickness of bone apposition by the periosteal bone neo-formation. The implant removal itself may have caused some interfacial damage, as indicated in previous studies [53,63]. Nevertheless, this minor damage should not be taken into account when interpreting the results, as healing processes in the surrounding bone and marrow cavity, and not at the interface itself, were described and qualified in this study.

3.2.2. The Effect of Immediate Implant Loading on Cortical Bone Healing

Well-controlled immediate implant loading did not cause large differences in the sequence of biological events leading to osseointegration in cortical bone, as there was the formation of a haematoma and an altered nuclear morphology of osteocytes surrounding the implantation site, followed by an intensive bone remodelling and the formation of new bone leading to the osseointegration of the implant. At early time-points, an endosteal and periosteal bone neo-formation was found, while the cortex itself contained damaged osteocytes. At later time-points, bone neo-formation was also found at the cortical level itself. Differences between loaded and unloaded implants were observed with larger reactions for the endosteal and periosteal bone for the immediately loaded implants after 28 and 42 days, respectively. At the end-point of the experiment, bone formation at the cortical level was reduced around the immediately loaded implants compared with the unloaded ones.

4. Mechanobiology at the Tissue-Implant Interface during Early Implant Loading

In early implant loading, an extended timeframe during which the bone is allowed to heal precedes the loading of the implant. For the animal studies presented below and classified as early implant loading conditions, a healing period of 7 days was respected prior to well-controlled loading of the implant by a custom-designed external loading device.

4.1. Animal Model: The Guinea Pig Cortical Tibia Model

The guinea pig animal model with percutaneous tibial implants as described by Jansen and de Groot [61] was used for investigating tissue regeneration and adaptation at the implant surface in response to early loading. Implants were installed bi-cortically with primary stability on the medial surface of the tibia near the ankle joint. The implants made direct contact with the host bone of the upper and lower cortices, while the mid part of the implant faced the bone marrow cavity. The design and the dimensions of the implant varied slightly between the three performed studies (see legends of Tables 7, 8 and 9). The percutaneous part of the implant fitted the mechanical stimulation lever. Mechanical stimulation was started 1 week after implant installation. In each animal, one implant was stimulated while the contralateral implant served as the unloaded control. The same set-up as described in paragraph 3.2.1 was used (Figure 5).

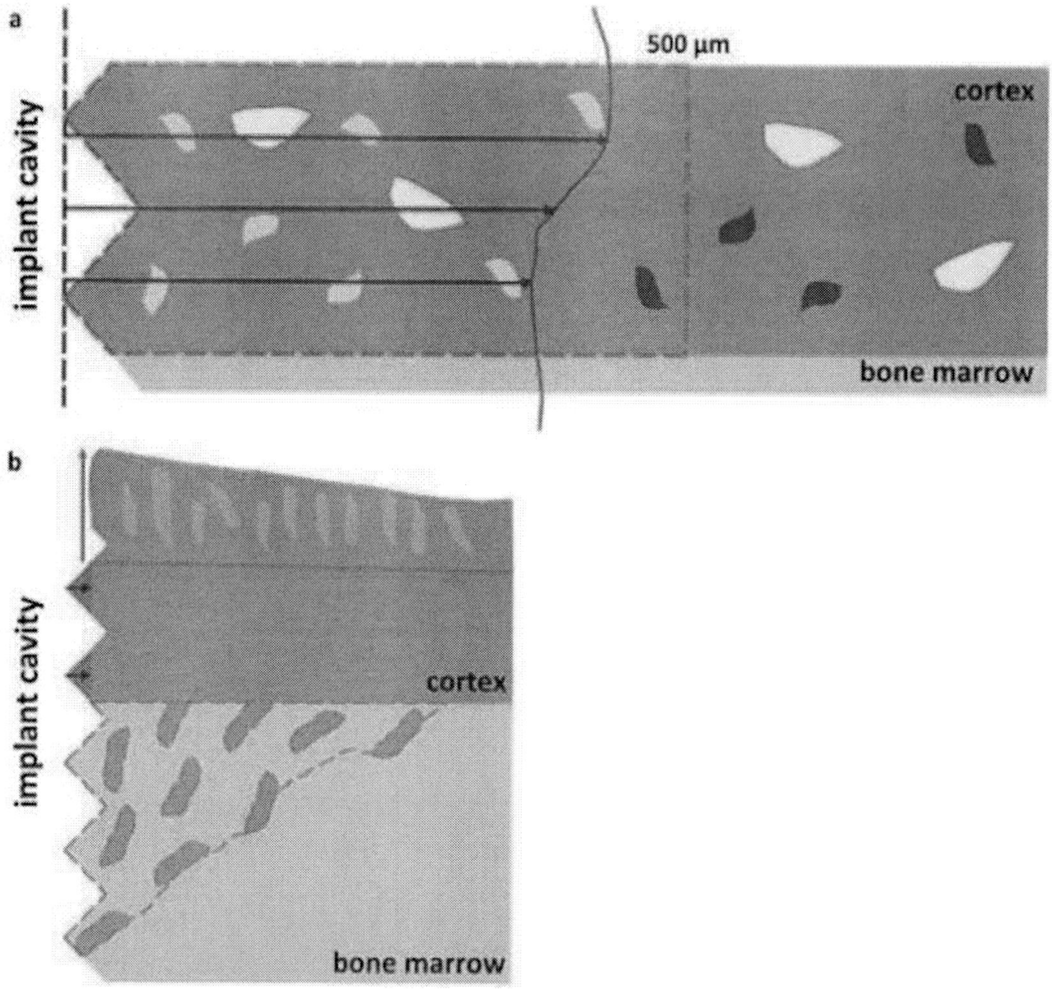

Figure 4. Schematic representation of the histomorphometric measurements. (a) The region of osteocytic cell damage (altered appearance: light blue, normal appearance: darker blue) in the cortical bone was assessed by measuring the distance (blue arrows) between implant cavity (blue dotted line) and the end of this region (full blue line). Quantification of the relative surface of the basic multicellular units (BMUs; indicated in yellow) occurred in a region of 500 µm from the implant (red dotted line). (b) Quantification of the new bone formation at the cortex was performed by measuring (black arrows) the average distance of new bone (in green) per indentation. The endosteal response was quantified by measuring the total surface of the reaction (green dotted line) and expressed as percentage of a control surface. The bone density in the endosteal response was defined as the trabecular surface (green dotted line around trabeculae) relative to the total surface of the endosteal reaction (green dotted line). The thickness of bone apposition by the periosteal response was quantified adjacent to the implant (green arrow).

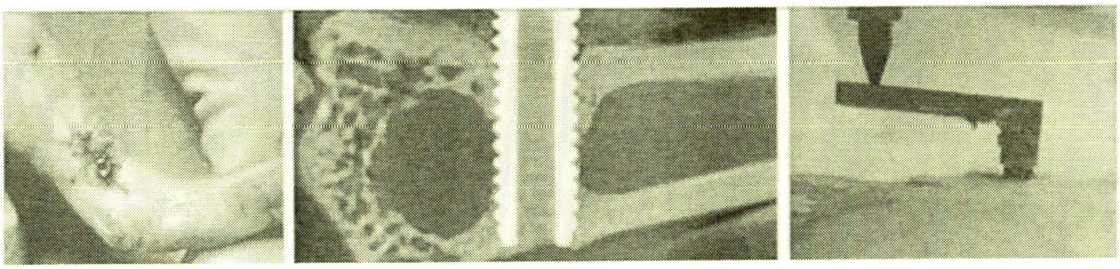

Figure 5. Anatomical location of the implant installation in the guinea pig model (left); X-ray image of an implant in situ in the tibia, with bicortical fixation (middle); External loading device, screw-retained onto the implant with the horizontal lever parallel to the tibia axis. Implant stimulation was performed at a distance of 20mm distally from the attachment point implant-loading device.

The influence of specific mechanical parameters, i.e. the duration, amplitude, rate and frequency of loading, of the early implant loading regime on peri-implant bone adaptation in the cortical host bone and on peri-implant bone formation in the marrow zone was investigated. The exact strain at the upper cortical implant interface in response to loading was experimentally calculated from cadaver strain gauge measurements. A strain gauge was glued on the outer surface of the tibial bone in the direction of the stimulation lever arm, at a distance of 1.3 mm from the implant surface. It was observed that the measured strains increased linearly with the forces applied, independently of the loading frequency. Bone strains of 1600 μstrain for loading amplitude of 6 N were measured. Based on these experimental results, a simplified finite element model of the guinea pig tibia with implant was built to extrapolate the strain distribution at the implant interface during stimulation. Bone strains in the cortical bone in the immediate vicinity of the implant were estimated to be in the order of magnitude of 2000-3000 μstrain for a force amplitude of 6 N. This is 1.3-2 times the strains quantified at a 1.3 mm distance from the implant.

Post-mortem 25μm-thick MMA-sections were cut by means of a modified inner circular saw microtome parallel to the longitudinal direction of the tibia and to the long axis of the implant. Three sections per implant were stained with methylene blue and basic fuchsin. The histomorphometrical analyses were carried out by means of a light microscope (Leica Microsystems GmbH, Wetzlar, Germany) connected to a PC, equipped with a video and image analysis system (Leica Q-wins Pro-image analysis system, Wetzlar, Germany). Digital images were made of the proximal and distal half of the histological section. A routine was written in the program for standardized segmentation of the bone. The same methodology for evaluation of the bone-to-implant contact and of the bone formation in the vicinity of the implant was used in all three studies, and is illustrated in Figure 6. Besides the well-known parameter bone-to-implant contact (BIC; %), the parameter 'delta bone mass' (Δ BM; %) was recorded. The bone mass (BM) was defined as the percentage of a defined area occupied by bone tissue. BM was performed for both cortices and for the marrow cavity, at a 500 μm, 1000 μm and 1500 μm distance from the implant surface, proximally and distally. The recorded values of BM for the unloaded implants were subsequently subtracted from the BM values of the loaded implants, resulting in the Δ BM value. A positive value indicates bone gain through early loading; inversely, a negative value reflects bone loss. Additionally to these histomorphometrical analyses, *in vivo* μCT-analysis (HOMX 161, Philips X-ray, Hamburg, Germany) for tracking the progress of peri-implant bone formation was performed in the 1st study (see 4.2.1) and vibration analysis (Osstell® device, Integration Diagnostics,

Savedalen, Sweden) for investigating the evolution of the implant stability in the 2nd and 3rd study (see 4.2.2 and 4.2.3).

4.2. Experimental Findings

Three guinea pig studies, each of them consisting of several experimental series, were performed, exploring the effect of early implant loading besides the separate mechanical parameters composing the early loading regime on the peri-implant tissue response. The investigated variable(s) in each study are highlighted in bold in the column headings.

4.2.1. Effect of Early Implant Loading on Peri-Implant Bone Formation and Adaptation

The first study was conducted to explore whether the mechanobiological rule of stimulation of bone tissue formation by moderate mechanical loading also holds true for the regenerating tissue surrounding implants in direct contact with the bone and with the bone marrow cavity [64]. Early loading experiments were designed for screw-shaped titanium implants. The implants in all series were loaded for 1800 cycles per day, 5 days/week for 4 weeks.

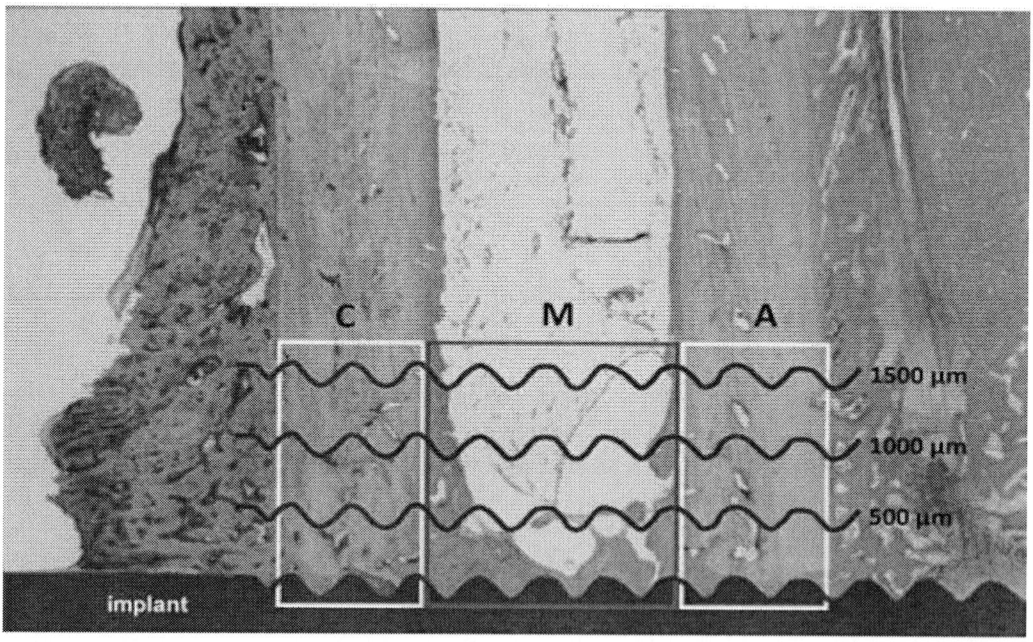

Figure 6. Quantification of the bone mass was performed on digital images of the proximal and distal half of the histological slices, at 3 x 2 x 3 regions of interest (ROI): at 500, 1000 and 1500 μm from the proximal and distal implant surfaces for the cervical cortex (C500, C1000, C1500), the medullar cavity (M500, M1000, M1500) and the apical cortex (A500, A1000, A1500).

Three loading series were characterized by low-amplitude/high-frequency stimulation, while in the other three series high amplitude/low-frequency stimulation was applied (Table 7). It was hypothesized that controlled early loading of the implants positively affects the tissue response around permucosal implants in contact with the host bone. Isolation of the loading regime evoking an optimal peri-implant tissue response was aimed for in this study.

Table 7. Loading parameters for 6 experimental series with percutaneous tibial implants (acid-etched; Ra-value 0.75-0.90 μm) in the guinea pig (n=55). Implants were stimulated 5 days/week for 4 weeks. The parameter kept constant is highlighted in bold; the varying parameters are highlighted in *italic*

Series number	Implant	# cycles/ loading session	*Force amplitude (N)*	Strain amplitude (με)	*Frequency (Hz)*	Strain rate (με/s)
6	IPI	1800	2	540	3	1620
5	IPI¶	1800	3	800	3	2400
4	MTI	1800	4	1000	3	3000
1	MTI	1800	0.5	160	30	4800
2	MTI	1800	1	260	30	7800
3	MTI	1800	1.5	400	30	12000

The mechanical parameters were arranged in order of increasing strain rate (= frequency x strain). The mentioned strains are the peak strains for the sinusoidally varying mechanical stimulation and have been calculated based on cadaver strain gauge measurements and on numerical simulations.

IPI, Steri-Oss implant (Ti6Al4V; 2.0 x 5 mm), Nobel Biocare, Göteborg, Sweden; IPI¶, Steri-Oss implant (Ti6Al4V; 2.0 x 5 mm) with a wider neck thread, manually adjusted; MTI implant (c.p. titanium; 1.8 x 7 mm), Dentatus, Hägersten, Sweden.

The histomorphometrical results revealed that none of the early loading regimes applied for 4 weeks to an implant bicortically fixated in the guinea pig tibia, provoked a significantly different effect on the degree of BIC compared to unloaded implants. However, bone tissue formation was enhanced (positive Δ BM) in response to early loading in the marrow cavity closest to the implant surface at the compressive side. It was possible to track the progress of bone formation in response to loading in time by *in vivo* μCT. The gain in bone mass was inversely correlated with the strain rate, with a significant result for the loading regimes with calculated cortical strain rate of up to 2400 με/s (series 5 and 6) at low-frequency stimulation (3 Hz). Based on these results, the guinea pig model as a model for cortical bone adaptation and medullar tissue differentiation to early implant loading was validated as being sensitive for assessment of the impact of early implant loading on the peri-implant endosteal tissue differentiation and bone formation. An improved bone reaction in the marrow cavity was achieved through early loading, confirming the study hypothesis.

4.2.2. Effect of Strain Rate of Loading on Peri-Implant Bone Formation and Adaptation

The role of the mechanical parameter strain rate (= frequency x strain) in the anabolic response of the implant surrounding tissues to early implant loading was investigated in a 2nd study with the same model [65]. Early loading experiments were designed for cylindrical threaded titanium implants. The implants in all series were loaded for 5 days/week for 4 weeks. The strain rate was kept constant for all groups (1600 με/s), while the amplitude and frequency varied in an inversely proportional manner (Table 8). It was hypothesized that the strain rate is the determining factor of the peri-implant tissue response to controlled early loading. No differences in BIC and Δ BM should thus be expected between the 3 experimental groups.

Table 8. Loading parameters for 3 series of experiments with percutaneous tibial implants (1.8 x 5 mm; TiO_2-blasted Ti6Al4V; Ra-value 1.74 μm; Astra Tech, Mölndal, Sweden) in the guinea pig (n=39). Implants were stimulated 5 days/week for 4 weeks. The parameter kept constant is highlighted in bold; the varying parameters are highlighted in *italic*

Series number	# cycles/ loading session	*Force amplitude (N)*	Strain amplitude (με)	*Frequency (Hz)*	Strain rate (με/s)
1	1800	2	533	3	1600
2	6000	0.6	160	10	1600
3	18000	0.2	53	30	1600

The mentioned strains are the peak strains for the sinusoidally varying mechanical stimulation and have been calculated based on cadaver strain gauge measurements and on numerical simulations.

Histomorphometry revealed that none of the early implant loading regimes had a significantly positive effect on BIC compared to unloaded implants. The resonance frequency analysis could not detect any differences in implant stability between loaded or unloaded implants, nor between the different series or over time. Furthermore, an anabolic effect of early mechanical loading on the peri-implant tissue response at the implant side under compression was observed, thereby confirming the results of previous study. Significant differences in stimulation of bone formation in the marrow cavity (Δ BM) for the different loading regimes were observed: the positive effect of early mechanical stimulation increased with increasing force amplitude and decreasing frequency. Based on these results, not the strain rate as such but its composing factors, the loading amplitude and frequency, are determinant factors for the positive effect of early implant loading on the peri-implant tissue mineralization. The hypothesis of the study could not be confirmed. Furthermore, this animal implant model has been shown to be most nociceptive to a low-frequency and high-amplitude early stimulation (see series 1).

4.2.3. Effect of Amplitude of Loading during Low-Frequency Loading on Peri-Implant Bone Formation and Adaptation

Based on the results of previous study, a low-frequency early loading protocol was selected for investigation of the effect of the amplitude of loading on peri-implant bone formation and adaptation [66]. Early loading experiments were designed for cylindrical threaded titanium implants. The implants in all series were loaded for 1800 cycles per day, 5 days/week for 4 weeks. Varying force amplitudes were combined with a frequency of stimulation of 3 Hz. For the 1st experiment, stimulation parameters of 2 N at 3 Hz were chosen, as these parameters provoked an optimal response in the previous study. For the 2 other experiments, lower force amplitudes were applied. It was hypothesized that strain levels of 700-1100 με at the implant interface (strain amplitudes values listed in Table 9, multiplied by 1.3-2; see paragraph 4.1) provoke the optimal peri-implant tissue response to early implant loading.

Table 9. Loading parameters for 3 series of experiments with percutaneous tibial implants (1.8 x 5 mm; TiO_2-blasted Ti6Al4Va; Ra-value 1.74 μm; Astra Tech, Mölndal, Sweden) in the guinea pig (n=30). Implants were stimulated 5 days/week for 4 weeks. The parameters kept constant are highlighted in bold; the varying parameter is highlighted in *italic*

Series number	# cycles/ loading session	*Force amplitude (N)*	Strain amplitude (με)	Frequency (Hz)	Strain rate (με/s)
1	1800	2	533	3	1600
2	1800	1	267	3	800
3	1800	0.5	133	3	400

The mentioned strains are the peak strains for the sinusoidally varying mechanical stimulation and have been calculated based on cadaver strain gauge measurements and on numerical simulations.

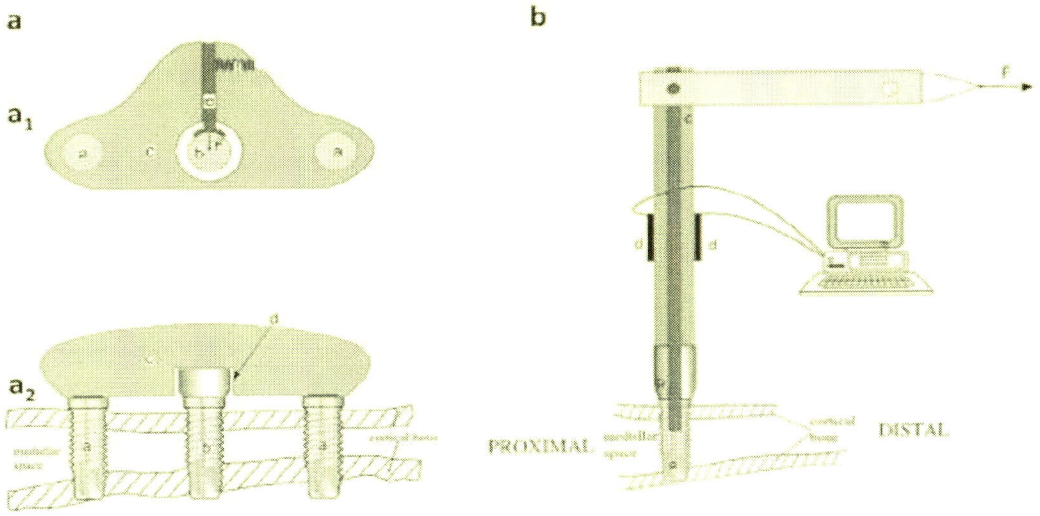

Figure 7. a. Schematic drawing of the static loading device. a_1: occlusal view, a_2: lateral view. a) implants supporting the static loading device, b) loaded implant, c) static loading device, d) central hole in the loading device to allow passage for the loaded implant, e) press plunger, f) screw to stabilise the press plunger after activation. F indicates the applied force; b. Schematic drawing of the dynamic loading device: a) loaded implant, b) abutment, c) metal beam mounted onto the abutment, d) strain gauges which record the deformation of the beam. F indicates the applied force.

Again, histomorphometry displayed no significantly different effect of early loading on BIC compared to unloaded implants. The resonance frequency analysis could not detect any differences in implant stability between loaded or unloaded implants, nor between the different series or over time. Furthermore, the anabolic effect of early implant loading on the peri-implant tissue response at the implant side under compression was confirmed. The most abundant bone formation response to early loading in the peri-implant marrow cavity was observed for the loading regime of series 2, and was significantly different with the tissue reaction in series 1 and 3. The relatively low (it was hypothesized that higher strains would be most beneficial in the tissue response) estimated strains of 350-540 με provoked the most pronounced anabolic process in the peri-implant tissues.

5. Mechanobiology at the Tissue-Implant Interface during Delayed Implant Loading

Being aware of the fact that interfacial micromotion immediately or early after implantation interferes with the local bone healing process, the delayed implant loading protocol was generally accepted since 1977. In the clinic, this protocol implies a load-free healing time for 3 months in mandible and 6 months in maxilla to allow uncompromised osseointegration and hence minimize the risk of soft tissue encapsulation. With respect to the bone turnover of animal species used (rabbit – 6 weeks for the femur), an unloaded integration period of the implant of 6 weeks was respected.

5.1. Animal Model: The Rabbit Cortical Tibia Model

Threaded 10 mm long implants (Brånemark SystemA, MKII, Nobel Biocare, Sweden) were placed bicortically in the tibiae and were allowed to heal for 6 weeks [67]. The osseointegrated implants were then subjected to controlled static loading, controlled dynamic loading or unloading. Static loading was performed by means of a press plunger and spring in contact with the abutment (Figure 7a). Dynamic loading was performed once a day through a beam mounted onto the abutment (Figure 7b). Recorded bone strains during implant pull-out tests (performed prior to the main experiment) were implemented in a 3D finite element model in order to determine the stresses induced in the implant surrounding bone and to use as a guideline for the dynamic loading protocol. The dynamic loading protocol applied excessive implant loading. Histomorphometrical quantifications of BIC and the bone density lateral to the implant at the marginal portion (BD; the percentage of the area between the lowest part of the implant shoulder and the third screw thread tip occupied by bone tissue) were performed on 25 μm thick undecalcified and toluidine blue stained sections, cut parallel to the loading direction.

5.2. The Effect of Static and Dynamic Loading on Marginal Bone Reactions around Osseointegrated Implants

The statically loaded implant was stimulated with a transverse force of 29.4 N applied on a distance of 1.5 mm from the top of the implant, resulting in a bending moment of 4.4 Ncm; the dynamically one with a transverse force of 14.7 N applied on a distance of 50 mm from the top of the implant, resulting in a bending moment of 73.5 Ncm, for a total of 2520 cycles at 1 Hz frequency for 14 days. The hypothesis was tested that an excessive load can lead to marginal bone loss and eventually to loss of osseointegration.

The histological picture was similar for statically loaded and control implants. Despite the applied static load, the bone around the implants was maintained. Crater-shaped bone defects in the marginal bone area around the dynamically loaded implants were observed, but without lower BIC values compared to the statically loaded and unloaded implants. However, BD of the marginal bone was significantly lower for the dynamically loaded implants, and signs of ongoing bone resorption were seen. This study confirmed the biological tolerance against static loads (induced by e.g. prosthesis misfit) but demonstrated that applications of well-calibrated dynamic loads that meet the limits of the bone strength are detrimental for the peri-implant marginal bone.

6. Principal Findings and Clinical Relevance

6.1. Principal Findings of the Different Animal Models

6.1.2. The Trabecular Bone Model

Bone formation in the bone chamber occurred exclusively via the intramembranous ossification pathway, as a consequence of the stable mechanical environment within the chamber and the highly vascularised site of implantation. The histological character of the bone that had formed remained consistent throughout all experiments. Inside an unloaded chamber the tissue consisted primarily of highly cellular fibrous tissue. Regions of a finely trabeculated structure were also evident, especially near the infiltration portals (perforations). On the contrary, the biopsies of loaded chambers were composed of bone marrow and a composite of woven and lamellar trabecular bone microstructure up to the implant. The trabecular bone neoformation may be the result not only of remodelling and appositional growth of existing lamellar trabeculae but also of the recruitment of new populations of osteogenic cells derived from the endosteal surfaces and the marrow. We have not yet specifically addressed the question whether the imposed loads "turned on" pre-existing osteoblasts or bone-lining cells (no increase in bone-forming cells) or whether the mechanical stimulus accelerated the recruitment and differentiation of pluripotential cells to matrix-producing osteoblasts (resulting in more bone-forming cells).

A significant stimulation of healing and bone formation associated with the application of controlled axial micro-movement was evidenced, provided that there exists a good biomechanical coupling. An adequate force transfer from the loaded implant to the surrounding tissues was promoted for implant designs favouring a good tissue interlocking (screw-shaped, micro-roughened implants), resulting in a stimulation of bone formation. Micro-motion up to 90 μm at the interface of a screw-shaped roughened implant positively influenced osseointegration. As the highest applied micro-motion (90 μm) for a screw-shaped roughened implant still possessed a bone inductive capacity, we could not define the threshold of micro-motion for that specific implant design. Based on the experimental findings, an attempt was made to conceptualize the establishment of osseointegration within a defined window of tolerated micro-motion for screw-shaped implants with a turned *versus* a rough micro-design (Figure 8).

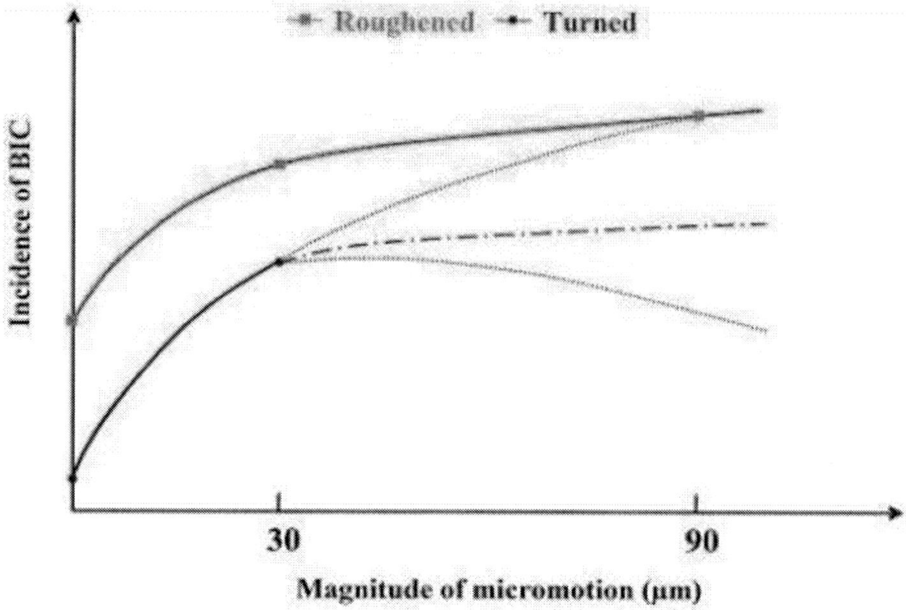

Figure 8. Relationship between the magnitude of the applied interfacial micro-motion during the healing phase and the establishment of osseointegration for a threaded implant with two different surface properties. Enhanced cell migration and adhesion around a roughened implant resulted in a more extensive mineralization of the interfacial tissues compared to a turned implant surface in the absence of loading. Peri-implant osteogenesis was enhanced at the same rate with increasing micromotion for both implant micro-designs, with micromotions up to 30 µm and 90 µm for a turned and roughened implant respectively. The ranges defined for a threaded turned implant subjected to a micromovement above 30 µm are estimates and may vary (dotted lines). The optimum value beyond which bone formation will decline or fibrous implant encapsulation is initiated, is undefined for both situations.

6.1.2. The Cortical Bone Model

In contrast to the trabecular model where the implant was axially loaded, a non-axial loading resulting in bending moments was applied to the implant in the cortical bone model. The resulting amount and distribution of the strains in the bone is dependent on the type of loading. With axial loading, the stress is more evenly distributed in the surrounding tissues, whereas non-axial loading causes stress gradients with concentrated high strains at the cortex and at the apex of the implant. Indeed, it was observed that excessive load negatively influenced the amount of bone in the marginal part of osseointegrated implants (delayed loading). Application of moderate loads during implant osseointegration (immediate or early loading) promoted bone formation in the adjacent bone marrow cavity, but did not affect the cortical bone. The most pronounced stimulation of bone formation in the marrow surrounding the implant was observed for low frequency stimulation (3 Hz) with low-level peri-implant strains (350-540 µε). These strain levels lie far beneath the level of minimum effective strain resulting in bone formation, as proposed by Frost [17]. This finding confirms that extrapolation of the gathered knowledge on the parameters of mechanically mediated bone modelling to the bone biology around osseointegrating implants should be performed with caution. Furthermore, no differences in bone-to-implant contact between the loaded and unloaded implants installed in this cortical bone model could be observed for none of the loading regimes. For the applied mechanical stimuli, bone formation in the cortical regions could not be elicited.

6.2. Clinical Relevance

The cells at the interface of a newly inserted implant are subjected to different stresses and strains. Because the loads are usually intermittent and uncontrolled, the biomechanical state of the peri-implant tissue varies continuously. Correlation between occlusal loading and implant success/failure is difficult to establish because of the problem of clinically quantifying the mechanical properties (magnitude, direction, etc.) of naturally occurring forces. In order not to be affected by this limitation, the animal models used allow well-controlled and well-defined mechanical stimulation of the implant. However, important differences with the clinical situation should be taken into account.

The osseointegration healing periods were individualized taking into account the bone turnover cycle for the animal species, the peri-implant environment (gap interface *versus* direct contact) and the anatomical location of implant installation (trabecular bone *versus* cortical bone). Whether and which factor could be applied when extrapolating the experimental findings to the clinical situation (e.g. the duration of the remodeling cycle in the rabbit is 6 weeks *versus* about 4 months in humans; i.e. a factor of 3), is not straightforward. Clinical research remains the ultimate test to validate the experimental results. Secondly, the implant installation in the animal studies was performed in a trabecular bone environment (bone chamber) or in tibial cortical bone (guinea pig and rabbit cortical model). The jawbones however are composed of both bone types. While cortical bone is mainly found in the mandible, trabecular bone is more abundant in the maxilla. Reports [53,63] comparing the healing process in these two bone types revealed faster healing of the peri-implant trabecular bone compared with cortical bone. However, cortical bone provides a firm mechanical interlocking of the implant immediately after installation (primary stability), which is also important when discussing the process of osseointegration. The differences in healing pattern and the level of primary stability should be taken into account when osseointegration of (immediately or early loaded) implants is anticipated in patients.

In the bone chamber model, a distance of 2.5 mm between the implant and the parent bone has to be bridged. Additionally, the recruitment of the bone forming cells from the parent bone and the mesenchymal cells from the bone marrow is partly obstructed by the bone chamber walls during tissue regeneration. The cells can only migrate towards the implant through the three perforations in the walls. Secondly, the exerted shear forces are not directly transferred to the surrounding bone in the first weeks – as it is for implants in direct contact with the native bone – but to the differentiating tissue in the gap interface. The bone chamber experimental set-up mimics more the immediate tooth replacement clinical situation where, mostly at the apex, the implant is in direct contact with the host bone, whereas more coronally only a blood cloth surrounds the implant. Primary mechanical implant stability – a prerequisite for successful osseointegration – was obtained in the model through the device itself through the central gliding bearing and was only interrupted during the loading sessions. Finally, as the bone chamber is a bone- and bone-marrow-containing defect, fibrous tissue does not originate from an extrabony site like the periosteum or the periodontal tissues in the clinical situation or in the classic animal models. The nature of the tissue can thus be exclusively related to the mechanical loading history of the implant (supposing no infectious agents penetrated into the bone chamber).

In the guinea pig and rabbit cortical model, the implant is installed bicortically in the tibia, while the mid-portion of the implant is surrounded by bone marrow. The clinical cases with implants installed in the interforaminal region of the mandible resemble most this situation. Due to the high primary stability of newly inserted implants offered through a bicortical fixation, micromotion at the implant's interface under early or immediate loading is levelled. Marginal cortical bone loss to mechanical loading can only be observed when excessive loading is applied. Investigation of newly formed bone in response to mechanical stimulation along the implant surface at its mid-portion (cancellous architecture) has not been reported yet in the clinical literature. Implant investigation carried out in the cortical bone tibia model may be invalid for predicting healing outcomes in the maxilla.

CONCLUSION

Detailed information about the quantification of the mechanical parameters from the results of clinical studies on immediate and early loading is lacking, because none of these studies have actually targeted those factors in a quantitative manner. Therefore, the results of this series of controlled and quantified implant loading studies contribute to the advanced understanding of the mechanobiology of implants subjected to immediate or early loading. Evidence was provided for the sensitivity of bone formation and resorption to the mechanical conditions at the peri-implant site. Implant loading can be performed immediately or early after insertion without disturbing the biological osseointegration process and can be beneficial for peri-implant bone formation. An optimal bone response to immediately or early loaded implants may thus not only be determined by the primary stability of the implant and by the host bone characteristics, but also by the individual loading parameters and by an optimized load transfer through an appropriate implant design. These findings should reinforce continued well-designed and fundamental research on immediate and early implant loading, as this loading protocol benefits many patients to a great extent.

REFERENCES

[1] Avila, G., Galindo, P., Rios, H. and Wang, H.L. (2007) Immediate implant loading: current status from available literature. *Implant Dent*,16,235-245.

[2] Esposito, M., Grusovin, M.G., Chew, Y.S., Coulthard, P. and Worthington, H.V. (2009a) Interventions for replacing missing teeth: 1- versus 2-stage implant placement. *Cochrane Database of Systematic Reviews*,8, CD006698.

[3] Esposito, M., Grusovin, M.G., Achille, H., Coulthard, P., Worthington, H.V. (2009b) Interventions for replacing missing teeth: different times for loading dental implants. *Cochrane Database of Systematic Reviews*,1,CD003878.

[4] Gapski, R., Wang, H.L., Mascarenhas, P. and Lang, N.P. (2003) Critical review of immediate implant loading. *Clinical Oral Implants Research*,14,515-527.

[5] Pilliar, R.M., Lee, J.M. and Maniatopoulos, C. (1986) Observations on the effect of movement on bone ingrowth into porous-surfaced implants. *Clinical Orthopaedics and Related Research*,208,108-113.

[6] Cameron, H.U., Pilliar, R.M. and MacNab, I. (1973)The effect of movement on the bonding of porous metal to bone. *Journal of Biomedical Materials Research*,7,301-311.

[7] Søballe, K., Hansen, E.S., B-Rasmussen, H., Jørgensen, P.H. and Bünger, C. (1992) Tissue ingrowth into titanium and hydroxyapatite-coated implants during stable and unstable mechanical conditions. *Journal of Orthopaedic Research*,10,285-299.

[8] Szmukler-Moncler, S., Salama, H., Reingewirtz, Y. and Dubruille, J.H. (1998) Timing of loading and effect of micromotion on bone-dental implant interface: review of experimental literature. *Journal of Biomedical Materials Research*,43,192-203.

[9] Goodship, A.E. and Kenwright, J. (1985) The influence of induced micromovement upon the healing of experimental tibial fractures. *Journal of Bone and Joint* Surgery. *British Volume*,67,650-655.

[10] Kenwright, J., Richardson, J.B., Cunningham, J.L., White, S.H., Goodship, A.E., Adams, M.A., Magnussen, P.A. and Newman, J.H. (1991) Axial movement and tibial fractures. A controlled randomised trial of treatment. *Journal of Bone and Joint Surgery. British Volume*,73,654-659.

[11] Goodship, A.E., Cunningham, J.L. and Kenwright, J. (1998) Strain rate and timing of stimulation in mechanical modulation of fracture healing. *Clinical Orthopaedics and Related Research*,355,S105-S115.

[12] Klein, P., Schell, H., Streitparth, F., Heller, M., Kassi, J.P., Kandziora, F., Bragulla, H., Haas, N.P. and Duda, G.N. (2003) The initial phase of fracture healing is specifically sensitive to mechanical conditions. *Journal of Orthopaedic Research*,21,662-669.

[13] Shi, H.F., Cheung, W.H., Qin, L., Leung, A.H. and Leung, K.S. (2009) Low-magnitude high-frequency vibration treatment augments fracture healing in ovariectomy-induced osteoporotic bone. *Bone*, Dec 2.

[14] Albrektsson, T. and Wennerberg, A. (2004) Oral implant surfaces: Part 1. Review focusing on topographic and chemical properties of different surfaces and in vivo responses to them. *International Journal of Prosthodontics*,17,536-543.

[15] Roux, W. (1881) *Der Kampf der Teile im Organismus.* Leipzig, Germany: Engelman.

[16] Wolff, J. (1892) *Das Gesetz der Transformation der Knochen.* Berlin, Germany: Springer.

[17] Frost, H.M. (1987) Bone "mass" and the "mechanostat": a proposal. *Anatomical Record*,219,1-9.

[18] Pauwels, F. Eine neue Theorie über den Einfluss mechanischer Reize auf die Differenzierung der Stützgewebe. *Zeitschrift für Anatomie und Entwicklungsgeschichte* 1960; 121:478-515. [Translated by Maquet, P, Furlong, R. A new theory on the influence of mechanical stimuli on the differentiation of supporting tissue. In: Pauwels F, ed. *Biomechanics of the Locomotor Apparatus*. Berlin: Springer; 1980; 375-407.]

[19] Carter, D.R., Blenman, P.R. and Beaupré, G.S. (1988) Correlations between mechanical stress history and tissue differentiation in initial fracture healing. *Journal of Orthopaedic Research*,6,736-748.

[20] Giori, N.J., Beaupré, G.S. and Carter, D.R. (1993) Cellular shape and pressure may mediate mechanical control of tissue composition in tendons. *Journal of Orthopaedic Research*,11,581-591.

[21] Prendergast, P.J., Huiskes, R. and Søballe, K. (1997) Biophysical stimuli on cells during tissue differentiation at implant interfaces. *Journal of Biomechanics*,30,539-548.
[22] Carter, D.R., Beaupré, G.S., Giori, N.J. and Helms, J.A. (1998) Mechanobiology of skeletal regeneration. *Clinical Orthopaedics and Related Research*,355,S41-S55.
[23] Claes, L.E. and Heigele, C.A. (1999) Magnitudes of local stress and strain along bony surfaces predict the course and type of fracture healing. *Journal of Biomechanics*,32,255-266.
[24] Geris, L., Gerisch, A., Maes, C., Carmeliet, G., Weiner, R., Vander Sloten, J. and Van Oosterwyck, H. (2006) Mathematical modeling of fracture healing in mice: Comparison between experimental data and numerical simulation results. *Medical and Biological Engineering and Computing*,44, 280-289.
[25] Lacroix, D. and Prendergast, P.J. (2002) A mechano-regulation model for tissue differentiation during fracture healing: analysis of gap size and loading. *Journal of Biomechanics*,35,1163-1171.
[26] 26. Isaksson, H., van Donkelaar, C., Huiskes, R. and Ito, K. (2006) Corroboration of mechanoregulatory algorithms for tissue differentiation during fracture healing: comparison with in vivo results. *Journal of Orthopaedic Research*,24,898-907.
[27] Isaksson, H., Comas, O., van Donkelaar, C.C., Mediavilla, J., Wilson, W., Huiskes, R., Ito, K. (2007) Bone regeneration during distraction osteogenesis: mechano-regulation by shear strain and fluid velocity. *Journal of Biomechanics*,40,2002-2011.
[28] Bailon-Plaza, A, van der Meulen, MC. (2001) A mathematical framework to study the effects of growth factor influences on fracture healing. *Journal of Theoretical Biology*, 212,191-209.
[29] Bailon-Plaza, A. and van der Meulen, M.C. (2003) Beneficial effects of moderate, early loading and adverse effects of delayed or excessive loading on bone healing. *Journal of Biomechanics*,36,1069-1077.
[30] Geris, L. (2007) *Mathematical modelling of bone regeneration during fracture healing and implant osseointegration.* Heverlee (Leuven), Belgium; 1-204. Thesis.
[31] Wiskott, H.W. and Belser, U.C. (1999) Lack of integration of smooth titanium surfaces: a working hypothesis based on strains generated in the surrounding bone. *Clinical Oral Implants Research*,10,429-444.
[32] Hsieh, Y.F., Wang, T. and Turner, C.H. (1999) Viscoelastic response of the rat loading model: implications for studies of strain-adaptive bone formation. *Bone*,25,379-382.
[33] Cullen, D.M., Smith, R.T. and Akhter, M.P. (2001)Bone-loading response varies with strain magnitude and cycle number. *Journal of Applied Physiology*,91,1971-1976.
[34] Mosley, J.E. and Lanyon, L.E. (1998) Strain rate as a controlling influence on adaptive modeling in response to dynamic loading of the ulna in growing male rats. *Bone*, 23,313-318.
[35] Skerry, T.M. and Lanyon, L.E. (1995) Interruption of disuse by short duration walking exercise does not prevent bone loss in the sheep calcaneus. *Bone*,16,269-274.
[36] Robling, A.G., Burr, D.B. and Turner, C.H. (2000) Partitioning a daily mechanical stimulus into discrete loading bouts improves the osteogenic response to loading. *Journal of Bone and Mineral Research*,15,1596-1602.
[37] Robling, A.G., Burr, D.B. and Turner, C.H. (2001) Recovery periods restore mechanosensitivity to dynamically loaded bone. *The Journal of Experimental Biology*, 204,3389-3399.

[38] Robling, A.G., Hinant, F.M., Burr, D.B. and Turner, C.H. (2002) Improved bone structure and strength after long-term mechanical loading is greatest if loading is separated into short bouts. *Journal of Bone and Mineral Research*,17,1545-1554.

[39] Srinivasan, S., Agans, S.C., King, K.A., Moy, N.Y., Poliachik, S.L. and Gross, T.S. (2003) Enabling bone formation in the aged skeleton via rest-inserted mechanical loading. *Bone*,33,946-955.

[40] Srinivasan, S., Weimer, D.A., Agans, S.C., Bain, S.D. and Gross, T.S. (2002) Low-magnitude mechanical loading becomes osteogenic when rest is inserted between each load cycle. *Journal of Bone and Mineral Research*,17,1613-1620.

[41] Rubin, C.T. and McLeod, K. (1994) Promotion of bony ingrowth by frequency-specific, low-amplitude mechanical strain. *Clinical Orthopaedics and Related Research*,298,165-174.

[42] Hsieh, Y.F. and Turner, C.H. (2001)Effects of loading frequency on mechanically induced bone formation. *Journal of Bone and Mineral Research*,16,918-924.

[43] Warden, S.J. 6 Turner, C.H. (2004) Mechanotransduction in the cortical bone is most efficient at loading frequencies of 5-10 Hz. *Bone*,34,261-270.

[44] Judex, S., Lei, X., Han, D. and Rubin, C. (2007) Low-magnitude mechanical signals that stimulate bone formation in the ovariectomized rat are dependent on the applied frequency but not on the strain magnitude. *Journal of Biomechanics*,40,1333-1339.

[45] Christiansen, B.A., Kotiya, A.A., Silva, M.J. (2009) Constrained tibial vibration does not produce an anabolic bone response in adult mice. *Bone*,45,750-759.,

[46] Turner, C.H., Akhter, M.P., Raab, D.M., Kimmel, D.B. and Recker, R.R. (1991)A non-invasive in vivo model for studying strain adaptive bone modeling. *Bone*,12,73-79.

[47] Kaspar, D., Seidl, W., Neidlinger-Wilke, C., Beck, A., Claes, L. and Ignatius, A. (2002) Proliferation of human-derived osteoblast-like cells depends on the cycle number and frequency of uniaxial strain. *Journal of Biomechanics*,35,873-880.

[48] Rubin, C.T., Lanyon, L.E. (1985) Regulation of bone mass by mechanical strain magnitude. *Calcified Tissue International*,37,411-417.

[49] Nkenke, E. and Fenner, M. (2006) Indications for immediate loading of implants and implant success. *Clinical Oral Implants Research*,17,19-34.

[50] Duyck, J., De Cooman, M., Puers, R., Van Oosterwyck, H., Vander Sloten, J. and Naert, I. (2004) A repeated sampling bone chamber methodology for the evaluation of tissue differentiation and bone adaptation around titanium implants under controlled mechanical conditions. *Journal of Biomechanics*,37,1819-1822.

[51] Vandamme, K., Naert, I., Geris, L., Vander Sloten, J., Puers, R. and Duyck, J. (2007c) The effect of micro-motion on the tissue response around immediately loaded roughened titanium implants in the rabbit. *European Journal of Oral Sciences*,115,21-29.

[52] Duyck, J., Slaets, E., Sasaguri, K., Vandamme, K. and Naert, I. (2007) Effect of intermittent loading and surface roughness on peri-implant bone formation in a bone chamber model. *Journal of Clinical Periodontology*,34,998-1006.

[53] Slaets, E., Carmeliet, G., Naert, I. and Duyck, J. (2006) Early cellular responses in cortical bone healing around unloaded titanium implants: an animal study. *Journal of Periodontology*,77,1015-1024.

[54] Vandamme, K., Naert, I., Geris, L., Vander Sloten, J., Puers, R. and Duyck, J. (2007a) Histodynamics of bone tissue formation around immediately loaded cylindrical implants in the rabbit. *Clinical Oral Implants Research*,18,471-480. ,

[55] Vandamme, K., Naert, I., Geris, L., Vander Sloten, J., Puers, R. and Duyck, J. (2007b) Influence of controlled immediate loading and implant design on peri-implant bone formation. *Journal of Clinical Periodontology*,34,172-181.

[56] Vandamme, K., Naert, I., Vander Sloten, J., Puers, R. and Duyck, J. (2008) Effect of implant surface roughness and loading on peri-implant bone formation. *Journal of Periodontology*,79,150-157.

[57] Duyck, J., Vandamme, K., Geris, L., Van Oosterwyck, H., De Cooman, M., Vandersloten, J., Puers, R. and Naert, I. (2006)The influence of micro-motion on the tissue differentiation around immediately loaded cylindrical turned titanium implants. *Archives of Oral Biology*,51,1-9.

[58] Geris, L., Vandamme, K., Naert, I., Vander Sloten, J., Duyck, J. and Van Oosterwyck, H. (2008) Application of mechanoregulatory models to simulate peri-implant tissue formation in an in vivo bone chamber. *Journal of Biomechanics*,41,145-154.

[59] Geris, L., Vandamme, K., Naert, I., Vander Sloten, J., Duyck, J. and Van Oosterwyck, H. (2009) Numerical simulation of bone regeneration in a bone chamber. *Journal of Dental Research*,88,158-163.

[60] Usui, Y., Zerwekh, J.E., Vanharanta, H., Ashman, R.B., Mooney, V. (1989) Different effects of mechanical vibration on bone ingrowth into porous hydroxyapatite and fracture healing in a rabbit model. *Journal of Orthopaedic Research*,7,559-567.

[61] Jansen, J.A. and de Groot, K. (1988) Guinea pig and rabbit model for the histological evaluation of permanent percutaneous implants. *Biomaterials*,9,268-272.

[62] Slaets, E., Naert, I., Carmeliet, G. and Duyck, J. (2009) Early cortical bone healing around loaded titanium implants: a histological study in the rabbit. *Clinical Oral Implants Research*,20,126-134.

[63] Slaets, E., Carmeliet, G., Naert, I. and Duyck, J. (2007) Early trabecular bone healing around titanium implants: a histologic study in rabbits. *Journal of Periodontology*, 78,510-517.

[64] De Smet, E., Jaecques, S.V., Wevers, M., Jansen, J.A., Jacobs, R., Vander Sloten, J., Naert, I.E. (2006) Effect of controlled early implant loading on bone healing and bone mass in guinea pigs, as assessed by micro-CT and histology. *European Journal of Oral Sciences*, 114,232-242.

[65] De Smet, E., Jaecques, S.V., Jansen, J.J., Walboomers, F., Vander Sloten, J, Naert, I.E. (2007) Effect of constant strain rate, composed of varying amplitude and frequency, of early loading on peri-implant bone (re)modelling. *Journal of Clinical Periodontology*, 34,618-624.

[66] De Smet, E., Jaecques, S.V., Jansen, J.J., Walboomers, F., Vander Sloten, J., Naert, I.E. (2008) Effect of strain at low-frequency loading on peri-implant bone (re)modelling: a guinea-pig experimental study. *Clinical Oral Implants Research*,19,733-739.

[67] Duyck, J., Rønold, H.J., Van Oosterwyck, H., Naert, I., Vander Sloten, J. and Ellingsen, JE. (2001) The influence of static and dynamic loading on marginal bone reactions around osseointegrated implants: an animal experimental study. *Clinical Oral Implants Research*,12,207-218..

In: Biomechanics of Dental Implants
Editor: Murat Çehreli

ISBN 978-1-62100-780-7

Chapter 6

Histologic and Histomorphometric Evaluation of Implants Retrieved from Humans

Giovanna Iezzi, Marco Degidi, Antonio Scarano, Jamil Awad Shibli, Carlo Mangano, Vittoria Perrotti and Adriano Piattelli

1. Introduction

Osseointegrated dental implants have been proven to be highly successful in several clinical indications with high survival and success rates, but some of them will still fail [1]. Many animal studies have been performed to evaluate, for example, different implant macrogeometries, different implant surface topographies, different loading conditions, different bone qualities and quantities, etc [2]. All these studies are extremely valuable; they are, however, of a low evidence quality and the results obtained from these studies could be automatically transposed to a human situation. Accordingly, it is very important to evaluate retrieved human implants. Implants can be removed due to problems (they are failing or are already failed), e.g. mobility [3], fracture [4-6], peri-implantitis, bone resorption, infection [7], etc. They can also be obtained for other reasons, psychological reasons, unrestorable prosthetics, misalignment, pain, dysesthesia, not optimal position from aesthetics and hygienic point of view, inability of an implant to meet changed prosthetic needs, or can be retrieved at autopsy [8-13]. In all these latter cases the retrieved implants continued to have an excellent bone anchorage. Implants from humans can also be obtained as part of a research protocol approved by an Ethical Committee. Also in these cases the bone anchorage is still present. The careful evaluation of all these different types of implants can be extremely useful to help in understanding the failure modalities or the reactions of the peri-implant tissues (soft tissues and bone). The concept of osseointegration was initially based on a histological definition but, however, there are no histomorphometrical diagnoses of osseointegration in that we don't know the precise values of the bone to implant contact percentages (BIC) that are needed by an implant to meet the requirement to be defined osseointegrated. A histological and histomorphometrical evaluation of a large series of retrieved implants could help to give some answers in this field [14-24]. The Implant Retrieval Center of the University of Chieti-Pescara, Italy, has been working continuously since mid 1988. In that time frame a large quantity of retrieved human implants have been treated to obtain thin ground sections and have been evaluated [25].

In this chapter, the main focus will be on implants with different surface characteristics, on implants retrieved after different loading periods, on implants inserted in poor bone sites, on the long term fate of the hydroxyapatite coating, on implants inserted in patients suffering from metabolic conditions as osteoporosis, on implants retrieved from post-extraction sites, on implants retrieved from grafted sites, on implants retrieved from smokers, on the peri-implant soft tissues.

2. Dental Implant Surfaces

The initial healing period of an implant is the phase of the osseointegration process that which is primarily affected by the surface condition of the implant. With roughened surfaces there is an increase of the BIC. When an implant is placed into a bone site, a cascade of biological events is initiated. There is a recruitment and migration of osteogenic cells to the implant surface. Then, new bone formation takes place, which results in the formation of a mineralized interfacial matrix, followed by a bone remodeling process [26, 27]. It is important to emphasize, on the other hand, that the surface characteristics of an implant are not the only requirement to obtain a long-lasting implant anchorage. The implant material, bone quality and quantity, surgical technique, surface characteristics, implant design, implant loading conditions, are all related to the implant long term success. The percentage of BIC may be employed to evaluate the stability of an implant, and values higher than 50% appear to be satisfactory. Torque removal force (RTV) has been used to describe the anchorage of an implant to the bone, and the higher the value, the greater the biomechanical strength of the bone-implant interface. We have already said that results obtained in studies performed in humans are more reliable than the findings obtained in studies performed in animals or *in vitro*. Some studies can, however, be performed using an animal model, e.g. RTV evaluations of implants with different macro-and microgeometries, and *in vitro* studies can be useful in helping to understand the biological response of different types of cultured cells in contact with different implant surface topographies. Each surface should be described by the combination of parameters representative of height and space [28, 29].

Machined surfaces were the most commonly used in the past. These surfaces were also called "turned" or "smooth", and microscopic observation under Scanning Electron Microscopy (SEM) revealed the presence of a slight roughness due to the grooves and ridges produced during the turning process. One of the main characteristics of the machined surfaces was that the bone growth pattern was characterized by "distance osteogenesis", i.e. bone growth toward the implant surface (implantopetal kind of bone growth).

Sandblasted surfaces were produced by blasting the metal with different types of blasting or gritting agents. This process was influenced by the number and the size of the particles used. The blasting procedure served to increase the irregularities of the implant surface of the implant, by using agents such as aluminum oxide (Al_2O_3) or titanium oxide (TiO_2). The large variability in surface appearance under SEM of different implant surfaces is due to the different techniques employed in the blasting procedure. The sandblasted surfaces have shown, in *in vitro* studies, a higher adhesion, proliferation, and differentiation of osteoblasts. Higher BIC values were found in histological studies that compared blasted and turned surfaces. Blasting procedures leave, however, blasting residual particles on the surface of the implant, and this fact could modify the bone healing process. Some researchers think that aluminum ions could impair bone formation by a possible competitive action to calcium, while others suggested that histological data did not provide evidence to support the hypothesis that residual aluminum oxide particles on the implant surface could affect the osseointegration of titanium dental implants. The bone growth pattern around blasted, rough surfaces is characterized by "contact osteogenesis", i.e. the osteoblasts start depositing osteoid matrix directly on the implant surface ("implantofugal type of growth"). This type of bone growth could produce an earlier and a higher quantity of bone at the interface with the implant (Figures 1 and 2) [30].

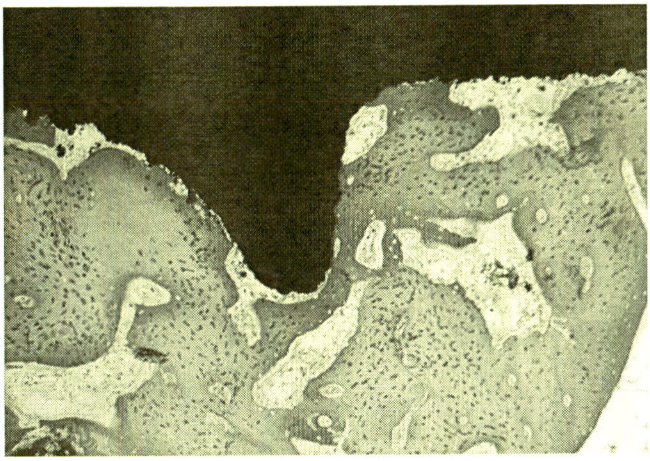

Figure 1. Sandblasted implant retrieved from the mandible. Trabecular bone is present around the implant (acid fuchsin-toluidine blue 40X).

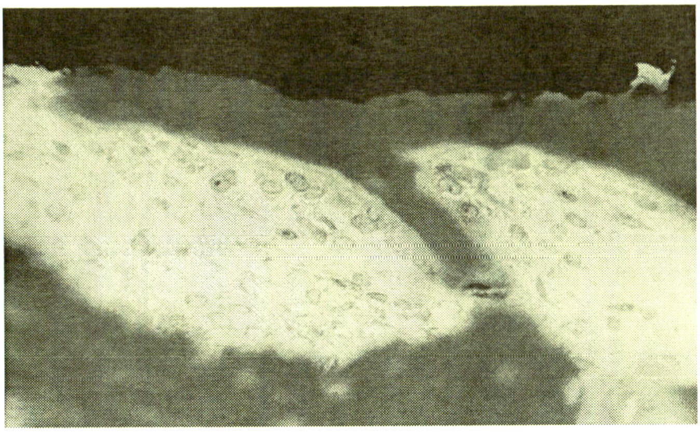

Figure 2. Sandblasted implant retrieved from the maxilla. Histologic ground section of sandblasted surface presenting osteoblasts apposing newly formed bone in close contact with the implant surface (acid fuchsin-toluidine blue 200X).

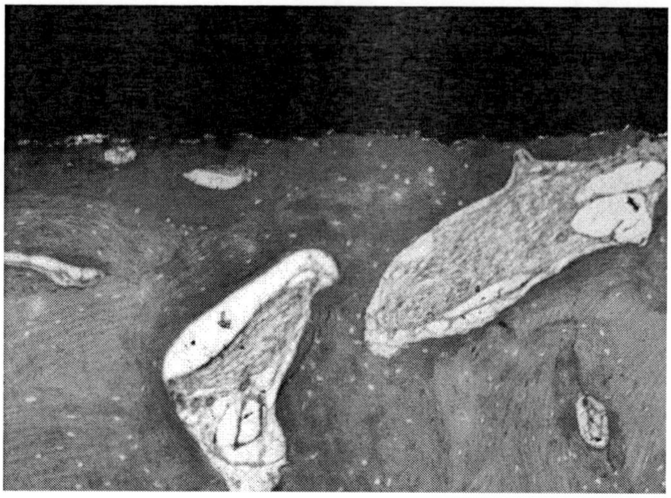

Figure 3. Plasma-sprayed surfaced implant retrieved from the mandible. It is possible to observe bone tissue in close contact with the implant surface and small marrow spaces in proximity of the surface (acid fuchsin-toluidine blue 100X).

Plasma sprayed surfaces have been used in orthopedics since many decades. These implants were prepared by spraying heat molten metal on the titanium base, which resulted in a surface with irregularly sized and shaped valleys and peaks, pores and cavities with an increase of the implant surface area by 6 to 10 times. This surface topography, in which it was possible to observe the formation of bone into the coating, improved the implant fixation in bone, by a biomechanical interlock. One disadvantage of this type of surface could be the detachment of titanium particles from the coating after implant insertion. The implications of this occurrence were, however, not clear (Figures 3 and 4) [31-36].

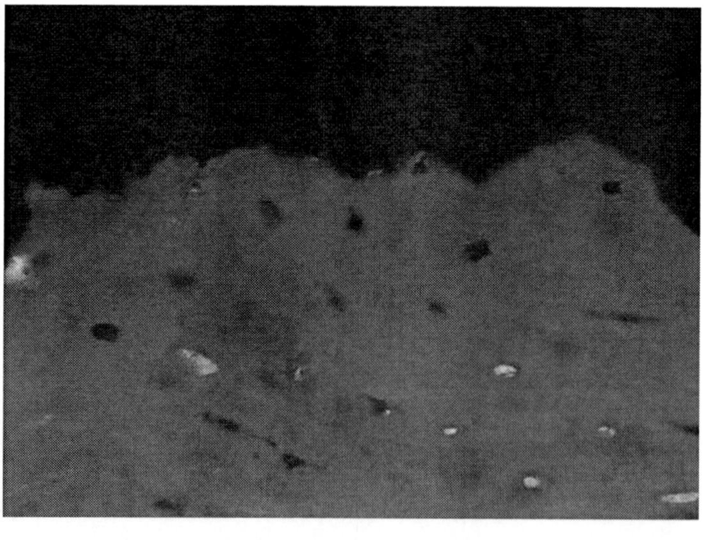

Figure 4. Plasma-sprayed surfaced implant retrieved from the mandible. Compact bone in close contact with the implant surface with no gaps at the interface (acid fuchsin-toluidine blue 200X).

Figure 5. Acid-etched surfaced implant retrieved from the mandible. Small marrow spaces can be observed in close proximity of the implant surface (acid fuchsin-toluidine blue 40X).

Acid-etched surfaces were introduced to modify the implant surfaces without the residues found after the blasting procedures, to have a more uniform surface treatment, and to control the loss of metallic substance. Baths using chlorides (HCl), sulfuric (H_2SO_4), hydrofluoric (HF), and nitric (HNO_3) acids, in different combinations, have been used. The acid-etching process was affected by the acid used, by the bath temperature, and by the etching time. The bone growth pattern was "contact osteogenesis" (Figures 5 and 6) [37-39].

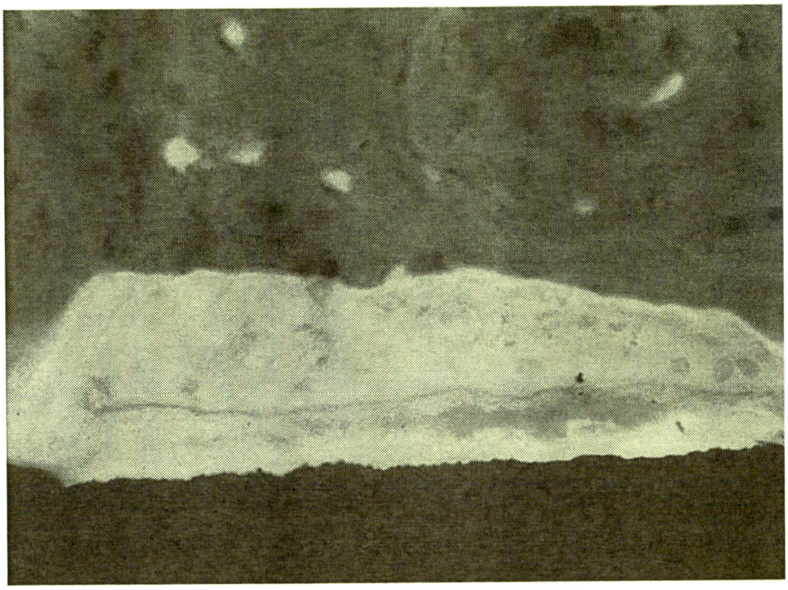

Figure 6. Acid-etched surfaced implant retrieved from the mandible. A osteoblast rim can be observed in the vicinity of implant surface. Osteoid matrix is present (acid fuchsin-toluidine blue 200X).

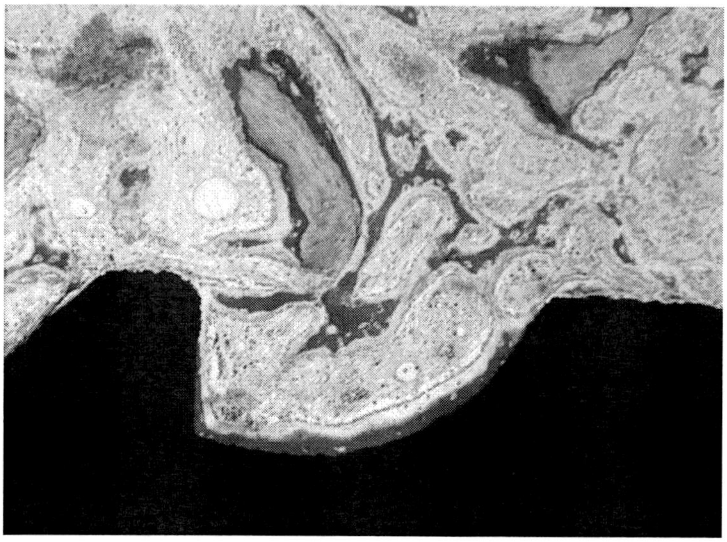

Figure 7. Sandblasted and acid etched implant retrieved from the maxilla. Trabecular bone with large marrow spaces is present. Newly formed bone can be observed inside the implant thread (acid fuchsin-toluidine blue 40X).

Sandblasted and acid etched surfaces were obtained with a combined procedure of blasting (to produce a macro-texture) followed by acid-etching (to produce a final micro-texture). Sandblasted and acid etched implants promoted a higher BIC at earlier time points compared to plasma-sprayed-coated implants. Sandblasted and acid-etched surfaces showed high osteoconductive properties and capabilities to induce cell proliferation (Figures 7 and 8) [40-42].

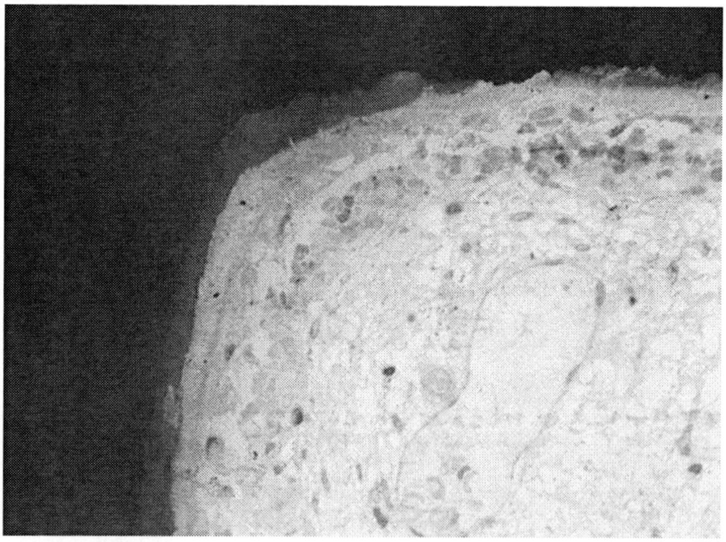

Figure 8. Sandblasted and acid etched surfaced implant retrieved from the maxilla after 4 weeks of healing. Newly formed bone inside the implant thread can be observed (contact osteogenesis) (acid fuchsin-toluidine blue 200X).

Figure 9. Ground section of hydroxyapatite (HA) coating implant surface. Inside the implant thread it is possible to observe the HA coating in close contact to newly formed bone (acid fuchsin-toluidine blue 200X).

Anodized surfaces were obtained by modifying the structure of the superficial oxide layer of the implant surface without depositing grit particles. Anodized surfaces were prepared by applying a voltage on the titanium specimen immersed in an electrolyte. The resultant surface presented micro-pores of variable diameters [43-45].

Hydroxyapatite (HA) coatings had a similar roughness and increase in surface area as that observed with titanium plasma spray surfaces. A direct bone bond was shown to be present with HA coated implants and the strength of the HA-to-bone interface was greater than that observed of titanium to bone and even greater than that seen in titanium plasma-sprayed surfaces and bone. In addition, accelerated interfacial bone formation and maturation have been observed in dogs. Gap healing, i.e. the healing in the space between the implant and bone, could be enhanced by the HA coating. The advantages of an HA coating are: increased surface area, increased roughness for initial stability, stronger bone-to-implant interface, faster healing at the interface, increased gap healing, less corrosion of metal. The coating may, however, flake, crack upon insertion, especially into dense bone. The increased surface roughness could increase the risk of bacterial contamination should the coating be found outside bone, e.g. in cases of peri-implant crestal resorption. There is also an increased cost of the coating, compared with uncoated implants (Figure 9) [46, 47].

Zirconia (zirconium oxide) is used in implantology for its biocompatibility, esthetics (its color is similar to tooth), mechanical properties. ZrO_2 implants are biocompatible, bioinert, radiopaque, and present a high resistance to corrosion, flexion and fracture. ZrO_2 implants have been reported to show a bone and soft tissue contact similar to that seen around titanium implantsZrO_2 can be used to produce a entire implant, or as a coating. Bioceramic molecular impregnation may modulate the characteristics of the protein layer adhesion in our body, the nanoscale structure of the extra-celullar matrix provides an essential and natural web of nanofibers to support cells and depict an instructive background to guide their behavior. Physical and bioceramic incorporation surface treatments at nanometer scale have shown higher means of BIC and torque values compared with rough implant surface topography at micrometer scale.

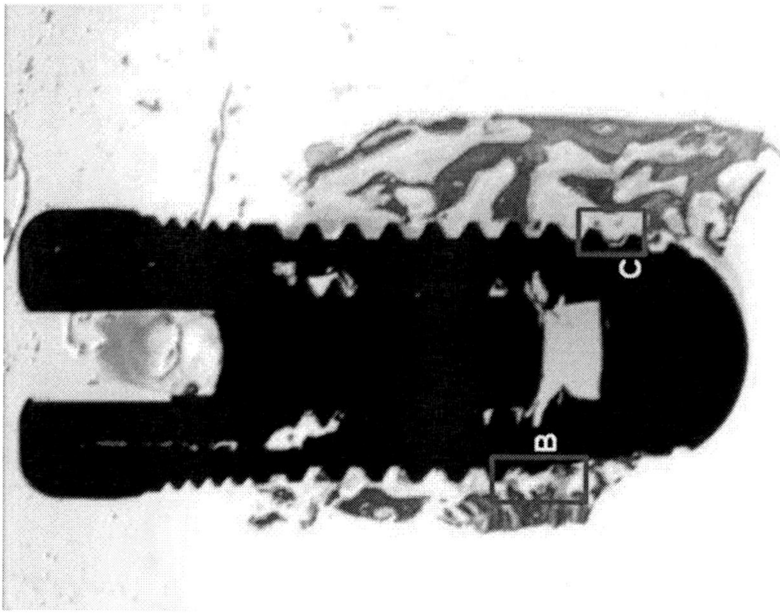

Figure 10. Histologic ground section of a biomolecular impregnated implant surface. Newly-formed bone with connecting bridges between the new bone trabeculae and the implant surface (Basic fuchsin and toluidine blue staining, original magnification x12).

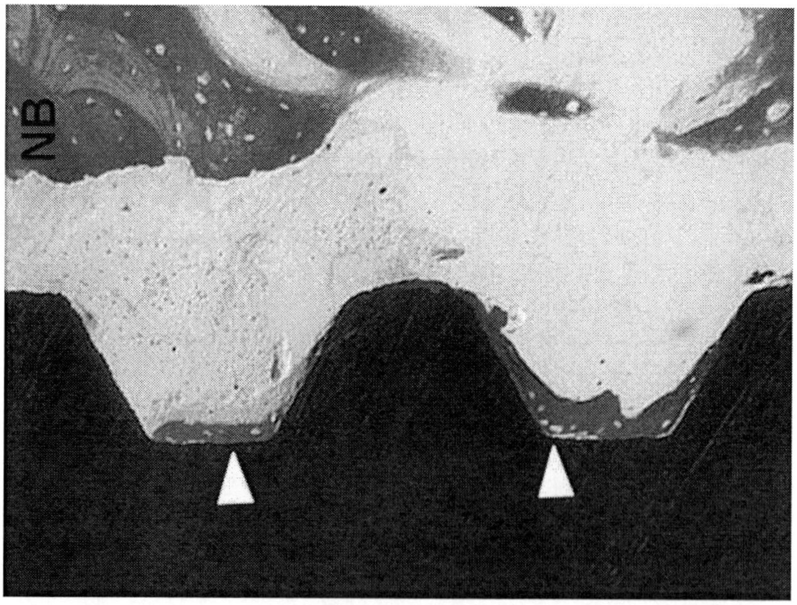

Figure 11. Ground section of a biomolecular impregnated implant surface presenting newly-formed bone at early stages of maturation and remodelling. The newly formed bone (NB) is close to the implant surface (arrows), suggesting contact osteogenesis (Basic fuchsin and toluidine blue staining, original magnification x100).

The application of nanotechnology for the alteration of texture and chemistry in dental implant topography may result in varied cell behavior, ranging from alterations in adhesion, orientation, mobility and surface antigen display of the pre-osteogenic and osteogenic cells. Complementary, nanoscale features may also affect the adsorption and conformation of integrin binding proteins, changing the availability of binding sites and modifying integrin signaling (Figures 10-12).

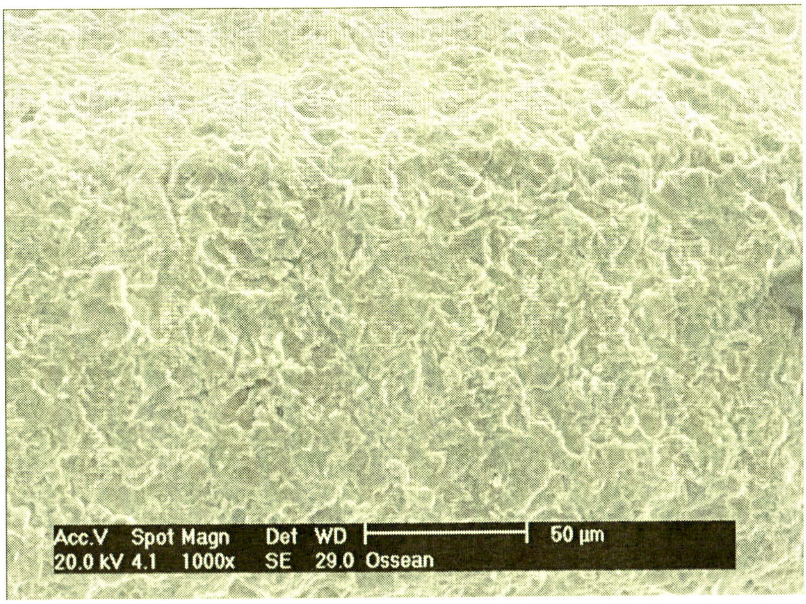

Figure 12. Scanning electron microscopy image showing the implant surface topography.

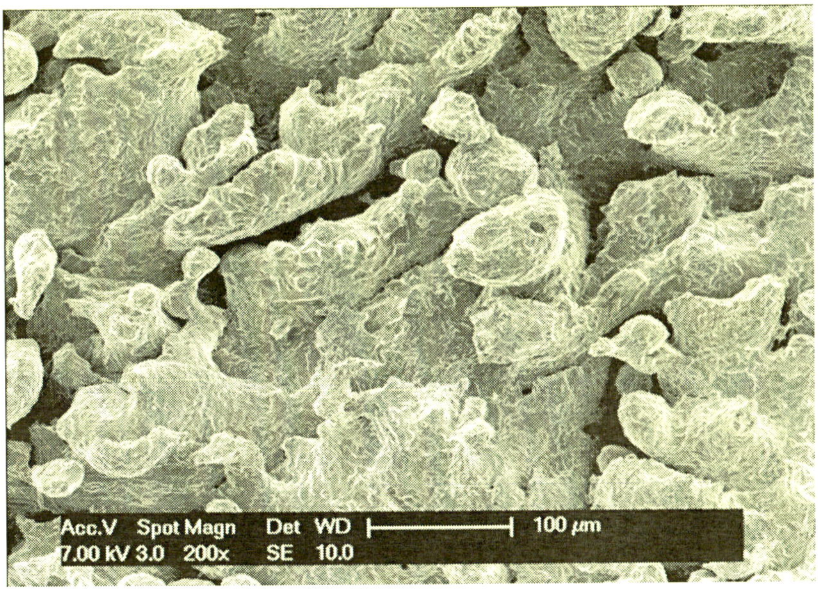

Figure 13. Scanning electron microphotograph of the direct laser fabrication surface.

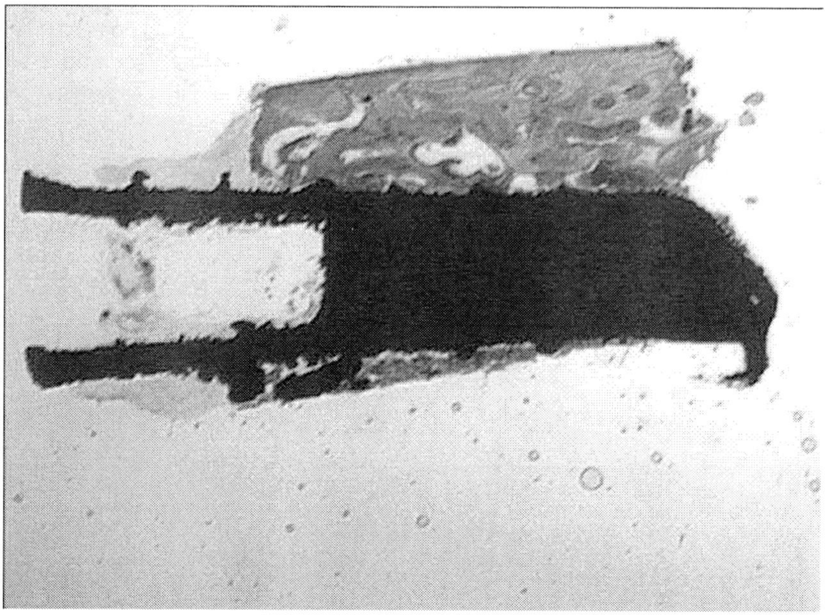

Figure 14. Histological ground section of the direct laser fabrication micro-implant surface after 2 months of healing showing the newly-formed bone at early maturing stages. There are connecting bridges between the new bone trabeculae and the implant surface (Basic fuchsin and toluidine blue staining, original x12 magnification).

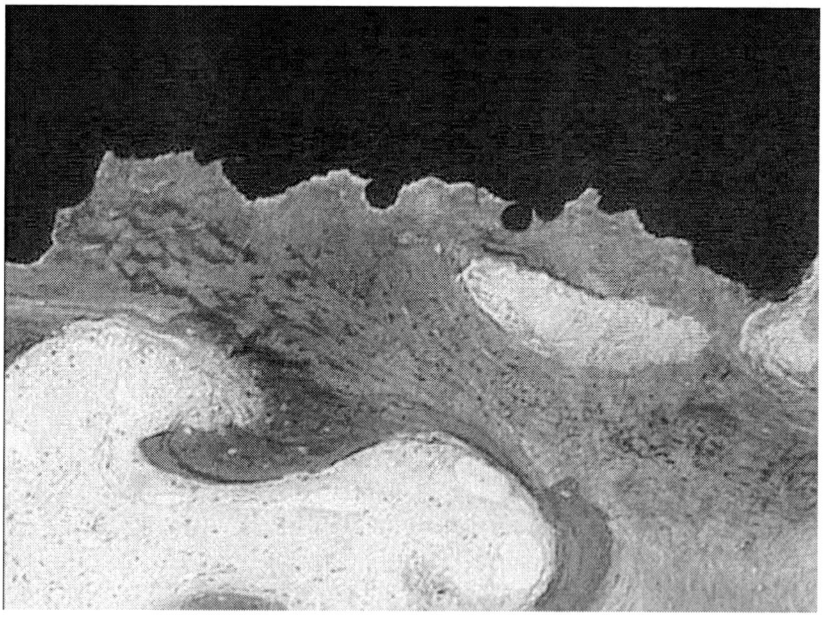

Figure 15. There are reversal lines between newly formed bone and the pre-existing bone. The newly formed bone tissue is in close contact with the implant surface (Basic fuchsin and toluidine blue staining, original 200X magnification).

Direct laser metal sintering implant surface previous studies have shown that direct laser metal sintering (DLMS) produces structures with complex geometry that allow better osteconductive properties. DLMS implant surface presents a similar cell density to that on rough surface but lower than on machined surfaces. Moreover, it was shown that implants obtained through laser metal sintering were better adapted to the elastic properties of bone. DLMS implant topography not only minimizes stress-shielding effects but also improves implants long-term success rates. These observations also suggested that DLMS technique is an economical method for producing implants from commercially pure titanium or alloys (Figures 13-15) [48].

3. Implants Retrieved after Different Time Periods

A submerged healing period of 3-4 months has been thought to be necessary to obtain mineralized bone at the interface of dental implants, and an earlier implant loading has been reported to determine the occurrence of fibrous tissue at the bone-implant interface [49-51]. On the other hand, several researchers have reported, in the last decade, that in early and immediately loaded (IL) implants, placed in good quality bone, it was possible to obtain a high level of osseointegration, clinically and radiographically similar to that of implants used with a standard submerged protocol. Moreover, very high implant survival rates for immediately loaded implants have been reported in the literature. Many patients found the wearing of provisional prostheses rather uncomfortable, and most certainly the possibility to shorten the healing time without jeopardizing the dental implants long-term success would be beneficial for most of them. Immediate loading has been reported to be a viable and successful treatment option. However, the generalization from the results of the clinical trials to everyday routine dental practice should always be made with extreme caution because in most trials the inclusion criteria were very strict, only very good candidates for the implant therapy were included and the clinicians were high skilled. Primary implant stability and lack of micromovement were considered to be the main factors involved in the success of IL implants. Macroretention offered by implant thread could reduce the risk of implant movements in the case of immediately loaded implants. Rigid splinting with minimal lateral forces decreased the amount of micromotion during the early healing phase, giving the implant a higher tolerance to deleterious micromotion. Healing processes were strongly influenced by the local mechanical loading history. Mechanical stimuli regulated cell division and differentiation and determined the tissue type and architecture. Well-controlled implant loading seemed to accelerate the formation of mineralized tissues at the interface. Histological evidence of clinical successfully osseointegrated implants was rare in the literature, especially after a period of functional loading of more than 1 year [52-59]. Moreover, it could also, perhaps, be useful to evaluate the healing events at the interface after different time periods. Histological data pertaining to IL implants demonstrate that IL didn't produce untoward effect in the bone healing. Histological evidence showed that, even with shorter healing periods (4, 6, and 8 weeks), it was possible to observe the formation of mineralized tissue at the interface (Table 1) [60-61].

Even in poor bone sites, a high bone-implant contact percentage were observed. In the spongious area the implant an almost continuous thin shell of newly formed bone usually covered surfaces. Mineralized tissues were found covering a large portion of the implant surface with no foreign body or inflammatory reactions visible. Bone remodeling was present in an area around the implant. The histological and histomorphometrical analysis on the interface of immediately loaded implants retrieved from humans showed a high percentage of BIC percentage even after a longer loading period.

Table 1. Histomorphometrical data (Bone-to-implant contact – BIC) of human immediately loaded implants

Implant system	Specifics	Loading (mo.)	No. of implants	Area	BIC (%)
3i®	Osseotite®	4	2	mandible	78-85 5.2 %
Nobel Biocare®	Ti-Unite®	6	1	maxilla	60 3.8 %
IMZ® Twin Plus	sandblasted/AE	10	2	mandible	64.5 4.7%
Maestro®	HA-coated/sandblasted	6	3	mandible	80.6 4.7 %
Ankylos®	sandblasted/AE	6	1	mandible	74 6 %
Frialit-2®	sandblasted/AE	10	7	2x maxilla, 7x mandible	66.8 3.8%
XiVE®	sandblasted/AE	6	2	mandible 2nd molar	61-72 3.8 %
XiVE® DPS	sandblasted/AE	14	1	mandible	72.6 2.7%
XiVE® plus	sandblasted/AE	2	1	mandible	71 3.2 %
XiVE® DPS	sandblasted/AE	6	16	mandible	67.4 2.1 %
LaserLok	sandblasted/AE	4	3	maxilla	68 5 %
PHI	TPS	8-9	2	maxilla/mandible	60-70 %
Blades	machined	2-21 years	72	maxilla/mandible	41.3 2.2 - 89.2 1.1 %
Screws	machined	3-16 years	31	maxilla/mandible	56.3 5.0 % - 78 2.3
MEAN					69.28 3.1 %

In conclusion, the data from the observations of the interface of retrieved, clinically stable, immediately loaded implants showed that, independent of where they were placed in the maxilla or the mandible and the implant design, the concept of immediate loading allows new bone formation at the interface of oral implants. High BIC percentages seemed to be possible (Figures 16-18).

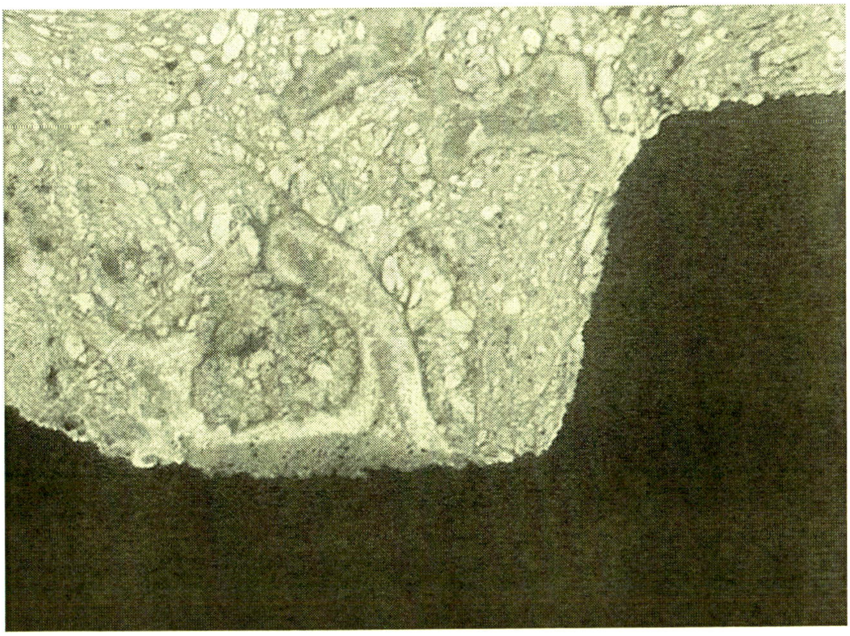

Figure 16. Implant retrieved after 4 weeks. Trabecular bone with large marrow spaces is present. Newly formed bone can be observed inside the implant thread (Basic fuchsin and toluidine blue staining, original 100X magnification).

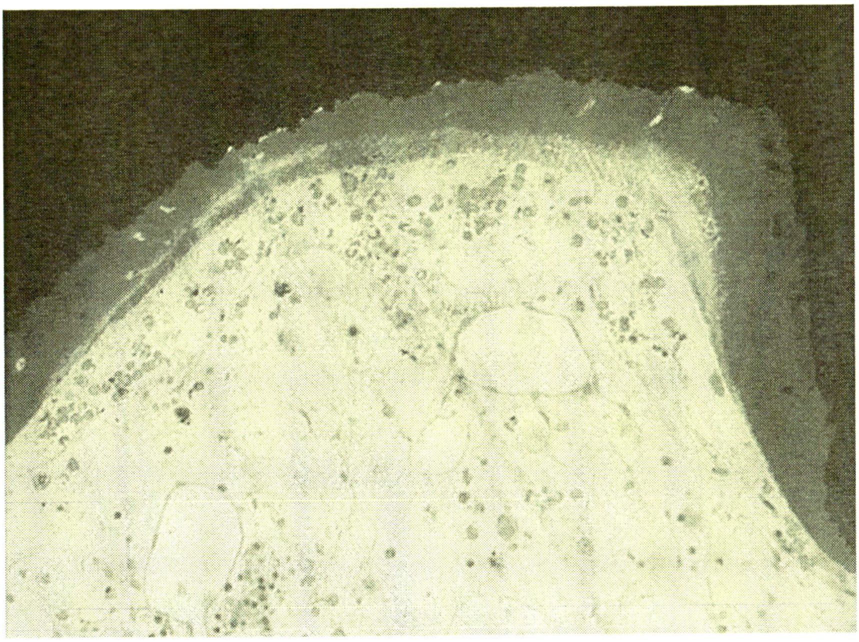

Figure 17. Implant retrieved after 8 weeks. Newly formed bone inside the implant thread can be observed (contact osteogenesis)(Basic fuchsin and toluidine blue staining, original 100X magnification).

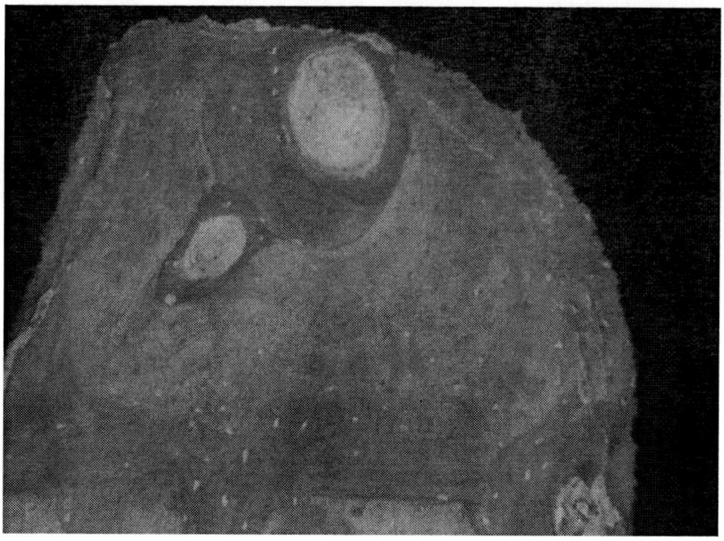

Figure 18. Implant retrieved after 7 years. Mature bone inside the implant thread can be observed. Remodelling areas are present (Basic fuchsin and toluidine blue staining, original 100X magnification).

4. Implants Inserted in Poor Bone Sites

An important parameter that influences the long-term success of oral implants is the bone quality of the implant bed. Posterior areas of the jaws have been avoided in implant dentistry due to their poor bone quality, higher chewing forces and presumed higher implant failure rates. Soft bone implant sites have been deemed by several researchers to be a great potential risk situation and most failures have been found in sites where the bone density from the start was low. Increasing the rate of early endosseous integration was a critical goal to achieve improved success rates.

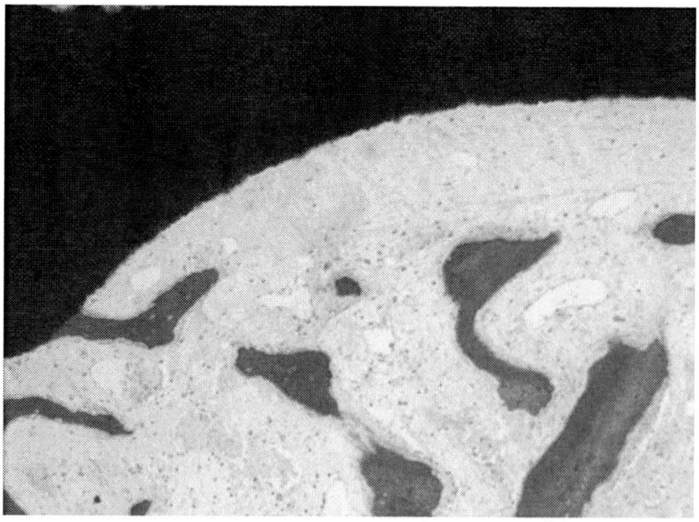

Figure 19. Implant retrieved from maxilla. Trabecular bone with large marrow spaces. (Basic fuchsin and toluidine blue staining, original 100X magnification).

Figure 20. Implant retrieved from maxilla. Small trabeculae in close proximity of the implant surface. (Basic fuchsin and toluidine blue staining, original 40X magnification).

The surface microtexture of the implants has been shown increasingly to be of relevant importance in the early stages of osseointegration. Some microstructured surfaces have an improved characteristic of contact osteogenesis even in soft bone, with coverage of the implant surface by a bone layer as a base for intensive bone formation and remodelling. A high BIC was observed even in implants inserted in poor bone sites (Figures 19 and 20) [62].

4.1. HA Coated Implants

A coating of the titanium surface with a layer of hydroxyapatite (HA) has been proposed to get a higher osseointegration rate, a faster bone attachment, a reduction in the healing time, a higher interfacial strength to bone, an enhancing of the load-stress distribution to the surrounding bone, better maintenance of the bone crest height, increased potential for guided bone regeneration. Concerns about the degradation of the coating over the years have been raised: it has been speculated that the resorption of the HA could produce a space between implant and bone with a resultant mechanical instability.

There is a risk for degradation of the coating, which can weaken the bone bond giving rise to implant failure. There have been some concerns about the long-term integrity of the coating *in vivo*, and the fact that the dissolution and detachment of the coating could expose the underlying metal surface. It has been hypothesized that this fact could have adverse effects on interfacial bone-implant apposition and on the bone-implant interface stability. Moreover, the breakdown of the coating could produce particulate material with phagocytic response by macrophages or a foreign body reaction. Furthermore, implant failure has been associated with the loss of coating integrity, and studies of failed HA-coated orthopaedic femoral stems have shown areas of coating degradation and separation. Moreover, the duration of the advantage of the HA coating is unclear.

Coating dissolution and detachment from the titanium surface have been described histologically. There is some controversy whether the loss of the HA coating could be detrimental to the integration of the implant. Porous HA is resorbed through physicochemical dissolution and cell-mediated phagocytosis. Additional dissolution may be due to the phagocytic and enzymatic action of macrophages recruited to the surface. However, histological studies have shown that after many years of function, HA-coated implants continued to demonstrate adequate BIC percentages. This fact seemed to support the view that an adequate stability of the HA-coated prostheses was maintained despite the coating loss. In some specimens, the almost complete resorption of the HA coating didn't appear to have interfered with the osseointegration processes. In these cases, the HA coating resorption, probably, did not have a great clinical significance because the implant was osseointegrated and was still providing an adequate function. Haversian systems were observed in close proximity to the implant surfaces in HA-coated implants and this fact pointed to a physiologic remodelling activity of the peri-implant bone. No foreign body reaction was observed associated with the HA particles that appeared to be detached from the coating. These particles were always surrounded by bone [63-68].

4.2. Implants in Patients with Osteoporosis

Osteoporosis is a disease that influences the quality of bone tissue so that it may become susceptible to fracture. While animal studies have described the deleterious effect of osteoporosis on osseointegration, no clinical studies showed a clear association between implant failure and osteoporosis. The mechanism by which osteoporosis acted on peri-implant bone was based on the decrease in both cancellous bone volume and BIC, consequently reducing bone tissue to support dental implants. However, in studies in humans, BIC was found to be similar for both osteoporosis and non-osteoporosis subjects. In conclusion, the results of the histomorphometrical studies, in implants retrieved from humans, suggested that osteoporosis might not present an absolute contra-indication for implant placement, at least, after osseointegration has been established (Figures 21 and 22) [69-72].

Figure 21. Implant retrieved in patients with osteoporosis. Mature cortical bone with no remodelling areas (Basic fuchsin and toluidine blue staining, original 200X magnification).

Figure 22. Implant retrieved in patients with osteoporosis. No osteoblastic activity can be detected. A gap is present at the bone-implant interface. (Basic fuchsin and toluidine blue staining, original 200X magnification).

4.3. Immediate Post-Extraction Implants

Subsequent to the removal of all teeth in the adult individual, the alveolar processes will undergo atrophy. Marked alterations of the height and width of the alveolar ridge will occur following single or multiple tooth extractions. The healing process following tooth removal apparently resulted in more pronounced resorption on the buccal than on the lingual/palatal aspects of the ridge.

So, after tooth extraction, the resorption and remodeling of the alveolar socket could result into a site that would be inadequate, from a dimensional point of view, for the implant placement. When an implant was placed into an extraction socket, osteogenic and osteoresorptive responses were already initiated following extraction, and this tissue could enhance the capacity for healing. An immediate implant is one implant that was placed into an extraction socket at the same time the tooth was extracted. Immediate post-extraction implants have several advantages such as fewer surgical procedures, preservation of bone volume, and shortening of the time needed until the implants could be restored. Additional advantages of the use of the immediate post-extraction implants were:

1. Shortening of the edentulous time period
2. Reducing the costs of treatment
3. Improve the psychological approach with the patient
4. Reduction of the comprehensive treatment time with less surgical procedures and morbidity
5. Optimal esthetic result, with an easier definition of the implant position as a consequence of correct fixture position and angulation
6. Improvement of biomechanics of the future restoration

Several different human clinical studies have demonstrated that with immediate post-extraction implants it was possible to obtain very high (more than 90%) long-term success percentages. Moreover, many experimental studies have confirmed that a high percentage of BIC could be achieved on light microscopic level in animals, when using immediate post-extraction implants. One major drawback in using immediately post-extraction implants was due to the lack of adaptation of the alveolar bone in the cervical portion of the implant. Soft tissues, creating problems in the osseointegration of the implant, could fill this space. Almost always, when using immediate post-extraction implants, it was necessary to resort to osteopromoting techniques, with the use of biomaterials and membranes. However, in a histological study aimed at evaluating the outcome of implantation in fresh extraction sockets without the use of membranes in humans in comparison with implants placed in healed, mature alveolar bone, no significant differences in the clinical and radiographic parameters were observed between the 2 experimental categories. Bone resorption was not present in any area of the histological sections [73-75].

5. Implants Inserted in Grafted Sites

5.1. Sinus Augmentation Procedures

The successful outcome of a sinus augmentation procedure can be evaluated best by a histological examination of the events at the bone-implant interface. A successful implant osseointegration in sinus augmentation procedures should be characterized by a high quantity of newly formed bone at the implant interface, to provide enough bone for mechanical support and integration of the implants. A bone substitute material should have the capability to allow the integration of loaded titanium implants. One of the most important questions about sinus augmentation procedures is if the regenerated bone obtained after the insertion of a graft is able to integrate dental implants. Other important questions are the extent of the surface of a dental implant placed into a grafted sinus which will be surrounded by bone in direct contact with the implant surface, and to what extent functional ankylosis will be present and if the obtained implant osseointegration will remain stable over the long period, after functional loading. Very high BIC was reported in implants retrieved from sinuses augmented after a period varying from some months to several years. All these implants had osseointegrated and had remained osseointegrated after over several years of loading. It has been reported that grafted particles in contact with the implant surface could lead to reduce mechanical support for the dental implants. No contact was, however, observed between grafted particles and implant surfaces in most of the reported implants. The continued presence of grafted particles in the peri-implant bone did not seem jeopardize the integration of the implant because no contact between the graft particles and the implant surface was observed, and a complete resorption of the grafted material did not seem to be a prerequisite needed to get formation of bone at the interface and implant osseointegration. On the other hand, the lack of complete resorption of the grafted material could even be advantageous in order to maintain the initial dimensions of the grafted area with time.

Other human histologic specimens retrieved from grafted sinuses after longer time periods will certainly help to clarify the question of the biomaterial resorption over time and of the potential of regenerated bone to achieve and maintain osseointegration with dental implants. Provisional implants are helpful in helping the patients to avoid the inconveniences of wearing a denture and they can, moreover, provide useful information after retrieval (Figures 23 and 24) [76, 77].

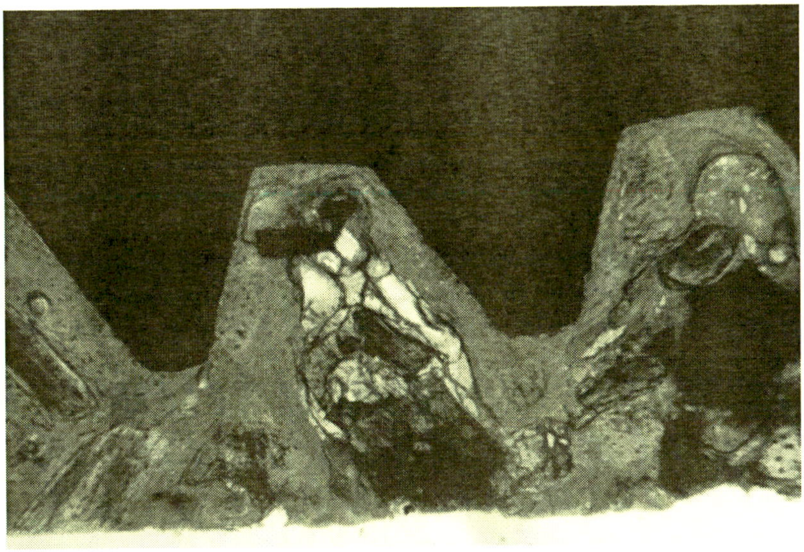

Figure 23. Implant retrieved from a site augmented with anorganic bovine bone and Pep-Gen P15 after 8 years. Compact bone in close contact to the implant surface is present. Residual biomaterial particles can be observed (Basic fuchsin and toluidine blue staining, original 40X magnification).

Figure 24. Implant retrieved from a site augmented with bovine bone and Pep-gen P15 after 8 years. No particles can be seen in contact with the implant surface. (Basic fuchsin and toluidine blue staining, original 40X magnification).

5.2. Iliac Grafts

It seemed that the characteristics of the regenerated bone issue and of the bone implant interface in implants inserted in grafted bone were similar to those seen in nongrafted situations.

6. Implants in Smokers

The influence of smoking on peri-implant bone has been evaluated in a several histologic animal models. The majority of these studies agree that smoking had a detrimental effect on bone healing, BIC and bone mineral density. Smoking delays the normal bone healing process by a mechanism that inhibits proliferation of precursor cells. Cigarette smoking is composed of over 4,000 toxins that had the potential to undermine the peri-implant bone healing. Toxins such as nicotine, carbon monoxide, nitrosamines, benzenes, aldehydes, and hydrogen cyanide have been shown to affect essential processes of bone healing. Nicotine was a potent vasoconstrictor that reduced not only blood flow and nutrient delivery to the surgical implant site but also inhibited the proliferation of fibroblasts, red blood cells and macrophages. Carbon monoxide decreased the oxygen carrying capacity of red blood cells, while hydrogen cyanide leads to hypoxia.

In human retrieved specimens, BIC% was found to be significantly lower in smokers. A tendency toward slower wound repair has been suggested. Moreover, cigarette smoking reduced the rate of bone formation and increased the rate of bone destruction in post-menopausal women. Cigarette smoking seemed to suppress osteoprotegerin levels and might contribute towards the decreased peri-implant bone formation. However, the precise mechanisms by which smoking exerted its deleterious effects on bone healing remained unclear (Figures 25 and 26) [78-81].

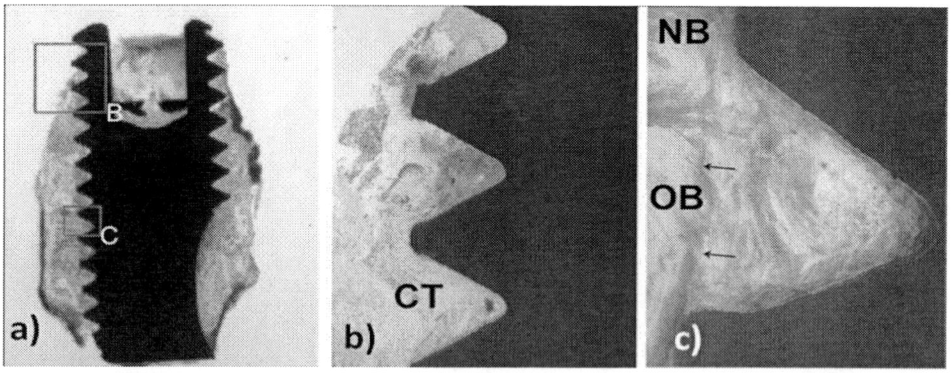

Figure 25. (A) Histological ground section of an oxidized micro-implant surface, showing the newly-formed bone at early maturing and remodelling stages, retrieved from a posterior mandible of smoker (Basic fuchsin and toluidine blue staining, original magnification 12x); (B) A Larger magnification of the lateral area in the section shown in (A). Note the presence of marginal bone loss and connective tissue (CT) in contact with the implant surface (Basic fuchsin and toluidine blue staining, original 40x magnification). C) A higher magnification of the lateral area in the section shown in (A). The arrow shows the reversal line between preexisting (OB) and a mixture of new bone (NB) and remaining bone from the drilling process. (Basic fuchsin and toluidine blue staining, original 100x magnification).

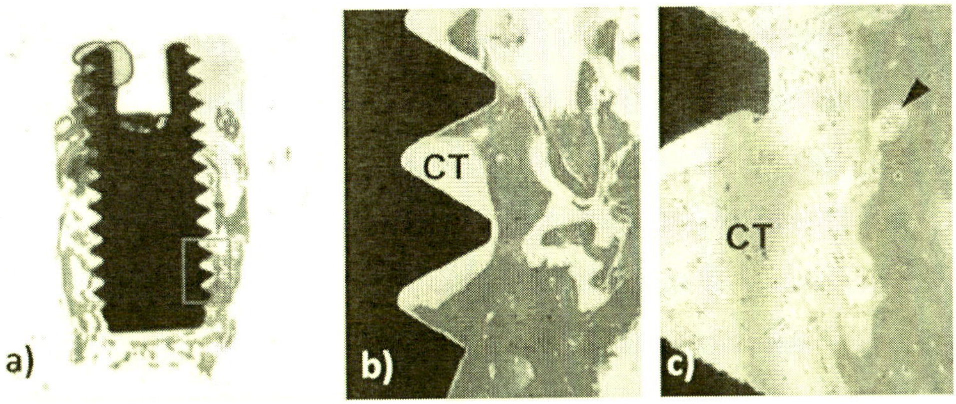

Figure 26. (A) Histological ground section of a micro-implant retrieved after 8 weeks of healing from a posterior maxilla of a smoker with newly-formed bone showing early maturing and remodelling stages. (Basic fuchsin and toluidine blue staining, original 12x magnification); (B) Higher power view of the lateral area in the section shown in (A). The newly formed bone tissue shows areas of direct contact with the oxidized implant surface, although in some areas there are also a lack of connecting bridges between the new bone and the implant surface (Basic fuchsin and toluidine blue staining, original 40x magnification); (C) Gap and connective tissue (CT) are present between newly formed bone and the implant surface. CT was loose with scattered inflammatory cells. The arrow shows an osteoclast. (Basic fuchsin and toluidine blue staining, original 100x magnification).

7. Peri-Implant Soft Tissues

Current knowledge of the histological and histomorphometrical features of the supracrestal peri-implant soft tissues was constituted, for the most part, by data obtained on studies in dogs or non-human primates. The distance from the mucosal margin to the alveolar bone was called biological width (BW). Around teeth the BW had a constant vertical dimension providing the gingival esthetics, and was composed by sulcus depth, junctional epithelium, and connective tissue attachment. The peri-implant soft tissues played a relevant role in the protection of the underlying bone from the invasion of oral bacteria. Around implants, the supracrestal soft tissues had many similarities to the dentogingival tissues around teeth and were composed by an epithelium and a connective tissue.

In human retrieved specimens, the sulcular epithelium (SE) was composed of about 4-5 layers of parakeratinized epithelial cells and had a length of about 1.2-1.3 mm. The junctional epithelium (JE) was composed of about 3-4 layers of epithelial cells and had a length of about 1.0-1.5 mm. Connective tissue attachment had a width of between 400 and 800 μm. Collagen fibers, in form of bundles, were oriented perpendicularly to the abutment surface, up to a distance of 200 μm from the surface, where they became parallel running in several directions. In some areas it was possible to see collagen fibers bundles oriented perpendicularly or obliquely to the section plane. In the area neighbouring to the abutment surface, the CT contained a few blood vessels, and dense collagen fibers, oriented parallel to the longitudinal axis of the abutment, were present. Collagen fibers oriented in a perpendicular way and inserting directly contacting on the abutment surface were not observed. Collagen fibers appeared to form a 3-dimensional network around the abutment. This differentiated network of fibers may have a clinical relevance as a mechanical protection of the underlying bone. The similarities between the dimension of the human peri-implant soft tissues and those described around teeth suggest that also the peri-implant BW is a physiologically formed and stable over time (Figures 27 and 28) [82].

Figure 27. Crestal bone with wide osteocyte lacunae, and a rim of osteoblasts deposing osteoid matrix. Dense connective tissue without inflammatory infiltrate. (Basic fuchsin and toluidine blue staining, original 20x magnification).

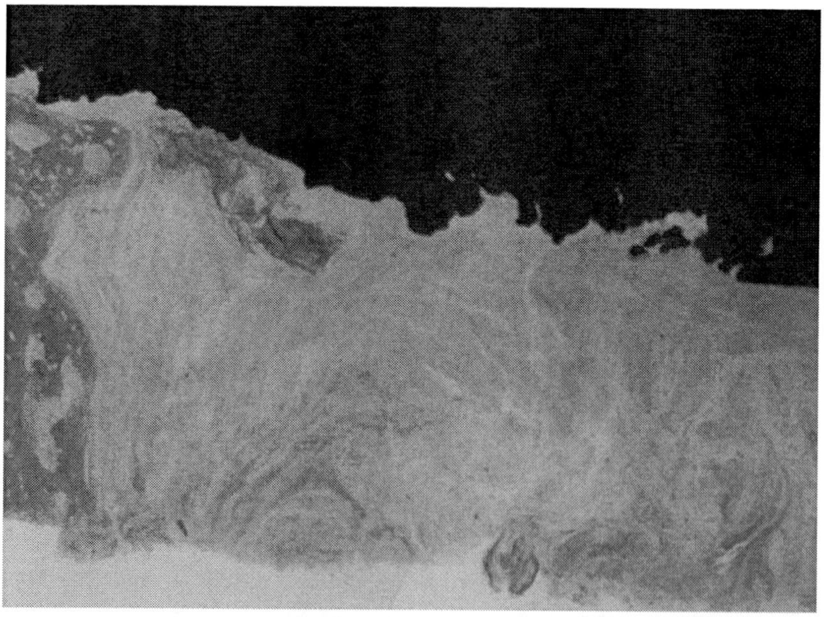

Figure 28. Higher magnification of the previous image (Basic fuchsin and toluidine blue staining, original 40x magnification).

REFERENCES

[1] Piattelli, A., Scarano, A. and Piattelli, M. (1998) Histologic observations on 230 retrieved dental implants: a 8 years experience (1989-1996). *Journal of Periodontology*, 69, 78-184.

[2] Mangano, C., Perrotti, V., Iezzi, G., Scarano, A., Mangano, F. and Piattelli A. (2008) Bone response to a new modified titanium surface implants in non-human primates (papio ursinus) and humans. Histological evaluations. *Journal of Oral Implantology*,34,17-24.

[3] Piattelli, A., Scarano, A., Favero, L., Iezzi, G., Petrone, G. and Favero G. (2003) Clinical and histological aspects of dental implants removed because of mobility. *Journal of Periodontology*, 74, 85-390.

[4] Piattelli, A., Scarano, A., Piattelli, M., Vaia, E. and Matarasso, S. (1998) Hollow implants retrieved for fracture: a light and scanning electron microscope analysis of 4 cases. *Journal of Periodontology*, 69, 185-189.

[5] Piattelli, A., Piattelli, M., Scarano, A. and Montesani, L. (1998) Light and scanning electron microscopic study of 4 fractured implants. *International Journal of Oral and Maxillofacial Implants*, 13,561-564.

[6] Sbordone, L., Traini, T., Caputi, S., Scarano, A., Bortolaia, C. and Piattelli, A. (2010) Scanning electron microscopy fractography analysis of fractured hollow implants. *Journal of Oral Implantology* ,36,105-111.

[7] Orsini, G., Fanali, S., Scarano, A., Petrone, G., Di Silvestro, S. and Piattelli, A. (2000) Tissue reactions, fluids and bacterial infiltration in implants retrieved at autopsy. *International Journal of Oral and Maxillofacial Implants*,15,283-286.

[8] Piattelli, A., Piattelli, M., Mangano, C. and Scarano, A. (1998) A histologic evaluation of eight cases of failed dental implants: is bone overheating the most probable cause? *Biomaterials*, 19,683-690.

[9] Piattelli, A., Scarano, A., Dalla Nora, A., De Bona, G. and Favero, G.A. (1998) Microscopical features in retrieved human Branemark implants: a report of 19 cases. *Biomaterials*, 19,643-649.

[10] Piattelli, A., Scarano, A., Piattelli, M., Vaia, E. and Matarasso, S. (1999) A microscopical evaluation of 24 retrieved failed hollow implants. *Biomaterials*, 20,485-489.

[11] Degidi, M., Scarano, A., Petrone, G. and Piattelli A. (2003) Histological analysis of clinically retrieved immediately loaded implants: a report of 11 cases. *Clinical Implant Dentistry and Related Research*,5,89-94.

[12] Romanos, G., Degidi, M., Testori, T. and Piattelli, A. (2005) Histological and histomorphometrical findings from human retrieved immediately functional loaded implants. *Journal of Periodontology*, 76,1823-1832.

[13] Degidi, M., Scarano, A., Iezzi, G. and Piattelli, A. (2003) Peri-implant bone in immediately loaded titanium implants: an histologic and histomorphometric evaluation in man. A report of 2 cases *Clinical Implant Dentistry and Related Research*,5,170-175.

[14] Degidi, M., Scarano, A., Piattelli, M., Perrotti, V. and Piattelli, A. (2005) Bone remodeling in immediately loaded and unloaded titanium implants: a histologic and histomorphometric study in man. *Journal of Oral Implantology*, 31,18-24.

[15] Traini, T., Degidi, M., Caputi, S., Strocchi, R., Di Iorio, D. and Piattelli, A. (2005) Collagen fiber orientation in human peri-implant bone of immediately loaded titanium dental implants. *Journal of Periodontology*, 76,83-89.

[16] Traini, T., Degidi, M., Strocchi, R., Caputi, S. and Piattelli, A. (2005) Collagen fiber orientation near dental implants in human bone: do their organization reflect differences in loading? *Journal of Biomedical Materials Research Part B: Applied Biomaterials*, 74B,538-546.

[17] Piattelli, A., Scarano, A. and Piattelli, M. (1996) Microscopical aspects of failure in osseointegrated dental implants: a report of five cases. *Biomaterials*, 17,1235-1241.

[18] Traini, T., Iezzi, G., Pecora, G. and Piattelli, A. (2006) Preferred collagen fiber orientation in the human peri-implant bone after short and long-term loading periods: a case. *Journal of Oral Implantology*,32,177-181.

[19] Traini, T., Degidi, M., Iezzi, G., Artese, L. and Piattelli, A. (2007) Comparative evaluation of the peri-implant bone tissue mineral density around unloaded titanium dental implants. *Journal of Dentistry*,35,84-92.

[20] Traini, T., Degidi, M., Murmura, G., Piattelli, A. and Caputi, S. (2007) Bone microstructure evaluation near unloaded dental implants combining confocal scanning electron microscopy and SEM backscattered electrons imaging. *International Journal of Immunopathology and Immunopharmacology*, 20:(S1) 37-41.

[21] Barros, R.R.M., Degidi, M., Novaes, A.B. Jr, Piattelli, A., Shibli, J.A. and Iezzi, G. (2009) Osteocyte density in the peri-implant bone of immediately loaded and submerged dental implants *Journal of Periodontology*, 80,499-504 .

[22] Degidi, M., Piattelli, A., Shibli, J.A., Perrotti, V. and Iezzi, G. (2009) Early bone formation around immediately restored implants with and without occlusal contact: a histologic and histomorphometric evaluation in man. A case report. *International Journal of Oral and Maxillofacial Implants*, 24,734-739.

[23] Degidi, M., Perrotti, V., Strocchi, R., Piattelli, A. and Iezzi, G. (2009) Is insertion torque (IT) correlated to bone-implant contact percentage (BIC) in the early healing period? A histological and histomorphometrical evaluation of 17 human retrieved dental implants *Clinical Oral Implants Research*,20,778-781 .

[24] Degidi, M., Perrotti, V., Piattelli, A. and Iezzi, G. (2009) Mineralized bone implant contact (MBIC) and Implant Stability Quotient (ISQ) in 16 human implants retrieved after early healing periods: a histological and histomorphometrical evaluation. *International Journal of Oral and Maxillofacial Implants*, 24,45-48.

[25] Piattelli, A., Scarano, A. and Quaranta, M. (1997) High-precision, cost-effective cutting system for producing thin sections of oral tissues containing dental implants. *Biomaterials*, 18,577-579.

[26] Grassi, S., Piattelli, A., de Figuereido, L.C., Feres, M., Iezzi, G. and Shibli, J.A. (2006) Histologic evaluation of early human bone response to different surfaces. *Journal of Periodontology*, 77,1736-43.

[27] Shibli, J.A., Grassi, S., de Figuereido, L.C., Feres, M., Marcantonio, E., Iezzi, G. and Piattelli, A. (2007) Influence of implant surface topography on early osseointegration. A histological study in human jaws. *Journal of Biomedical Materials Research Part B: Applied Biomaterials*, 80B,377-385.

[28] Degidi, M., Piattelli, A., Shibli, J.A., Strocchi, R. and Iezzi, G. Bone formation around a dental implant with a platform switching and another with a TissueCare connection. A histologic and histomorphometric evaluation in man .

[29] Shibli, J.A., Mangano, C., D'Avila, S., Piattelli, A., Pecora, G., Mangano, F., Onuma, T., Cardoso, L.A., Ferrari, D.S., Aguiar, K.C. and Iezzi, G. (2010) Influence of direct laser fabrication implant topography on type IV bone : a histomorphometric study in humans *Journal of Biomedical Materials Research*,93A,607-614.

[30] Piattelli, A., Podda, G. and Scarano, A. (1996) Histological evaluation of bone reactions to Al_2O_3 dental implants in man: a case report. *Biomaterials*, 17,711-714.

[31] Piattelli, A., Emanuelli, M., Scarano, A. and Trisi, P. (1996) A histological study of nonsubmerged titanium plasma-sprayed screw implants retrieved from a patient: a case report. *International Journal of Periodontics and Restorative Dentistry*,6,138-147.

[32] Piattelli, A., Scarano, A., Vaia, E. and Matarasso, S. (1996) Histologic evaluation of the peri-implant bone around plasma-sprayed non-submerged titanium implants retrieved from man: a report of two cases. *Biomaterials*, 17,2219-2224.

[33] Piattelli, A., Corigliano, M., Scarano, A. and Quaranta, M. (1997) Bone reactions to early occlusal loading of two-stage titanium plasma-sprayed implants: a pilot study in *International Journal of Periodontics and Restorative Dentistry*,17,163-169.

[34] Piattelli, A., Paolantonio, M., Corigliano, M. and Scarano, A. (1997) Immediate loading of titanium plasma-sprayed screw-shaped implants in man: a clinical and histological report of two cases. *Journal of Periodontology*,68,591-597.

[35] Piattelli, A., Scarano, A., Piattelli, M., Bertolai, R. and Panzoni, E. (1997) Histologic aspects of the bone and soft tissues surrounding three titanium non-submerged plasma-sprayed implants retrieved at autopsy: a case report . *Journal of Periodontology*,68,694-700.

[36] Piattelli, A., Corigliano, M. and Scarano, A. (1996) Microscopical observations of the osseous responses in early loaded human titanium implants: a report of two cases. *Biomaterials*,17,1333-1337.

[37] Testori, T., Szmukler-Moncler, S., Francetti, L., Del Fabbro, M., Scarano, A., Piattelli, A., Weinstein, R. (2001) Immediate loading of osseotite implants. A case report and histologic analysis after 4 months of occlusal loading. *International Journal of Periodontics and Restorative Dentistry*,21,451-459.

[38] Degidi, M., Petrone, G., Iezzi, G. and Piattelli, A. (2003) Bone contact around acid-etched implants: a histological and histomorphometrical evaluation of two retrieved implants. *Journal of Oral Implantology*,29,13-18.

[39] Orsini, G., Piattelli, M., Scarano, A., Petrone, G., Kenealy, J., Piattelli, A. and Caputi, S. (2007) Randomized-controlled histological and histomorphometric evaluation of implants with nanometer-scale calcium phosphate added to the dual acid-etched surface in the human posterior maxilla. *Journal of Periodontology*,78,209-218.

[40] Grassi, S., Piattelli, A., Sanchez Ferrari, D., de Figuereido, L.C., Feres, M., Iezzi, G., Shibli, J.A. (2007) Histologic evaluation of human bone integration on machined and sandblasted and acid-etched titanium surfaces after 2 months of healing in type IV bone *Journal of Oral Implantology*,33,8-12.

[41] Degidi, M., Piattelli, A., Shibli, J.A., Perrotti, V. and Iezzi, G. (2009) Bone formation after 4 and 8 weeks around immediately loaded and submerged dental implants with a modified sandblasted and acid-etched surface. A histologic and histomorphometric

analysis in man *International Journal of Oral and Maxillofacial Implants*, 2009;24:896-901

[42] Degidi, M., Piattelli, A., Shibli, J.A., Perrotti, V. and Iezzi, G. (2009) Bone formation around one-stage implants with a modified sandblasted and acid-etched surface. Human histological results at 4 weeks. *International Journal of Periodontics and Restorative Dentistry*.29,607-613.

[43] Degidi, M., Petrone, G., Iezzi, G. and Piattelli, A. (2002) Histologic evaluation of a human immediately loaded titanium implant with a porous anodized surface. *Clinical Implant Dentistry and Related Research*, 4, 110-114.

[44] Shibli, J., Feres, M., de Figueiredo, L.C., Iezzi, G., Piattelli, A. (2007) Histologic comparison of bone to implant contact in two types of dental implant surfaces: a single case study. *Journal of Contemporary Dental Practice*, 8,29-36.

[45] Shibli, J.A., Grassi, S., de Figueiredo, L.C., Feres, M., Iezzi, G. and Piattelli, A. (2007) Human peri-implant bone response to turned and oxidized titanium implants inserted and retrieved after 2 months. *Implant Dentistry*,16,252-259.

[46] Piattelli, A. and Trisi, P. (1993) Microscopic and chemical analysis of bone hydroxyapatite interface in a human retrieved implant. A case report. *Journal of Periodontology*,64,906-909.

[47] Iezzi, G., Scarano, A., Petrone, G. and Piattelli, A. (2007) Two human hydroxyapatite-coated dental implants retrieved after a 14 years loading period. A histologic and histomorphometric case report *Journal of Periodontology*,2007;78:940-947

[48] Mangano, C., Piattelli, A., d'Avila, S., Iezzi, G., Mangano, F., Onuma, T. and Shibli, J.A. (2010) Early human bone response to laser metal sintering surface topography: a histologic report. *Journal of Oral Implantology*,36,91-96.

[49] Degidi, M., Iezzi, G., Scarano, A. and Piattelli, A. (2008) Immediately loaded titanium implant with a tissue stabilizing/maintaining design ("beyond platform switching") retrieved from man after 4 weeks. A histological and histomorphometrical evaluation. A case report. *Clinical Oral Implants Research*, 19,276-282.

[50] Degidi, M., Scarano, A., Iezzi, G. and Piattelli, A. (2005) Histologic analysis in man of an immediately loaded implant retrieved after 8 weeks. *Journal of Oral Implantology*,31,247-254.

[51] Iezzi, G., Pecora, G., Scarano, A., Perrotti, V. and Piattelli, A. (2006) Histological evaluation of 3 immediately loaded implants retrieved from man after a 4 months loading period: a case report. *Implant Dentistry*, 15,305-312.

[52] Piattelli, A., Degidi, M., Marchetti, C. and Scarano, A. (1997) Histologic analysis of the interface of a titanium implant retrieved from a non vascularized mandible block graft after a 10 month loading period *International Journal of Oral and Maxillofacial Implants*, 12,840-843.

[53] Degidi, M., Scarano, A., Iezzi, G. and Piattelli, A. (2005) Histologic and histomorphometric analysis of an immediately loaded implant retrieved from man after 14 months of loading. *Journal of Long Term Effects of Medical Implants*,15,489-498.

[54] Vantaggiato, G., Iezzi, G., Fiera, E., Perrotti, V. and Piattelli, A. (2008) A histological and histomorphometrical case report of 3 immediately loaded screw implants retrieved from man after a 3 years loading period . *Implant Dentistry*,19,192-199.

[55] Traini, T., De Paoli, S., Caputi, S., Iezzi, G. and Piattelli, A. (2006) Collagen fibers orientation near a fractured dental implant after a 5 years loading period: a case report *Implant Dentistry*, 15,70-76.

[56] Iezzi, G., Pecora, G., Scarano, A., Perrotti, V. and Piattelli, A. Immediately loaded screw implant retrieved from man after a 12 years loading period: a histological and histomorphometrical case report . J Osseointegration

[57] Piattelli, A., Scarano, A. and Paolantonio, M. (1997) Immediately loaded screw implant removed for fracture after a 15-year loading period: histological and histochemical analysis *Journal of Oral Implantology* ,23,75-79.

[58] Trisi, P., Emanuelli, M., Quaranta, M. and Piattelli, A. (1993) A light microscopy, Scanning Electron Microscopy and Laser Scanning Microscopy Analysis of retrieved blade implants after 7 to 20 years of clinical function. *Journal of Periodontology*,64,374-378.

[59] Di Stefano, D., Iezzi, G., Scarano, A., Perrotti, V. and Piattelli, A. (2006) Immediately loaded blade implant retrieved from man after a 20 years loading period: a histological and histomorphometrical case report . *Journal of Oral Implantology*,32,171-176.

[60] Degidi, M., Petrone, G., Iezzi, G. and Piattelli, A. (2003) Histologic evaluation of 2 human immediately loaded and 1 submerged titanium implants inserted in the posterior mandible and retrieved after 6 months. *Journal of Oral Implantology*,29,223-229.

[61] Degidi, M., Scarano, A., Piattelli, M. and Piattelli, A. (2004) Histologic evaluation of an immediately loaded titanium implant retrieved from a human after 6 months in function. *Journal of Oral Implantology*,30,289-296.

[62] Iezzi, G., Degidi, M., Scarano, A., Perrotti, V. and Piattelli, A. (2005) Bone response around submerged unloaded implants inserted in poor bone sites: a retrospective histological and histomorphometrical study of 8 titanium implants retrieved from man. *Journal of Oral Implantology*,31,225-233. HA COATED IMPLANTS.

[63] Piattelli, A., Trisi, P. and Emanuelli, M. (1993) Bone reactions to hydroxyapatite coated dental implants in man: an histological study by means of SEM, light microscope and laser scanning microscope. *International Journal of Oral and Maxillofacial Implants*, 8,69-74.

[64] Piattelli, A., Trisi, P., Romasco, N. and Emanuelli, M. (1993) Histological analysis of a screw implant retrieved from man: influence of early loading and primary stability. *Journal of Oral Implantology*,19,303-306.

[65] Piattelli, A. and Trisi, P. (1994) Light and laser scanning microscopy study of bone/hydroxyapatite-coated titanium implants interface: Histochemical evidence of unmineralized material in humans. *Journal of Biomedical Materials Research*, 28,529-536.

[66] Piattelli, A., Cosci, F., Scarano, A. and Trisi, P. (1995) Localized chronic suppurative bone infection as a sequel of peri-implantitis in a hydroxyapatite-coated dental implant *Biomaterials*, 16,917-920.

[67] Piattelli, A., Scarano, A., Di Alberti, L. and Piattelli, M. (1999) Bone-hydroxyapatite interface in retrieved hydroxyapatite-coated titanium implants: a clinical and histologic *International Journal of Oral and Maxillofacial Implants*, 14,233-238.

[68] Iezzi, G., Orlandi, S., Pecora, G. and Piattelli, A. (2009) Histologic and histomorphometric evaluation of the bone response around an hydroxyapatite-coated

implant retrieved after 15 years. *International Journal of Periodontics and Restorative Dentistry*, 29,99-105.

[69] Shibli, J.A., Grande, P.A., d'Avila, S., Iezzi, G. and Piattelli, A. (2008) Evaluation of human bone around dental implant retrieved from subject with osteoporosis. *General Dentistry*, 56,4-67.

[70] De Melo, L., Piattelli, A., Iezzi, G., d'Avila, S., Zenobio, E.G. and Shibli, J.A. (2008) Human histologic evaluation of a six-year-old threaded implant retrieved from a subject with osteoporosis. *Journal of Contemporary Dental Practice*,9,99-105.

[71] Shibli, J.A., Aguiar, K.C., De Melo, L., Ferrari, D.S., d'Avila, S., Iezzi, G. and Piattelli, A. (2008) A histological analysis of human peri-implant bone in type 1 osteoporosis. *Journal of Oral Implantology*,34,12-16.

[72] Shibli, J.A., Melo, L., d'Avila, S., Zenobio, E.G., Faveri, M., Iezzi, G. and Piattelli, A. (2008) Histological comparisons on implants retrieved from subjects with and without osteoporosis. *International Journal of Oral and Maxillofacial Surgery*,37,321-327.

[73] Cornelini, R., Scarano, A., Covani, U., Petrone, G. and Piattelli, A. (2000) Immediate one-stage post-extraction implant. A clinical and histologic case report in man. *International Journal of Oral and Maxillofacial Implants*,15,432-437.

[74] Paolantonio, M., Dolci, M., Scarano, A., D'Archivio, D., Di Placido, G., Tumini, V. and Piattelli A. (2001) Immediate implantation in fresh extraction sockets. A controlled clinical and histological study in man . *Journal of Periodontology*,72,1560-1571.

[75] Guida, L., Iezzi, G., Annunziata, M., Iuorio, G., Costigliola, G., Salierno, A. and Piattelli, A. (2008) Immediate placement and loading of dental implants: a human histologic case report. *Journal of Periodontology*,79,575-581.

[76] Iezzi, G., Fiera, E., Scarano, A., Pecora, G. and Piattelli, A. (2007) Histological evaluation of a provisional implant retrieved from man after 7 months after insertion in a sinus augmented with calcium sulphate. A case report. *Journal of Oral Implantology*,33,89-95.

[77] Iezzi, G., Scarano, A., Mangano, C., Cirotti, Piattelli, A. (2008) Histologic results in a human implant retrieved for fracture 5 years after insertion in a sinus augmented with anorganic bovine bone. A case report. *Journal of Periodontology*,79,192-198.

[78] Shibli. J.A., Vitussi, T.R.C., Garcia, R.V., Zenobio, E.G., Ota-Tsuzuki, C., Piattelli, A. and d'Avila, S. Implant surface analysis and microbiologic evaluation of failed implants retrieved from smokers. *Journal of Oral Implantology,*33,232-238.

[79] Shibli, J.A., Vitussi, T.R.C., Garcia, R.V., Zenobio, E.G., Ota-Tsuzuki, C., Cassoni, A., Piattelli, A. and d'Avila, S. (2007) Implant surface analysis and microbiologic evaluation of failed implants retrieved from smokers. *Journal of Oral Implantology*,33,232-238.

[80] Shibli, J.A., Piattelli, A., Iezzi, G., Cardoso, L.A., Onuma, T., Perri de Carvalho, P.S., d'Avila, S., Ferrari, D.S., Mangano, C., Zenobio, E.G. (2010) Effect of smoking on early bone healing around oxidized surfaces: a prospective controlled study in human jaws. *Journal of Periodontology*,81,575-583.

[81] D'Avila, S., Delfino Dos Reis, L., Piattelli, A., Aguiar, K.C.S., de Faveri, M., Borges, F.L., Iezzi, G., Oliveira, N.T.C., Cardoso, L.A., Shibli, J.A. (2010) Impact of smoking on human bone apposition at different dental implant surfaces: a histologic study in type IV bone. *Journal of Oral Implantology*, 36,85-90.

[82] Romanos, G., Traini, T., Johansson, C. and Piattelli, A. (2010) Biological width and morphologic characteristics of soft tissues around immediately loaded implants. Studies performed on human autopsy specimens. *Journal of Periodontology*,81,70-78.

In: Biomechanics of Dental Implants
Editor: Murat Çehreli

ISBN 978-1-62100-780-7

Chapter 7

Finite Element Analysis in Dental Implant Biomechanics

Roberto S. Pessoa and Siegfried V. N. Jaecques

1. Introduction

The replacement of lost teeth by dental implants has proven to be a reliable treatment modality in a variety of clinical scenarios. However, regardless of the rehabilitation protocol, the predictability and long-term success of implant treatment are greatly influenced by the biomechanical environment. The intimate bone-implant contact at the interface allows the direct transmission of the loads applied over the implant-supported prostheses to the surrounding bone. The stress concentration can exceed bone's tolerance level, cause microdamage accumulation and induce bone resorption [1-3]. Under certain conditions, this excessive occlusal loading may cause the failure of successfully osseointegrated implants [4,5]. In the immediate loading protocols, the overall requirement is to control interfacial movement between the implant and the surrounding bone. Micromovements that exceed 150 μm can induce fibrous connective tissue formation instead of the desirable bone regeneration [6-8]. Hence, adverse forces over the implant-supported prosthetic device could not only cause abutment screw loosening and mechanical failures, but could also impair osseointegration.

Several factors are recognized to influence the biomechanical environment which the implants are exposed to, such as: bone quality in the insertion area, the nature of the bone-implant interface, the materials' properties of the implants and prosthesis, the surface roughness of the implant material, the occlusal condition (i.e., magnitude, direction and frequency of the loading), and the design of the implant [3,9-12]. Thus, from a biomechanical point of view, a key factor for the predictability of implant protocols is the development of implants and prosthesis designs capable of providing some degree of stability, under masticatory standard loading. However, the elaborate design of the implants and their relationship with the supporting tissues and prosthetic restoration prevent the use of simple analytical formulas for the evaluation of the effect of external loading on the internal stresses and displacements. In these analysis types, the finite element method (FEM) has provided valuable data, for a relatively low operational cost and time investment [13-15]. This technique implies the solution of a complex mechanical problem, by dividing an intricate geometry into smaller and simpler geometric domains called elements [16]. These elements are interpolated by nodes, which allow the determination of displacement, stress and strain resulting from external force, not only in each element, but also in the whole analyzed structure. In implantology, finite element analysis (FEA) has been applied to predict the biomechanical behavior of various implant designs, clinical scenarios and prosthesis designs [17-21].

However, the success of a FEA study depends on the proper simulation of geometries, interface characteristics, material properties, loading and support conditions of the implant and surrounding bone, as well as on the correct interpretation of the obtained results. In this way, the present chapter aims to discuss the methodology, applications and limitations of finite element techniques in implant dentistry.

2. FEA Basic Concepts

The finite element analysis (FEA) is a technique by which a physical prototype can be evaluated through a detailed mathematical model. This method uses a computer to solve a large set of mathematical equations, which simulate the physical properties of the evaluated structure [16].

The FEA is essentially a discretization process, in which an infinity-dimensional problem is transformed into a finite-dimensional problem, with a finite number of unknown quantities. The method consists in dividing the domain over which the problem is studied into many smaller interrelated regions, so called elements. Thus, the finite elements are subdivisions of a model, sufficiently small to justify an analytical approximation in each of these elements and also in the combination of its effects. These elements are interconnected through union points (inside and/or at the border) called nodal points or nodes (Figure 1). The interpolation functions allow, once the displacement in each node is determined, to interpolate displacements and to calculate the stress and strains in any point of the structure. The set of elements used in the discretization is called a mesh [16]. The type, arrangement and total number of elements, as well as the interpolation function, are important factors affecting the accuracy of the results.

Although FEA is an indisputably valuable and efficient tool for the analysis of diverse Biomechanical problems, it is important to understand the limitations of the method. FEA uses approximated mathematical models to represent the behavior of physical systems. In addition, the elaboration of a finite element model for an implant evaluation involves the acquisition of the implant system and bone geometries, the assignment of appropriate materials properties and interface conditions, and the determination of proper boundary conditions, loading magnitude and direction. Simplifications in any of these stages will have a direct effect on the model's precision. Thus, to validate a finite element model, it is recommended to confront its previsions with data derived from other analysis types, in particular, from experimental measurements.

2.1. Individualized Finite Element Modeling

In one of the first publications using FEA in implant dentistry, Atmaram and colleagues [22] evaluated five different materials and three different geometries of implants. The authors reported that the optimal combination of implant material and geometry may reduce the stress in the implant system and surrounding tissues threefold, compared with an arbitrary design. From those initial studies on, new implant designs and materials have been developed and the indications for dental implants have been amplified. Likewise, once the importance of the Biomechanics on the predictability and long term success of implant treatment had been confirmed, more researchers have become interested in this assessment modality.

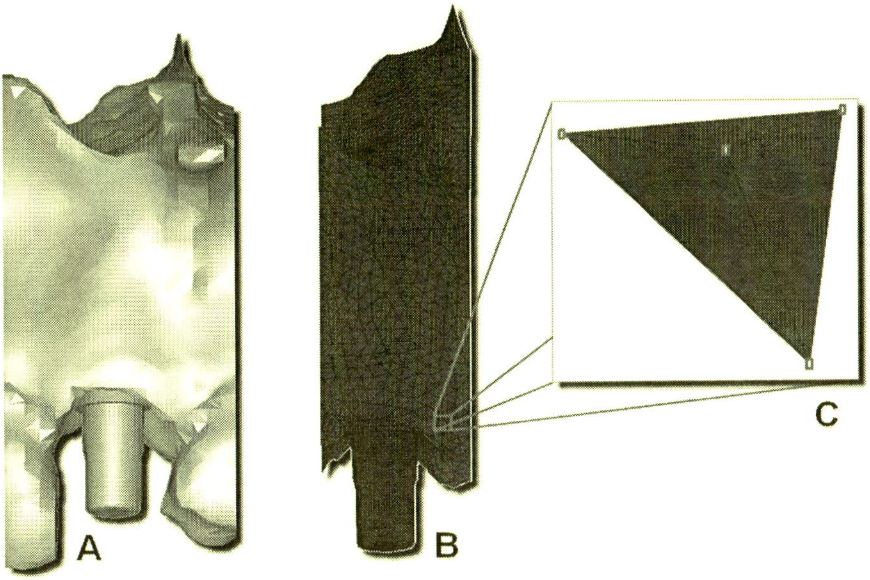

Figure 1. A – CAD solid model. B – Finite element model. C – Tetrahedral element used in the discretization of the model. Note the nodes at the element vertices.

However, the finite element modeling of valid and precise three-dimensional (3D) maxillary bones and implant systems is difficult to accomplish using traditional modeling techniques. Early studies frequently assumed two-dimensional (2D) representation of implants and jawbone structures. In addition, some studies even failed to recognize the difference between cortical and trabecular bone, and failed to accurately describe the implant's geometry. At that time, the authors claimed that, in a comparative analysis, complex reality could be simplified, assuming that proportions and relative effects would reflect the actual clinical situation with sufficient accuracy. As those studies were merely concentrated on the comparison of different conditions, the relative values were supposed to still lead to a better qualitative understanding of the biomechanics around implants. Furthermore, 3D models would lead to a huge number of nodes and, thus to unacceptably high computational costs, considering the limited performance of the hardware available at the time. Hence, little attention was paid to the extent to which the modeling method significantly influences the results. Considering these limitations, and although those pioneer studies were extremely important for the development of FEA in the implantology field, 2D FEA can no longer be accepted as a clinically useful guidance to the implant treatment. As a matter of fact, the in vivo stress and strain state is a three-dimensional problem and the meticulous representation of the object to be modeled greatly influences the accuracy of a FEA.

On the other hand, the evaluation of the local mechanical loading stimulus determinant in the processes of tissue differentiation and bone formation/resorption around a given implant involves a detailed knowledge of the stress and strain state in the peri-implant bone [23]. The finite element (FE) model used to address this aim necessarily should incorporate the site-specific bone anatomy, bone density distribution, implant position and in vivo measured implant loads [24]. In addition, the assumptions made during the process of developing the numerical model, especially regarding the assignment of material properties and interfaces conditions, entail a rigorous experimental validation. However, once a relationship between the numerical results and the real in vivo biomechanical environment of an implant is established, it is possible to determine the specific conditions that accelerate or consolidate osseointegration, as well as the aspects related to implant failure. This information may be further applied in the optimization of implant designs as a function of the biomechanical parameters beneficial to the peri-implant bone. In this way, especially considering the recent changes in osseointegrated implant usage clinical protocols, individualized FEA can contribute to more accurate treatment decisions, diminishing the risks of implant failure.

2.2. Bone Geometry

One major difficulty in simulating the mechanical behavior of dental implants is the modeling of human bone tissue. The substantial complexity of the mechanical characteristics of bone and its interaction with implant systems has led to major and often incorrect simplifications made in early FEA. However, for an accurate individualized FE model, the correct bone and implant geometry as well as the correct position of the implant with respect to the bone must be incorporated in the model. A state of the art method to perform such a task involves X-ray computer tomography (CT) images and computer image processing technique. CT images have become a valuable tool to create anatomical, patient-specific finite element models of bony parts, either or not in combination with implants [23].

Once a set of digital CT images of the jawbone are obtained, an accurate 3D geometric model of jawbone and surrounding tissue can be readily established using image processing and reverse engineering approaches. Figure 2 shows the acquisition of CT images of an extraction socket of a dry maxilla, provided by the Department of Anatomy of the Faculty of Odontology at Araraquara (São Paulo State University, Brazil), taken by a Picker UltraZ CT scanner with a gantry tilt of 0°, at 120 kV acceleration voltage and 1mA current. The projection data were exported using the DICOM (Digital Imaging and Communication in Medicine) file format. The data set had a voxel size of 0.391 x 0.391 x 1.000 mm and consisted of contiguous slices with respect to the Z-axis. The higher the quality of the CT images, the better is the geometric representation of the bone derived from it. In some studies, because of the high resolution needed to visualize the bone geometry and the bone adaptive response to the loading regime, micro computer tomography (μCT) instead of conventional CT was used to create individualized, animal-specific FE models [1,10,11,23].

In the present example, bone segmentation and reconstruction of the bone geometry were accomplished by thresholding within an image-processing software (MIMICS®, Materialise,

Haasrode, Belgium) (Figure 3). Subsequently, in order to obtain the FE model, a computer-aided design (CAD) approach was followed. This means that, a geometrical model (CAD model), consisting of analytical surfaces and solids, was first created. If the interest in the study is on the biomechanical environment in the vicinity of the implant, only the relevant part of the bone must be reconstructed. A reconstruction of the entire maxilla, for instance, would have led to a larger FE model, with unnecessary higher time processing.

An analytical surface was fitted to the outer surface of the segmented bone. Based on this analytical surface, the bone solid model (volumetric model) was designed in a CAD system (Unigraphics). This CAD model was then imported to MSC.Patran® 2005r2 (MSC.Software, Gouda, the Netherlands) and meshed (Figure 4). Tetrahedral elements were used for both bone and implant system to ensure smooth contact at the interfaces, which would not have been possible when using hexahedral elements [25]. The volumetric meshes of bone and implant were generated based on the surfaces' standard triangulated language (STL) descriptions.

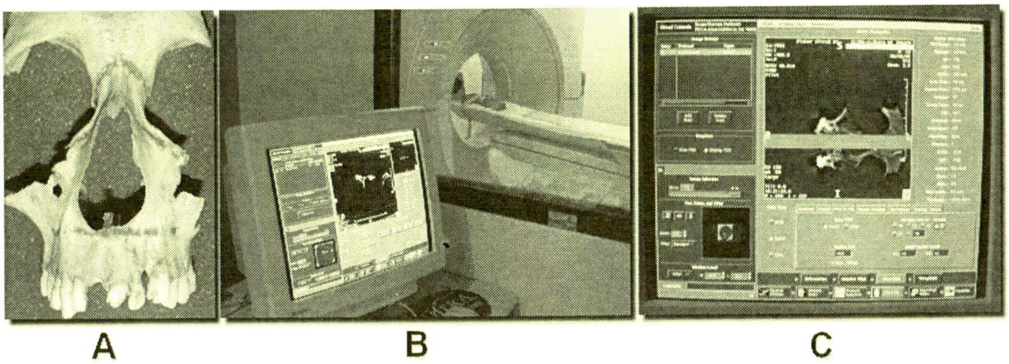

Figure 2. A – Dry maxilla. B, C – Acquisition of CT images of an extraction socket of the dry maxilla.

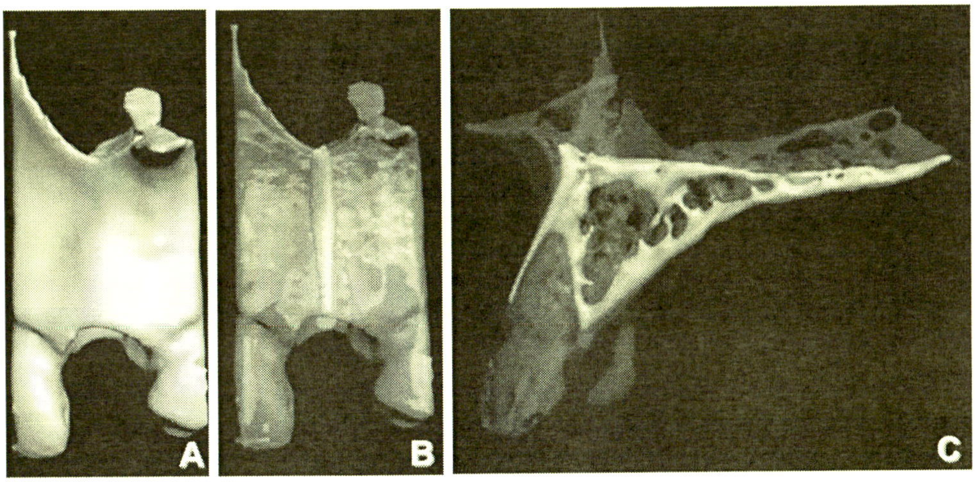

Figure 3. Extraction socket solid model. A – Buccopalatal view. B – Buccopalatal view, evidencingthe internal cortical bone structure. C – Lateral view, evidencing the internal cortical bone structure.

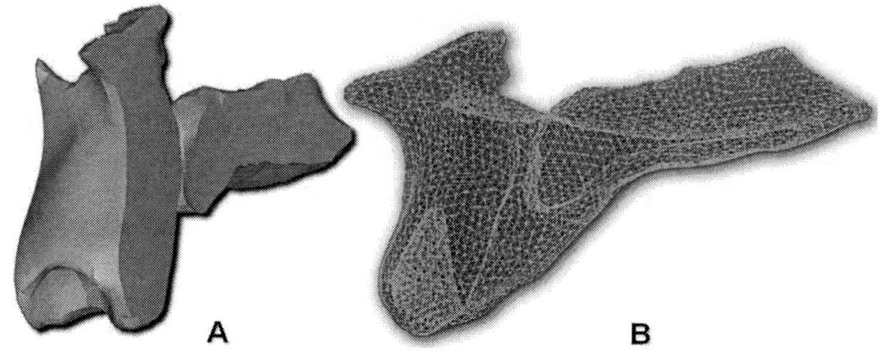

Figure 4. A – Extraction socket solid model. B – Extraction socket mesh.

2.3. Implant System Geometry and Positioning

Creating an accurate analytical model of a dental implant involves the modeling of all the possible aspects that may exert an influence within the region to be investigated. In producing realistic and reliable solutions, the modeling of the whole implant design is desirable (i.e. implant shape, length and diameter, thread design, implant-abutment connection design, abutment design and abutment screw design). Modeling assumptions and software limitations might lead to a number of inaccuracies within the obtained results. Figure 5 shows 3 examples of detailed implant system CADs (Ø 4.5 x 13 mm SIN SW® [SIN Sistema de Implante, São Paulo, Brazil], Ø 4.1 x 12 mm RN synOcta® ITI Standard [Institut Straumann AG, Basel, Switzerland] and Ø 4.3 x 13 mm Nobel Replace™ [Nobel Biocare AB, Göteborg, Sweden]) made by reverse engineering and exported as CAD models. The implant system CAD models were then separately meshed in MSC.Patran® 2005r2 (MSC.Software, Gouda, the Netherlands). Different degrees of mesh refinement were used for feature recognizing (i.e., at threads). In addition, as the results of FEA to some extent depend upon the size of the elements, the mesh was refined at locations with large stress gradients. The smaller elements used in this instance were about 50 μm.

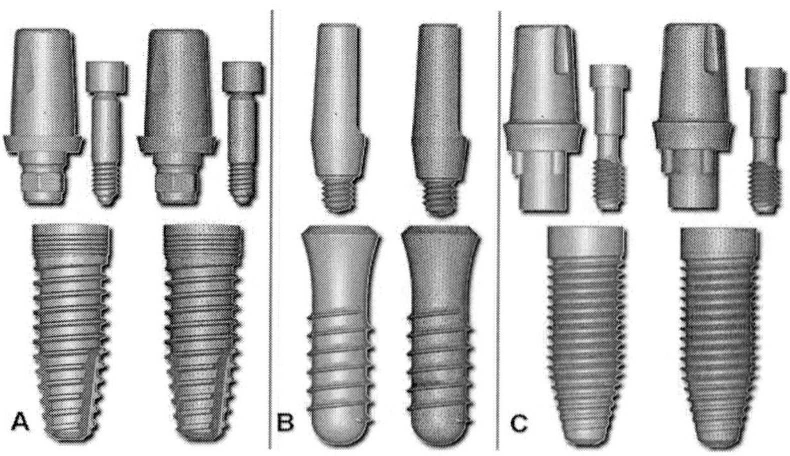

Figure 5. Implant, abutment and abutment screw solid models (left) and meshes (right). A – SIN SW®. B - ITI® Standard. C - Nobel Replace™.

This process allows the construction of high quality structured meshes in a relatively simple and effective way. Figure 6A shows a STL triangulation of a 3i® implant (3i, Palm Beach Gardens, USA) exported by SolidWorks® (Solidworks Corporation, Concord, MA, USA). The STL triangulation was then worked in Patran® and the resultant mesh is presented in figure 6B. Although a considerable improvement in STL tessellation can be noted in 6B implant mesh, an even higher quality could be reached when the mesh was constructed in Patran® based on the implant CAD models (Figure 6C). Furthermore, as can be clearly noted in figure 6A, STL triangulations cannot be directly used to generate FE meshes. In FEA, the elements must have a specific shape and a proper quality factor to achieve accurate predictions [26].

Moreover, no simplifications should be made regarding the implant external and internal threads (i.e., the spiral characteristic of the threads was maintained). There is a large body of biomechanical research, however, in which to simplify the modeling, the threads of the implant and the abutment were not represented in their spiral characteristics, but as symmetric rings. Huang and colleagues [27], comparing spiral and nonspiral threaded implants by FEA, found differences in stress values and distribution between the evaluated models. The author argued that a nonspiral threaded implant does not represent actual implant design and, consequently, the information obtained by concentric circle threads should be interpreted with some care.

After having created the bone solid model, the implant solid model must be correctly positioned with respect to the bone structures. In Figure 7, the implant was positioned 1 mm deep inside the extraction socket, in a central position, to a palatal direction. The contra-lateral teeth, as well as the socket anatomy, were used as a reference. Another option, when the CT images already contain a previously placed implant, is to align the implant solid model with implant contours extracted from the CT images [23]. A first approximation of the correct position is performed in Unigraphics. The final—more accurate—position of the implant is determined by means of a matching algorithm. This may be accomplished on the STL files within 3Matic® software (Materialise, Haasrode, Belgium) [28].

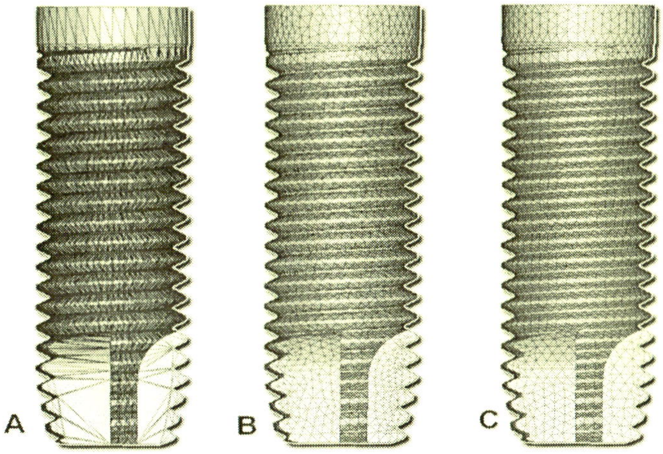

Figure 6. Different implant (3i®) mesh qualities. A – STL triangulation exported by SolidWorks®. B - Tetrahedral mesh constructed in Patran® based on the STL triangulation (A). C – Tetrahedral mesh constructed in Patran® based on the implant CAD model. Note the higher quality of 6C implant mesh.

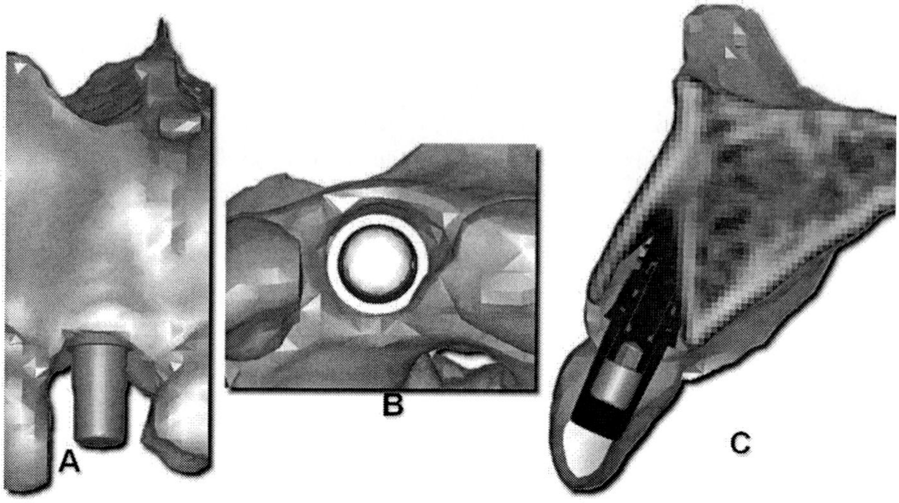

Figure 7. 3D CT-based model of an upper central incisor extraction socket and an implant model positioned inside of the alveolus. A – Frontal view. B – Occusal view. C – Proximal view.

After having obtained the correct position of the implant, the implant insertion hole in the bone solid model may be obtained by means of a Boolean subtraction. The abutment and abutment screw must be subsequently aligned to the implant, following the instructions by the implant producer.

2.4. Materials Properties of Bone and Implant Components

The quality and quantity of bone surrounding an implant greatly influences the load transfer from implant to the jawbone. Hence, it is important for the correct evaluation of the biomechanical environment around a given implant to implement in the FE model the local tissue heterogeneity in a patient-dependent manner. However, in the example given in figure 4, during meshing of the bone solid model, the entire volume that is contained within the outer bone surface is meshed. This means that the mesh consists of tetrahedral elements located in either cortical or trabecular bone. To discriminate between both tissues, different elastic properties can be assigned, based on the grey values in the CT images [20,21,23]. In this way, the information in the CT images may be used not only to extract the patient's bone geometry and but also to assign patient-specific bone mechanical properties (Figure 8).

Nevertheless, trabecular bone is an extremely porous structure, especially in this region of the maxilla. It consists of interconnected plates and rods of bone surrounded by marrow. At a microscopic level, the tissue elastic properties of each bone trabecula could be assigned. But in this case, once again the mesh density and, afterward, the computational costs would be unacceptable (unless high performace computing (HPC) resources such as a computing cluster are available). In the presented FE bone model the mechanical behavior of trabecular bone should be studied at the macroscopic level, by considering it as a continuum. Therefore, the stresses and strains calculated for trabecular bone must be interpreted in a qualitative rather that a quantitative way. On the other hand, the stresses and strains that were calculated in the cortical bone can also be interpreted quantitatively.

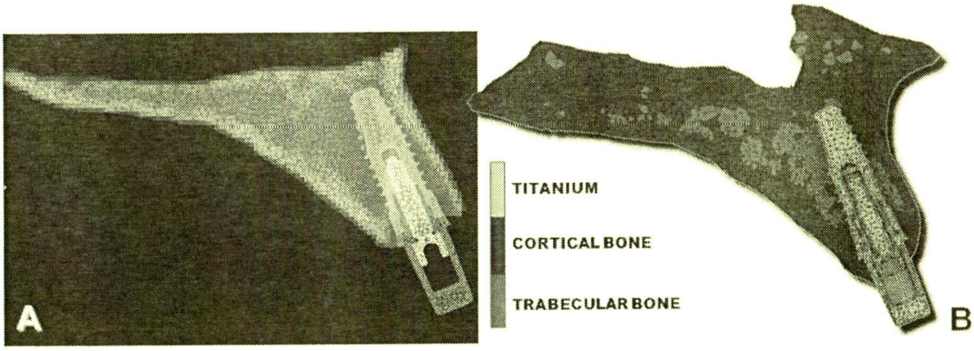

Figure 8. Assignment of CT based properties to the bone: buccopalatal medial plane view. A – Bone and implant system solid meshes superposed to the CT image. B – 3D FE model with the assigned material properties. Note the coincidence between the grey values in CT scan (A) and the elements assigned as cortical bone (B). Also note that the CT is a 2D slice, while the 3D FE model is shown as a sectional view starting in the plane of the CT slice and showing also the bone behind this plane.

The living cortical as well as trabecular bone tissues exhibit a certain degree of anisotropy, to some extent they are viscoelastic and contain voids [29]. However, as data on measurements of anisotropic properties from the human mandible and maxilla are still lacking today, even some of the more recent FEA studies considered the cortical and trabecular jawbones as isotropic, homogeneous, and linearly elastic [17,19,20,21,27]. The mechanical parameters necessary for the characterization of the mechanical behavior of the elastic materials are the Young's modulus and Poisson ratio. The values of these parameters for the materials used in different FEA may be found in the relevant literature [13,14]. Further investigations on the measurement of anisotropic properties and its influence on the stress pattern around oral implants are required.

On the other hand, the mechanical properties of an implant are very different to the original tissue. In general, dental implants are made from titanium and bio-ceramic materials, such as hydroxyapatite (HA). These materials present the advance of high compatibility with hard tissue and living bone [30]. Titanium has reasonable stiffness and strength, while HA has low stiffness, low strength and high ability to reach full integration with living bone. The majority of the previous and current FEA used titanium implants and components in their simulations. A high number of material properties have been adopted for titanium in these investigations, based on different material testing [13,14].

2.5. Interface Conditions

Linear static models have been employed extensively in previous FEA studies. These analyses usually assumed that all modeled volumes were bonded as one unit. However, the validity of a linear static analysis may be questionable when the investigation aims to explore more realistic situations that are generally encountered in the dental implant field. Some actual implant clinical situations will give rise to nonlinearities, principally related to the changing of interrelations between the simulated constituents of a FE model [15]. Moreover, frictional contact mode provides a greater fidelity with respect to the relative inter-components micro-motion within the implant system, and, therefore, a more reasonable representation of the real implant condition [31]. This configuration allows minor displacements between all components of the model without interpenetration. Under these conditions, the contact zones transfer pressure and tangential forces (i.e. friction), but not tension. Some FE analyses have shown remarkable differences in the values and even in the distribution of stresses between "fixed bond" and "non-linear contact" interface conditions [6,27,32]. Figure 9 presents a comparison between these two conditions of bone to implant interface, by a 3D FEA. The same interface condition (contact) was assumed for the implant system components in both situations. The results presented in figure 9 corroborates that not only the stress and strain levels but also the stress and strain distribution are highly affected by the interface state. Van Oosterwyck et al. [32] argued that through the bonded interface the force was dissipated evenly in both the compressive site and the tension site. However, on the contact interfaces, tensions are not transferred and force is only passed on through the compressive site, which results in excessive stresses. Additionally, note that the condition of the bone to implant interface also influences the strain distribution and level even inside of the implant system (Figure 9).

Conventionally, the osseointegrated bone to implant interface is treated as fully bonded. This assumption is supported by experimental investigations in which removal of rough implant implants frequently resulted in fractures in bone distant from the implant surface [33], suggesting the existence of an implant-bone "bond". On the other hand, frictional contact elements are used to simulate a nonintegrated bone to implant interface (i.e. in immediately loaded protocols), which allows minor displacements between the implant and the bone. The occurrence of relative motion between implant and bone introduces a source of non-linearity in the FEA, since the contact conditions will change during load application.

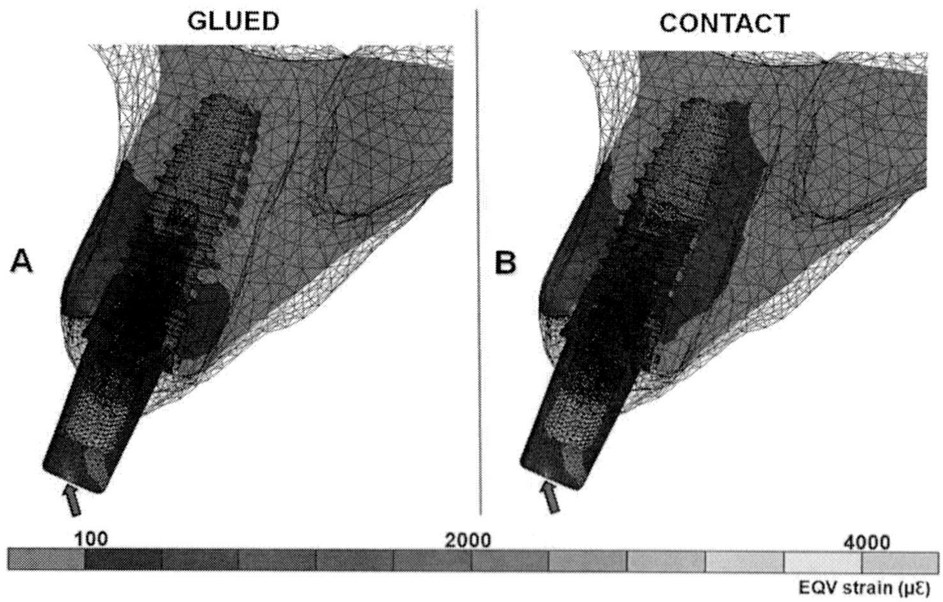

Figure 9. Equivalent strain (μ□) distribution for 100N loaded implant in a superior central incisor region, in a median buccopalatal plane. The bone to implant interfaces were assumed as fixed bond ("glued") (A) and frictional contact (B). The arrows indicate the loading direction for clarity.

The friction coefficient (μ) to be used in such simulations depends on many factors including mechanical properties and the roughness of the contact interface, exposure to interfacial contaminants [34] and in some cases the normal load [35]. A $\mu = 0.3$ was measured for interfaces between a smooth metal surface and bone, while a $\mu = 0.45$, for interfaces between a rough metal surface and bone [36]. Frequently, the frictional coefficient between bone and implant is assumed as being $\mu = 0.3$. In this direction, Huang et al. [27] investigated the effects of different frictional coefficients ($\mu = 0.3$, 0.45 and 1) on the stress and displacement of an immediately loaded implant simulation. The authors demonstrated that enlarging the value of μ shows no significant influence for increasing or decreasing the tensile and compressive stresses of bone. Nevertheless, increasing μ from 0.3 to 1, the interfacial sliding between implant and bone was mainly reduced from 20% to 30–60%, depending on the implant design [27].

Moreover, when an implant is surgically placed into the jawbone, the implant is mechanically screwed into a drilled hole of a smaller diameter. Large stresses will occur due to the torque applied in the process of implant insertion. As the implant stability and stress state around an immediately loaded implant may be influenced by such conditions, this should be also considered in a FE simulation of immediately loaded implants. However, this phenomenon has not been researched adequately yet. The implementation of such implant insertion stresses in a FEA is still unclear and should therefore be a matter of further investigations.

In addition, some earlier FEAs have undertaken linear solutions, underestimating the friction and torque between implant components. The solution of such linear FEA is simple, with a very low computational cost. However, perfect bonding between implant, abutment and abutment screw is not the actual scenario for dental implants. Non-linear contact analysis was proven to be the most effective interface condition for realistically simulating the relative micro-motions occurring between different components within the implant system [20,21,31]. For lateral or oblique loading conditions, specific parts can separate, or new parts that were initially not in contact can come into contact (Figure 10). Consequently, higher stress levels may be expected to occur in an implant-abutment connection simulated with contact interfaces, compared to a glued connection. In this regard, the pattern and magnitude of deformation will be influenced by the implant connection design [20,21,31].

Therefore, for correct simulation of an implant-abutment connection, frictional contact should be defined between the implant components. For the friction coefficients, Abkowitz et al. [37] reported a value of 0.5 for dry titanium to titanium friction and Steinemann et al. [38] found values of 0.43 to 0.53 for titanium in sodium chloride solution. Accordingly, between implant, abutment and abutment screw regions in contact, a frictional coefficient of 0.5 was generally assumed in non-linear simulations of implant-abutment connection [17,20,21,31].

Furthermore, besides the implant-abutment connection design, preload applied to the abutment screw was also considered to be a factor influencing the abutment stability [31,39]. The tightening process causes interference in the abutment screw, which in turn causes the threads of the abutment screw and the implant to engage with a positive force. This resulting tensile load in the abutment screw is known as the preload [40]. The achieved preload is influenced by the geometry of the screw, the contact between the screw head and abutment screw bore, the screw threads and the screw bore internal to the implant, friction between the various implant parts, and the material properties of the screws [41-43].

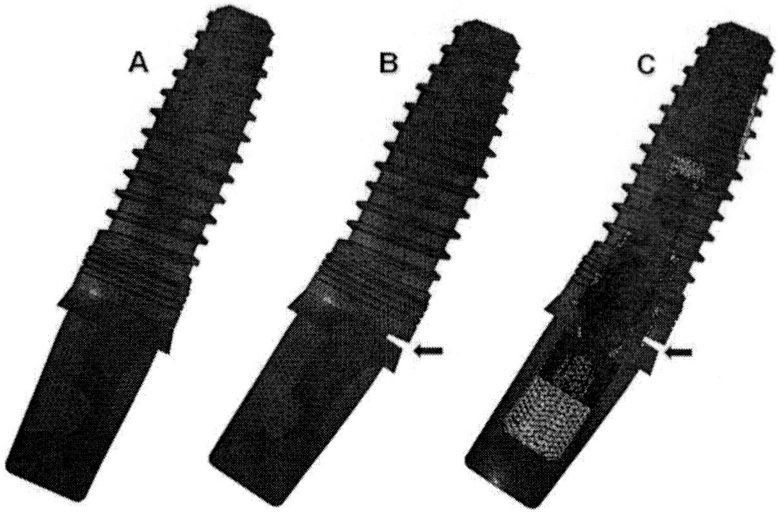

Figure 10. A – Stress distribution on implant-abutment surface, lateral view. B – The deformation is 10-fold magnified (note the abutment displacement; arrow indicates where contact is lost and non-linearity is introduced). C - The deformation is 10-fold, medial slice.

The implant-abutment joint efficiency, and therefore the strain state in the implant connection region, is considered a function of the design characteristics of the implant-abutment connection as well as, to some extent, of the preload stress achieved in the abutment screw when the suggested tightening torque is applied. Iplikçioglu et al. [44] compared the type and magnitude of strain on the implant collar assessed by FEA and strain gauge analysis (SGA). The authors found the same quality of strains on the implant collar, for lateral loading. However, FEA showed almost 2-fold higher strain than in vitro SGA. They argued that this finding may be dependent on several factors in FEA, including assumptions made during the construction of the mathematical model, the contact phenomenon, number and type of elements, and number of nodes used for calculation. Another reason for these differences may be lack of preload application in the Iplikçioglu et al. [44] FE model. The absence of preload may not have any effect during vertical loading, but will result in more separation at the screw joint under lateral loading. Considering that this clamping force has a considerable effect on the maintenance of abutment complex stability, a decreased amount of strain and separation is likely to be observed when preload is incorporated in FEA [44].

In this way, preload in the abutment screw is desirable to be included for a realistic FEA evaluation of the implant-abutment connection. Merz et al [31] included, in a non-linear FEA study, a torque value of 53.2 N for the taper joint and 358.6 N for the butt joint, which was determined by preload testing and calculations. The determined axial preload was introduced into the model with the help of a layer of temperature sensitive elements between the implant body and the cone or butt joint. These were submitted to a negative temperature difference, which resulted in a contraction such that the required tension was generated. Superior stress-relieving mechanics was indicated for the taper joint connection compared to the butt joint [31].

2.6. Loading and Boundary Condition

In order to successfully replicate, by means of FEA, the clinical situation that an implant might encounter in the oral environment, it is important to understand and correctly reproduce the natural forces that are exerted against the implants. These forces are mainly the result of the masticatory muscles action, and are related to the amount, frequency and duration of the masticatory function.

Forces acting on dental implants possess both magnitude and direction, and are referred to as vector quantities (Figure 11B). For accurate predictions on the implant–jawbone behavior, it is essential to determine realistic in vivo loading magnitudes and directions. However, at each specific bite point, bite forces can be generated in a wide range of directions. Moreover, although bite forces are generally supposed as acting downward, toward the apex of the implant, therefore tending to compress the implant into the alveolar bone, tensile forces and, principally, bending moments, may also be present depending on where the bite force is applied relative to the implant-supported prosthesis. This fact is even more important when the investigation aims to simulate multiple-implant models, because of the geometric factors involving the restorations which are linking the implants, such as the existence of distal cantilevers [45]. Nevertheless, although for implants used in single-tooth replacement simulations, in vivo forces ought to replicate the forces exerted on natural teeth, factors such as the width of the crown occlusal table, the height of the abutment above the bone level, and the angulation of the implant with respect to the occlusal plane will affect the value of the moment on the single-tooth implant.

A significant amount of investigations have assumed the direction of the load applied to the implant to be horizontal, vertical and oblique. Figure 11B shows vertical and horizontal components of load applied on the top of an abutment, at the central region. The resultant force simulates a palato-buccal static point oblique loading, with 45 degrees of inclination in relation to the implant longitudinal axis.

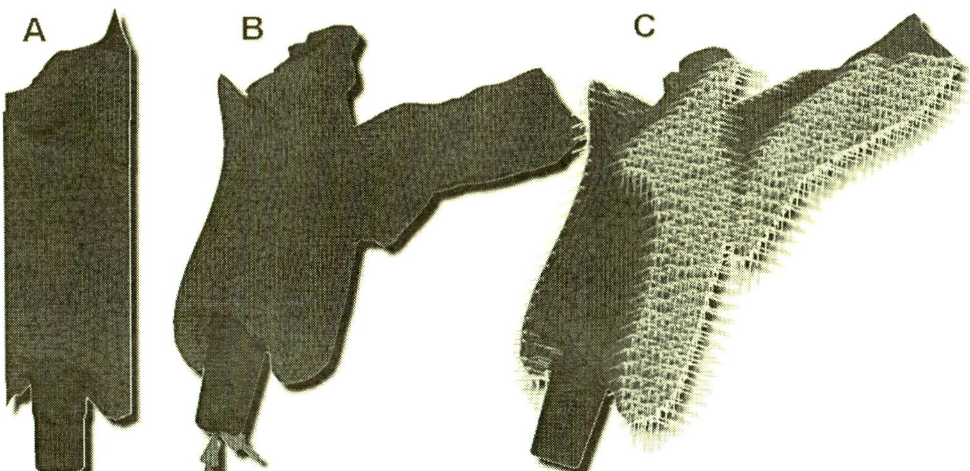

Figure 11. A – FE model of a superior central incisor extraction socket and the positioned implant. B – Loading applied on the abutment central region. C – Displacement constrain in all directions, at the nodes on mesial and distal borders.

The rationale for use of an oblique loading condition in this case was based on the finding that vertical (axial) forces directed to the implant system are relatively low and well tolerated in comparison to oblique forces, which generate bending moments, principally in the superior central incisor region. The selected location for loading was assumed to simulate the contact point with the antagonist tooth (Figure 11B).

Many investigators have tried to gain insight into implant loading magnitudes by performing tests using experimental, analytic, and computer-based simulations of various implant-supported prosthesis types [45-47]. Bite forces ranging from 50 to 400 N in the molar regions and 25 to 170 N in the incisor areas have been reported. These variations are influenced by patient's gender, muscle mass, exercise, diet, bite location, parafunction, number of teeth and implants, type of implant-supported prosthesis, physical status and age [46,47].

Ideally the entire jawbone structure should be evaluated for its contribution to the force exerted unto the dental implant. However, since the simulation of the whole mandibular and maxillary bone is very elaborate, smaller models have been proposed for parameter studies [17,19-21,27]. As already discussed above, if the interest in the study is on the biomechanical peri-implant environment, the modeling of no more than the relevant segment of the bone is required. This procedure allows saving computing and modeling time. In addition, Teixeira et al. [48] demonstrated, by a 3D FEA, that modeling the mandible at a distance greater than 4.2 mm mesially or distally from the implant did not result in any significant improvement in accuracy. Hence, besides the application of a proper implant loading, it is also essential for a reasonable FE modeling the determination of restrictions to the model displacement compatible with the anatomic segment to be simulated. In figure 11C, the model of a central incisor extraction socket was fully constrained in all directions at the nodes on mesial and distal borders, preventing the rotation of the maxillary section. Obviously, if necessary, expanding the domain of the model could reduce the effect of inaccurate modeling of the boundary condition [49].

2.7. Convergence Study

In a FEA, the results are an approximation rather than an exact solution. Under considerations of computational resource cost and limitations, the target to reach is the establishment of a compromise between the complexity of the model and the accuracy that is considered satisfactory.

Although in a comparative FEA study, to have similar element size meshes would be enough to eliminate the effect of different mesh density on the stress/strain values and distribution, the mesh refinement is a major factor in the achievement of an accurate model [50]. In addition, an accurate FE model would allow comparison with in vivo available data and further validation of the model. Therefore, following FEA best practice, a convergence test of the FE models should be performed to verify the mesh quality, ensuring that the numerical results are mesh-independent.

This requires comparison of the numerical results yielded by some different element size models, which have a difference in excess of 10% between the numbers of nodes of the meshes (Figure 12), thus ensuring that numerical convergence will be achieved without further mesh refinement. Figure 13 exemplifies a convergence study, in which the convergence criterion was considered as being less than 5% change of the peak von Mises stress (EQV Stress) at the bone-implant edge. In this region the highest stress levels were encountered in the first test analysis. These peaks were found to be in the same coordinates for all 5 mesh element sizes tested. Obviously, the time of processing was also considered as an exclusion criterion in the study. Based on the result of the convergence study, the optimal global element size for the bone mesh was 0.75 mm (Figure 13).

A FEA study can be used to gain insight into the biomechanics of oral implants and to verify some of the hypotheses that relate mechanical loading to peri-implant bone responses [1,7,10,23]. This can be achieved by a combination of the FEA and animal experiments. An individualized FE model, which incorporates specific bone geometry, implant position relative to the bone geometry and peri-implant bone quality may be created, in order to calculate the bone stresses and strains resulting from a loading experiment, and then relate them to the observed bone response of a given implant. In summary, animal experiments provide insights and measurements, which can then be interpreted within the context of FEA. FE simulations permit investigation of possible explanations that require in vivo validation and will suggest further experimental investigations. From this approach, important parameters for osteogenic, as well as for bone-resorptive mechanical stimulus can be identified.

On the other hand, even generic FE models, which intend to focus only on the relative influence of some implant parameter rather than to the absolute in vivo results, may be evaluated in respect to its coherence with biological available data [51]. Hence, it is possible to determine whether numerical models are consistent in their predictive capacity and whether the provided information could be extrapolated, or at least be useful, to the clinical context. Some criteria for adaptive bone modeling (bone gain and bone loss) have been proposed in relevant Biomechanical literature, and might be used as reference for FEA results.

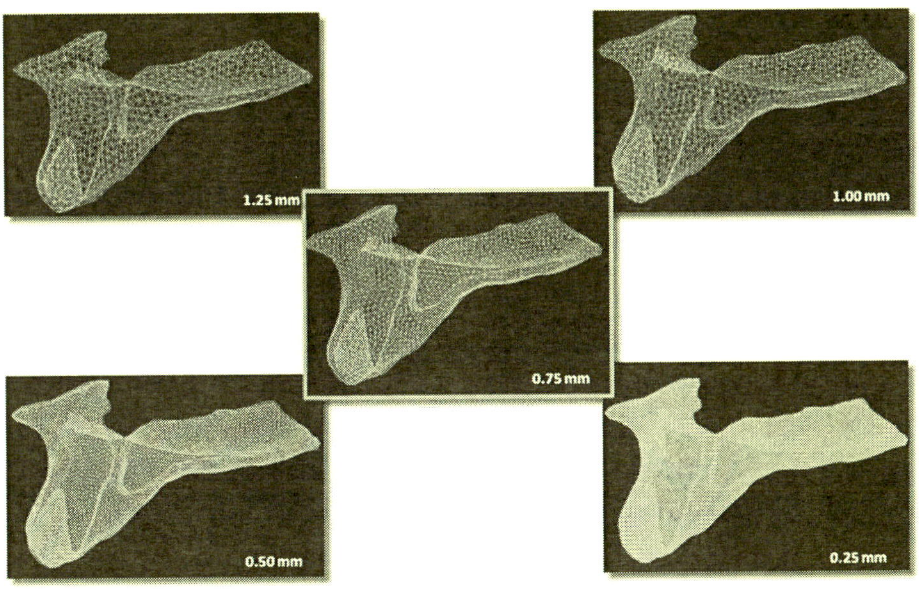

Figure 12. Different element size mesh models of a superior central incision extraction socket.

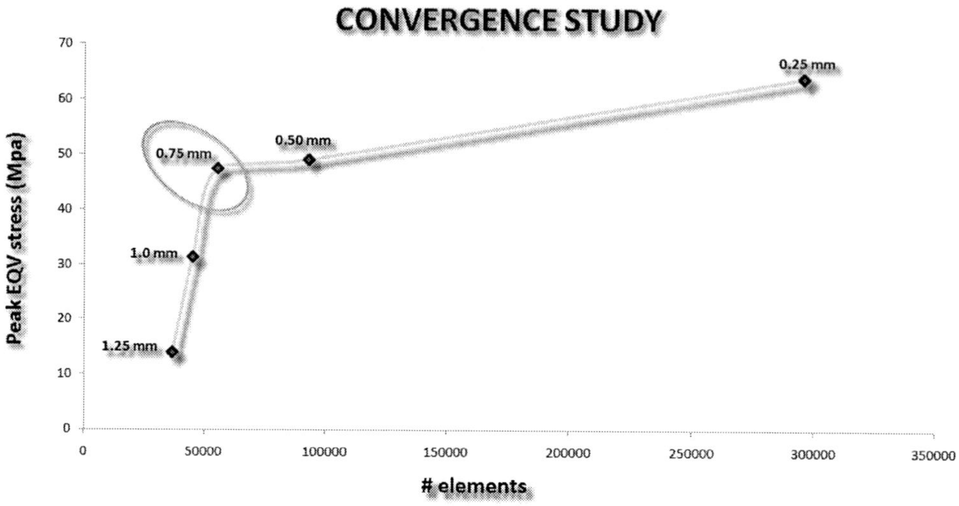

Figure 13. Influence of element size (1.25 mm, 1.0 mm, 0.75 mm, 0.50 mm and 0.25 mm) on bone mesh density and peak EQV stress in a bone model.

2.8. Reference Values for FEA Results

Although, precise determination of the loading level that separates mechanical loading into acceptable, osteogenic or failure-inducing levels is difficult and until now unresolved, some authors focused on the bone strain amplitudes as the mechanical stimulus determinant to bone adaptive process. It has been shown that above a certain strain threshold (1,500 μm), bone formation is initiated in cortical bone and that with increasing strain, bone responds with increasing formation activity [10,52-54. Underloading of the bone (< 100 μm) may lead to disuse atrophy and eventually to bone loss [55,56]. A possible threshold for pathological bone overload was considered by Frost (1992) as 4,000 μm. Also Duyck et al. [1], by FEA based on CT-images, estimated 4,200 μm as the value associated with overload-induced resorption.

Figures 14 and 15 show the strain distribution obtained by a FE model of an immediately placed and loaded implant. The model scale was set to range between 100-4,000μm to facilitate the visualization of the strain state in bone. Nevertheless, although the possible zones of bone underloading (< 100 μm, bone loss by disuse), normal load (100 – 1,500 μm, bone maintaining), mild overload (2,000 – 4,000 μm, bone gain) and pathologic overload (> 4000 μm, bone loss by microdamage accumulation) can be easily identified, this does not necessarily imply the bone remodeling in such areas. Rather than only strain amplitude, also loading frequency and number of loading cycles are parameters capable to greatly influence the cortical bone adaptive response [10, 57-60].

Furthermore, the loading applied in the presented simulation was static and bone responds to dynamic rather than to static loads [1,61-63]. In this way, it must be clear that the modeling of bone adaptive processes was not one of the aims in the given FEA example. In addition, the peaks of strain appeared locally in a minor part of the marginal bone, where indeed some localized bone resorption is likely to occur (Figure 15).

Likewise, principal stresses have been also used as a local risk indicator of physiological bone failure and of the activation of bone resorption at the bone-implant interface. Assuming ultimate bone strength as a physiological limit, local overloading at cortical bone arises in compression when the maximum compressive principal stress exceeds 170-190 MPa and in tension when the maximum tensile principal stress exceeds 100-130 MPa [64,65]. Ultimate compressive strength for cancellous bone was calculated as 5–6MPa [66]. Nevertheless, the strength of cortical bone varies between different patients, diverse anatomical locations and is also dependent upon the strain rate [67].

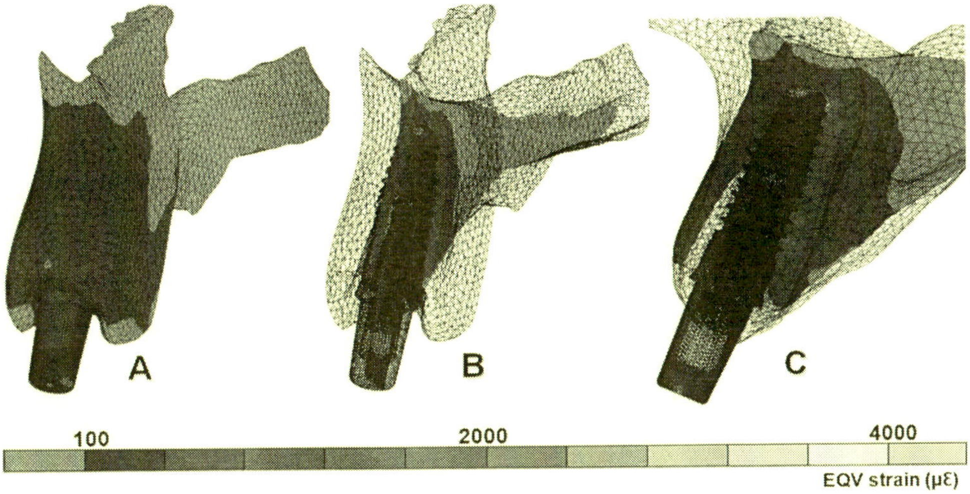

Figure 14. A – Strain distribution, latero-frontal view. B – bucco-palatal medial slice, lateral-frontal view. C – bucco-palatal medial slice, lateral view.

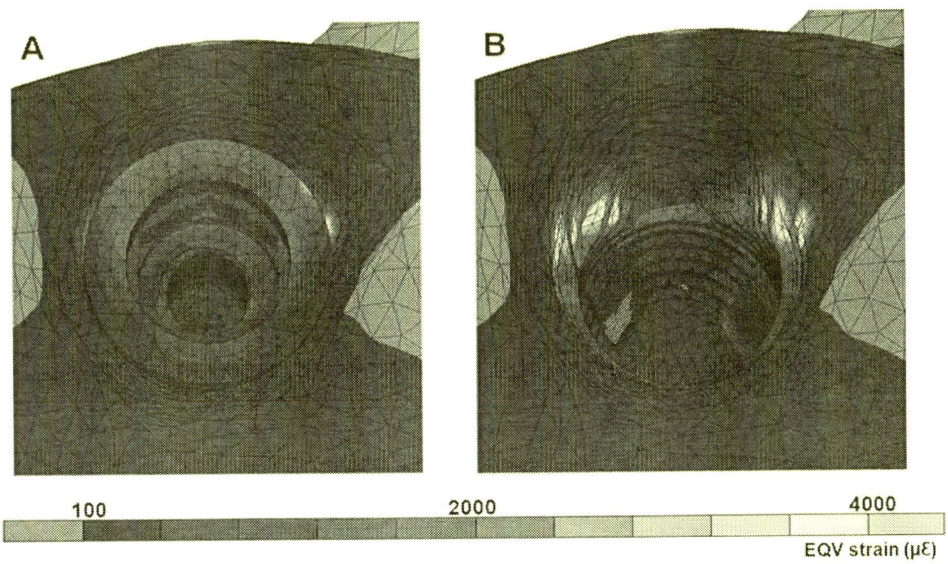

Figure 15. Occlusal view of the strain distribution in the alveolus. A – Implant positioned inside the alveolus. B – Strain distribution in bone (Note that the implants were removed for clarity).

On the other hand, a number of FEA studies have been using the von Mises stress to provide a global measure of load transfer mechanisms at the peri-implant region. However, considering the fact that bone responds differently to different types of stress [64], the Von Mises stress is not the most appropriate alternative to describe the stress state in the bone. It has been demonstrated that, cortical bone is strongest when loaded in compression, 30% weaker when subjected to tensile forces, and 65% weaker when loaded in shear [68].

On contrary, the implant, abutment and abutment-screw may be correctly evaluated by means of von Mises stress (Figure 16 – Neodent® Implant System, Curitiba, Brazil). The maximum admissible stress value for the titanium is about 900 MPa.

Regarding the immediately loading protocols, some recent FEA studies used the bone to implant relative displacement as a parameter for the evaluation of diverse implant aspects [20,21,27]. In fact, one of the most critical elements for the promotion of a safe biomechanical environment for an uneventful bone tissue formation around an immediately loaded implant is a stiff bone-implant interface, allowing low implant micro-movement in bone. Vandamme et al. [11] demonstrated that an implant displacement between 30 and 90 μm positively influenced osseointegration compared with no implant displacement. On the other hand, micromovement beyond 150 μm can induce fibro connective tissue formation, preventing immediately loaded implant osseointegration [7]. In addition, Duyck et al. [12] reported a negative influence of 30 μm micro-motion on osseointegration of cylindrical smooth-surfaced implants. The authors suggested that a good force transfer from the implant towards the surrounding tissues is mandatory and is dictated not only by the mechanical loading, as such, but also by the implant geometry and roughness. However, the above mentioned data were retrieved from studies in a bone chamber, in well-controlled mechanical conditions, protected from external influences [69]. Although implant installation in fresh extraction sockets mimics the bone chamber environment, at least in its neck part, direct extrapolation of the results of this in vivo animal study to the clinical context is not possible.

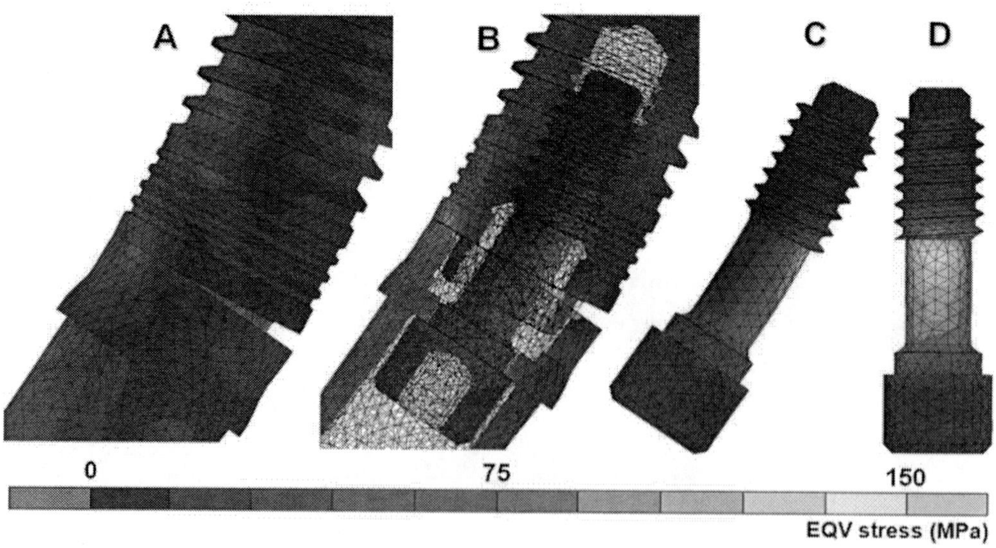

Figure 16. von Mises (EQV) stress evaluation of implant components (Neodent®). The deformation is 50-fold magnified. A – Implant-abutment surface, lateral view. B – Implant-abutment connection, medial slice. C – Abutment screw, lateral view. D – Abutment screw, frontal view.

2.9. Statistical Analysis in FEAs

Although it is an incontestably useful tool to obtain information that is difficult to acquire from laboratory experiments or clinical studies, the results obtained by FEA should be interpreted with some care. The assumptions made during the process of developing a FE model, principally regarding the material properties and the interface conditions, limit the validity of the absolute values of the stress/strain and displacement calculated in a model in which a rigorous experimental validation was not accomplished. In addition, few attempts have been made to evaluate model sensitivities to variations of input parameters (i.e. clinical situation, implant design, implant-abutment connection type, prosthesis design, implant position, bone quality, loading condition) or possible interactions between these parameters. On the contrary, most of the FEA studies have focused on only one or two parameters.

Table 1 exemplifies a possible inaccuracy that would occur on the interpretation of the results for the peak equivalent strain (EQV stain) obtained by FE simulations of implants with different implant-abutment connection types, immediately loaded into extraction socket [20]. If the connection type had been the only included parameter in the study, the conclusions would wrongly be that the Morse-taper connection presented the highest peak EQV strain in the bone for the immediately placed protocol.

Table 1. Results for the peak equivalent strain (EQV strain) in the bone (Pessoa et al. [20]). The results for the 100N loading models are highlighted in yellow. Note that, considering these models, the Morse-taper connection presented the highest EQV strain in the bone

Connection types	Clinical situations	Loading	Bone EQV strain (με)
Morse-taper	immediately placed	50N	4,997.8
	osseointegrated	50N	892.9
	immediately placed	100N	6,060.3
	osseointegrated	100N	1,817.0
	immediately placed	200N	9,910.1
	osseointegrated	200N	3,924.4
Internal hex	immediately placed	50N	3,762.4
	osseointegrated	50N	1,103.1
	immediately placed	100N	5,053.4
	osseointegrated	100N	2,593.5
	immediately placed	200N	9,048.2
	osseointegrated	200N	5,079.4
External hex	immediately placed	50N	4,013.7
	osseointegrated	50N	1,119.0
	immediately placed	100N	5,444.6
	osseointegrated	100N	2,232.8
	immediately placed	200N	10,209.8
	osseointegrated	200N	4,329.9

However, by modeling some other clinical situations and loading magnitudes, and applying a Analysis of Variance (ANOVA) on the results, Pessoa et al. [20] conclude that the connection type does not have a significant influence on the variation of EQV strain in the immediately placed situation (Table 2). Loading magnitude and the clinical situation are the main factors affecting the equivalent bone strain in this protocol.

In this way, the association of the FEA with a statistical analysis has been demonstrated as capable of accurately interpreting the relative influence that each of the input parameters have on the encountered results of implant FEAs [17-21,70].

Nevertheless, when applying the FE method to explore every possible combination of values for each investigated parameter, the total number of simulations required might be very high. In addition, the using of advanced engineering techniques, such as CT imaging, CAD systems and reasonable interface conditions, such as contact, to construct FE models capable to simulate the detailed mechanical characteristics of implant systems, may lead to complex models and relatively high processing time.

Table 2. Analysis of Variance for the peak equivalent strain in the bone. $P < 0.05$, * statistically significant (Pessoa et al. [20]). DF: degrees of freedom; SS: sum of squares; MS: mean square. Note that the connection type does not have a significant contribution to the results

Parameter	DF	SS	MS	*P*-Value	Contribution (%)
Connection Type	2	83004.42	41502.21	0.6799	0.06
Clinical Situation	1	69652650.49	69652650.49	<0.0001 *	49.94
Connection Type X Clinical Situation	2	2301719.20	1150859.60	0.0210 *	1.65
Loading Magnitude	2	63012208.21	31506104.10	<0.0001 *	45.18
Connection Type X Loading Magnitude	4	340997.31	85249.33	0.5503	0.24
Clinical Situation X Loading Magnitude	2	4068982.43	2034491.22	0.0077 *	2.92

Hence, instead of having a full factorial scheme, which must include all possible combination of parameters, Dar et al. [70] suggested combining FEA with the statistical Taguchi method, which utilizes an orthogonal array to reduce significantly the total number of required simulations and contains a well-chosen subset of all possible test condition combinations [70]. This procedure allows achieving a balanced comparison of the levels of any factor. Therefore, the Taguchi method can be combined with non-linear FEA to explore the sensitivity of a model to different input parameters and thereby reduce the experimental effort required to investigate multiple factors in FEA studies.

Conclusion

An in-depth understanding of the biomechanical environment of dental implants can be gained through the use of FEA. This increase in knowledge of stress/strain distributions and magnitudes within implant systems and surrounding jawbone may give support for the optimization of the implant designs and protocols of implant usage, as a function of the parameters beneficial to peri-implant bone, thereby diminishing the risks of implant failure. Therefore, it is of great importance that the clinician and researchers gain understanding on the methodology, applications and limitations of FEA in implant dentistry and become more confident to interpret the results of FEA studies.

REFERENCES

[1] Duyck, J., Ronald, H.J., Van Oosterwyck, H., Naert, I., Vander Sloten, J. and Ellingsen, J.E. (2001) The influence of static an dynamic loadinf on marginal bone reactions around osseointegrated implants: an animal experimental study. *Clinical Oral Implants Research*,12, 207-218.

[2] Hoshaw, S.J., Brunski, J.B. and Cochran, G.V.B. (1994)Mechanical loading of Brånemark implants affects interfacial bone modeling and remodeling. *International Journal of Oral Maxillofacial Implants*,9,345–360.

[3] Misch, C.E., Suzuki, J.B., Misch-Dietsh, F.M. and Bidez, M.W. (2005)A positive correlation between occlusal trauma and peri-implant bone loss: literature support. *Implant Dentistry*, 14,108-16.

[4] Isidor, F. (1996) Loss of osseointegration caused by occlusal load of oral implants. A clinical and radiographic study in monkeys. *Clinical Oral Implants Research*,7,143–152.

[5] Isidor, F. (1997) Histological evaluation of periimplant bone at implants subjected to occlusal overload or plaque accumulation. *Clinical Oral Implants Research*,8,1–9.

[6] Brunski, J.B. (1992) Biomechanical factors affecting the bone-dental implant interface. *Clinical Materials*,10, 53–201.

[7] Geris, L., Andreykiv, A., Van Oosterwyck, H., Vander Sloten, J., van Keulen, F., Duyck, J. and; Naert, I. (2004) Numerical simulation of tissue differentiation around loaded titanium implants in a bone chamber. *Journal of Biomechanics*,37,763–769.

[8] Søballe, K., Brockstedt-Rasmussen, H., Hansen, E.S. and Bünger, C. (1992) Hydroxyapatite coating modifies implant membrane formation. Controlled micromotion studied in dogs. *Acta Orthopeadica Scandinavica*,63,128–140.

[9] Bozkaya, D., Müftü, S. and Müftü, A. (2004) Evaluation of load transfer characteristics of five different implants in compact bone at different load levels by finite elements analysis. *Journal of Prosthetic Dentistry*,92,523-530.

[10] De Smet, E., Jaecques, S.V.N., Jansen, J.J., Walboomers, F., Vander Sloten, J. and Naert, I.E. (2007) Effect of constant strain rate, composed by varying amplitude and frequency, of early loading on peri-implant bone (re)modelling. *Journal of Clinical Periodontology*,34,618–624.

[11] Vandamme, K., Naert, I., Geris, L., Vander Sloten, J., Puers, R. and Duyck, J. (2007) The effect of micromotion on the tissue response around immediately loaded roughened titanium implants in the rabbit. *European Journal of Oral Science*, 115,21–29.

[12] Duyck, J., Vandamme, K., Geris, L., Van Oosterwyck, H., De Cooman, M., Vander Sloten, J., Puers, R. and Naert I. (2006) The influence of micro-motion on the tissue

differentiation around immediately loaded cylindrical turned titanium implants. *Archives of Oral Biology*,51,1–9.

[13] Geng, J.P., Tan, K.B. and Liu, G.R. (2001) Application of finite element analysis in implant dentistry: a review of the literature. *Journal of Prosthetic Dentistry*,85,585-98.

[14] Van Staden, R.C., Guan, H. and Loo, Y.C. (2006) Application of the finite element method in dental implant research. *Computer Methods Biomechanics Biomedical Engineering*,9,257-270.

[15] Wakabayashi, N., Ona, M., Suzuki, T. and Igarashi, Y. (2008) Nonlinear finite element analyses: Advances and challenges in dental applications. *Journal of Dentistry*,26,463-471.

[16] Zienkiewicz, O. C. and Taylor, R. L. (1989) *The Finite Element Method*. New York, McGraw-Hill.

[17] Lin, C-L., Chang, S-H., Chang, W-J. and Kuo, Y-C. (2007) Factorial analysis of variables influencing mechanical characteristics of a single tooth implant placed in the maxilla using finite element analysis and the statistics-based Taguchi method. *European Journal of Oral Science*,115,408–416.

[18] Ding, X., Liao, S-H., Zhu, X-H., Zhang, X-H., and Zhang, L. (2008) Effect of Diameter and Length on Stress Distribution of the Alveolar Crest around Immediate Loading Implants. *Clinical Implant Dentistry and Related Research*, 2008 DOI 10.1111/j.1708-8208.2008.00124.

[19] Lin, C-L., Wang, J-C. and Chang, W-J. (2008) Biomechanical interactions in tooth–implant-supported fixed partial dentures with variations in the number of splinted teeth and connector type: a finite element analysis. *Clinical Oral Implants Research*,19,107–117.

[20] Pessoa, R.S., Muraru, L., Marcantonio Jr, E., Vaz, L.G., Vander Sloten, J., Duyck, J., and Jaecques, S.V.N. (2009) Influence of implant connection type on the biomechanical environment of immediately placed implants – CT-based nonlinear, 3D finite element analysis. *Clinical Implant Dentistry and Related Research*, DOI 10.1111/j.1708-8208.2009.00155.x.

[21] Pessoa, R.S., Vaz, L.G., Marcantonio Jr, E., Vander Sloten, J., Duyck, J. and Jaecques, S.V.N. (2009) Biomechanical evaluation of platform switching in different implant protocols – CT based 3D finite element analysis. *International Journal of Oral Maxillofacial Implants*, accepted for publication.

[22] Atmaram, G.H., Mohammed, H. and Schoen, F.J. (1979) Stress analysis of single tooth implants. I. Effects of elastic parameters and geometry of implant. *Biomaterials, Medical Devices and Artificial Organs*, 7,99–104.

[23] Jaecques, S.V.N., Van Oosterwyck, H., Muraru, L., Van Cleynenbreugel, T., De Smet, E., Wevers, M., Naert, I. and Vander Sloten, J. (2004) Individualised, micro CT-based finite element modelling as a tool for biomechanical analysis related to tissue engineering of bone. *Biomaterials*, 25, 1683–1696.

[24] Van Oosterwyck, H., Duyck, J., Vander Sloten, J., Van der Perre, G., De Cooman, M., Puers, R. and Naert, I. (2001) Patient-dependent FE modeling as a tool for biomechanical optimization of oral reconstruction. In: Middleton, J; Jones, ML; Shrive, NG; Pande, GN; *Computer Methods in Biomechanics and Biomedical Engineering*. Amsterdam: Gordon and Breach Science Publishers;3, 559–564.

[25] Marks, L. W. and Gardner, T. N. (1993). The use of strain energy as a convergence criterion in the finite element modelling of bone and the effect of model geometry on stress convergence. *Journal of Biomedical Engineering*, 15, 474-476.
[26] Bechet, E., Cuilliere, JC. and Trochu, F. (2002) Generation of a finite element mesh from stereolithography (STL) files. *Computer-Aided Design*,34, 1-17.
[27] Huang, H.L., Hsu, J.T., Fuh, L.J., Tu, M.G., Ko, C.C. and Shen, Y.W. (2008) Bone stress and interfacial sliding analysis of implant designs on an immediately loaded maxillary implant: A non-linear finite element study. *Journal of Dentistry*,36, 409-417.
[28] Muraru, L., Van Lierde, C., Naert, I., Vander Sloten, J. and Jaecques, S.V.N. (2009) Three-dimensional finite element models based on in vivo microfocus computed tomography: Elimination of metal artefacts in a small laboratory animal model by registration with artefact-free reference images. *Advances in Engineering Software*,40,1207–1210.
[29] Katz, J.L. (1971) Hard tissue as a composite material. 1. Bounds on the elastic behaviour. *Journal of Biomechanics*,4,455–473.
[30] Hedia, H. S. and Mahmoud, N. A. (2004). Design optimization of functionally graded dental implant. *Biomedical Materials and Engineering*, 14, 133–143.
[31] Merz, B.R., Hunenbart, S. and Belser, U.C. (2000) Mechanics of the implant-abutment connection: An 8-degree taper compared to a butt joint connection. *International Journal of Oral Maxillofacial Implants*,15,519-526.
[32] Van Oosterwyck, H., Duyck, J., Vander Sloten, J., van der Perre, G., de Cooman, M., Lievens, S., Puers, R. and Naert, I. (1998) The influence of bone mechanical properties and implant fixation upon bone loading around oral implants. *Clinical Oral Implants Research*,9, 407–418.
[33] Gotfredsen, K., Berglundh, T. and Lindhe, J. (2000) Anchorage of titanium implants with different surface characteristics: An experimental study in rabbits. *Clinical Implant Dentistry Related Research*,2,120-128.
[34] Williams, J.A. (2000) *Engineering Tribology*. Oxford: Oxford University Press.
[35] Adams, G.G., Muftu, S. and Mohd Azar, N. (2003) A scale-dependent model for multi-asperity model for contact and friction. *Journal of Tribology*,125,700-708.
[36] Rancourt, D., Shirazi-Adl, A., Drouin, G. and Paiement, G. (1990) Friction properties of the interface between porous-surfaced metals and tibial cancellous bone. *Journal of Biomedical Materials Research*,24,1503-1519.
[37] Abkowitz, S., Burke, J.J. and Hiltz, R.H. (1955) *Titanium in Industry*. New York: Van Nostrand Co Inc.
[38] Steinemann, S.G., Mäusli, P.A., Szmukler-Moncler, S., Semlitzsch, M., Pohler, O., Hintermann, H.E. and Perren, S.M. Betatitanium alloy for surgical implants. In: Froes, F. H. and Caplan, I. (1993). *Titanium '92. Science and Technology*. The Minerals, Metals and Materials Society, 2689–2696.
[39] Schwarz, M.S. (2000) Mechanical complications of dental implants. *Clinical Oral Implants Research*, 11(Suppl.), 156–158.
[40] Bozkaya, D. and Müftü, S. (2005) Mechanics of the taper integrated screwed-in (TIS) abutments used in dental implants. *Journal of Biomechanics*, 38, 87–97.
[41] Lang, L.A., Wang, R.F. and May, K.B. (2002) The influence of abutment screw tightening on screw joint configuration. *Journal of Prosthetic Dentistry*,87,74-9.

[42] Burguete, R.L., Johns, R.B., King, T. and Patterson, E.A. (1994) Tightening characteristics for screwed joints in osseointegrated dental implants. *Journal of Prosthetic Dentistry*,71,592-599.

[43] Sakaguchi, R.L. and Borgersen, S.E. (1995). Nonlinear contact analysis of preload in dental implant screws. *International Journal of Oral Maxillofacial Implants*, 10, 295-302.

[44] Iplikçioglu, H., Akça, K., Çehreli, M.C. and Sahin, S. (2003) Comparison of Non-linear Finite Element Stress Analysis with In Vitro Strain Gauge Measurements on a Morse Taper Implant. International *Journal of Oral and Maxillofacial Implants*,18,258-265.

[45] Mericske-Stern, R., Piotti, M., and Sirtes, G. (1996) 3-D in vivo force measurements on mandibular implants supporting overdentures. A comparative study. *Clinical Oral Implants Research*, 7,387–396.

[46] Duyck, J., Van Oosterwyck, H., Vander Sloten, J., De Cooman, M., Puers, R. and Naert, I. (2000) Magnitude and distribution of occlusal forces on oral implants supporting fixed prostheses: An in vivo study. *Clinical Oral Implants Research*,11,465–475.

[47] Morneburg, T.R. and Proschel, P.A. (2002). Measurement of masticatory forces and implant loads: a methodologic clinical study. *International Journal of Prosthodontics*,15,20-27.

[48] Teixeira, E.R., Sato, Y., Akagawa, Y. and Shindoi, N. (1998) A comparative evaluation of mandibular finite element models with different lengths and elements for implant biomechanics. *Journal of Oral Rehabilitation*,25,299-303.

[49] Zhou, X., Zhao, Z., Zhao, M. and Fan, Y. (1999) The boundary design of the mandibular model by means of the three-dimensional finite element method. *West China Journal Stomatology*,17,1-6.

[50] Hart, R.T., Hennebel, V.V., Thongpreda, N., van Buskirk, W.C. and Anderson, R.C. (1992) Three-dimensional finite element study of the biomechanics of the mandible. *Journal of Biomechanics*,25,261-286.

[51] Mellal, A., Wiskott, H.W.A., Botsis, J., Scherrer, S.S. and Belser, U.C. (2004) Stimulating effect of implant loading on surrounding bone. Comparison of three numerical models and validation by in vivo data. *Clinical Oral Implants Research*,15,239-248.

[52] Rubin, C. and Lanyon, L. (1985). Regulation of bone mass by mechanical strain magnitude. *Calcified Tissue International*, 37, 411–417.

[53] Turner, C.H., Forwood, M.R., Rho, J.Y. and Yoshikawa, T. (1994) Mechanical loading thresholds for lamellar and woven bone formation. *Journal Bone and Mineral Research*,9, 87-97.

[54] Pilliar, R.M., Deporter, D.A., Watson, P.A. and Valiquette, N. (1991) Dental implant design– effect on bone remodelling. *Journal of Biomechanics and Materials Research*,25,467-483.

[55] Vaillancourt, H., Pilliar, R.M. and McCammond, D. Factors affecting crestal bone loss with dental implants partially covered with a porous coating: a finite element analysis. *International Journal Oral Maxillofacial Implants*, 1996 11, 351–359

[56] Frost, HM. (1992) Perspectives: bone's mechanical usage windows. *Bone Mineral*,19,257–271.

[57] Forwood, M.R. and Turner, C.H. (1994). The response of rat tibiae to incremental bouts of mechanical loading: a quantum concept for bone formation. *Bone*,15,603-609.

[58] Hsieh, Y.F. and Turner, C. . (2001) Effects of loading frequency on mechanically induced bone formation. *Journal of Bone and Mineral Research*, 16, 918–924.

[59] Robling, A.G., Hinant, F.M., Burr, D.B., Turner, C.H. (2002) Improved bone structure and strength after long-term mechanical loading is greatest if loading is separated into short bouts. *Journal of Bone and Mineral Research*,17,1545-1554.

[60] Rubin, C.T. and McLeod, K.J. (1984). Promotion of bony ingrowth by frequency-specific, low-amplitude mechanical strain. *Clinical Orthopaedics and Related Research*,298,165-174.

[61] Lanyon, L.E. and Rubin, C.T. (1984). Static vs dynamic loads as an influence on bone remodelling. *Journal of Biomechanics*, 17,897-905.

[62] Robling, A.G., Duijvelaar, K.M., Geevers, J.V., Ohashi, N., Turner, C.H. (2001) Modulation of appositional and longitudinal bone growth the rat ulna by applied static and dynamic force. *Bone*,29,105-113.

[63] Turner, CH. (1998) Three rules for bone adaptation to mechanical stimuli. *Bone*,23,399-407.

[64] Reilly, D.T. and Burstein, A.H. (1975). The elastic and ultimate properties of compact bone tissue. *Journal of Biomechanics*,8,393-405.

[65] Martin, R.B., Burr, D.B. and Sharkey, N.A. (1998). *Skeletal tissue mechanics*. New York: Springer, 127-78.

[66] Birnbaum, K., Sindelar, R., Gartner, J.R. and Wirtz, D.C. (2001) Material properties of trabecular bone structures. *Surgical and Radiological Anatomy*,23,399-407.

[67] Carter, D.R. and Hayes, W.C. (1977). The compressive behavior of bone as a two-phase porous structure. *Journal of Bone and Joint Surgery*, 59-A, 954–962.

[68] Cowin, SC. (1989)Bone mechanics. Boca Raton, FL: CRC Press.

[69] Duyck, J., De Cooman, M., Puers, R., Van Oosterwyck, H., Vander Sloten, J. and Naert, I. (2004) A repeated sampling bone chamber methodology for the evaluation of tissue differentiation and bone adaptation around titanium implants under controlled mechanical conditions. *Journal of Biomechanics*,37,1819-1822.

[70] Dar, F.H., Meakina, J.R. and Aspden, R.M. (2002) Statistical methods in finite element analysis. *Journal of Biomechanics*,35,1155–1161.

In: Biomechanics of Dental Implants
Editor: Murat Çehreli
ISBN 978-1-62100-780-7

Chapter 8

STRAIN MEASUREMENT AND ELECTRIC RESISTANCE STRAIN GAUGES

Ergin Tönük

1. INTRODUCTION

Classical mechanics is the branch of physical science that deals with the behavior of bodies under the action of forces. Bodies either move (without any shape change, like a rigid body) or deform (undergo a shape change) under the action of forces. In general, motion and deformation occur together if a body is under the action of forces.

Mechanical stress, which is the internal reaction force intensity to external forces, is an important parameter to decide whether the material can withstand the external loads safely or would it end up with a static or fatigue failure or a permanent shape change. Therefore stress analysis received attention in various disciplines including dentistry. Further, living biological materials undergo remodeling and adaptation under mechanical stress.

There exists three alternative methods for stress analysis:

- *Theoretical Analysis:* Mathematical model and analytic solution for the problem exists, therefore, for given external forces, stress at a desired point is available by the solution of either an equation or a set of equations.
- *Computational Analysis:* When analytical solution to the mathematical model of the problem is not available due to complicated geometrical shape, complicated external forces, contact, nonlinear material behavior, or, large strains and/or displacements, an approximate solution is obtained with the help of numerical methods (computational mechanics) so stress due to external forces may be estimated. One very common and powerful computational method capable of handling the mentioned problems in dentistry is the finite element analysis (FEA).
- *Experimental Analysis:* Although, the development in computational analysis reduced the need for experimental studies, because mathematical modeling is not an exact replica of the nature due to simplifications and idealizations, it is often necessary to perform experiments to check the validity of models. Establishment of

material laws (constitutive equations) and identification of the so-called material coefficients in these laws require experimental work as well. Stress, however, cannot be measured, but, it is relatively easy to measure normal strain. If mechanical behavior (*i.e.* stress-strain relationship or constitutive equation) is known for the material under consideration, stress can be estimated from experimental strain measurements.

2. Basic Definitions and Terminology

2.1. Continuum

The basic idealization in stress analysis is the continuum approach. Continuum approach idealizes the real materials with a continuous material model at every scale. This idealization assumes that there is material at every mathematical point in space or the real material is idealized with an infinite number of particles, therefore the abstract copy of a continuum is the real number set. One may always find a real number in between any two distinct real numbers. It is always possible to find another real number between the new number and the old one and this process may continue forever as one may find a material particle at every point in continuum (Malvern, 1969, Fung, 1994).

Continuum approach is an idealization because it is known for a long time that real materials are not continuum at atomic scale. For many biological materials, even at larger scales, the micro structure deviates considerably from a continuum. Therefore, continuum approximation may only be useful for predicting macro mechanics and averaged gross behavior. However continuum mechanics utilizes very powerful tools of calculus: limit, differentiation and integration to build up a mathematical theory which can accurately predict the observed macro mechanical behavior (i.e. motion and deformation) of various solids and fluids under the action of external forces.

2.2. Stress

In continuum mechanics, stress is defined as the force per unit area at the limit the area shrinks to zero. For practical materials, continuum stress may be interpreted as the average force per unit area for a sufficiently small area for which the continuum approximation is still valid.

Unlike static pressure, which is also force per unit area, stress is not a scalar quantity. Stress may be tensile, compressive or shear. For a rod of uniform cross section under the action of tensile or compressive forces, the internal force intensity at a cross section sufficiently away from geometric and material discontinuities is the normal stress which is equal to force divided by the cross sectional area (Figure 1). In normal stress the force intensity is perpendicular (*i.e.* normal) to the cross sectional area under consideration but it is also possible to have this force intensity within the area (Figure 2) which is called shear stress. Both definitions discussed are based on the assumption that the internal force intensity is constant over the cross section. However for a general body under the action of forces, stress (sometimes referred as Cauchy stress) may be visualized as in Figure 3.

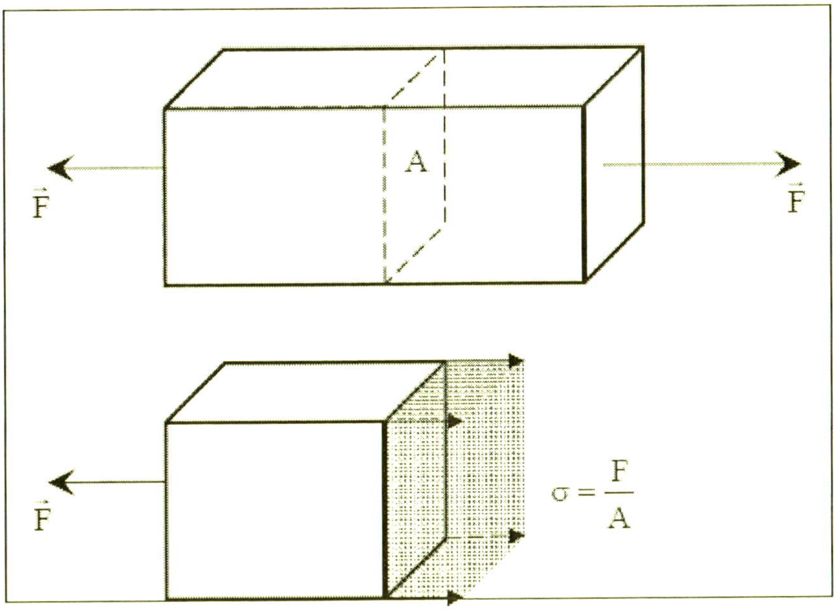

Figure 1. Normal stress.

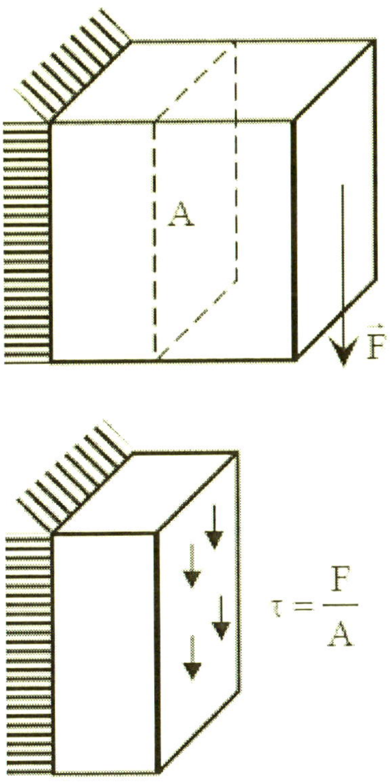

Figure 2. Shear stress.

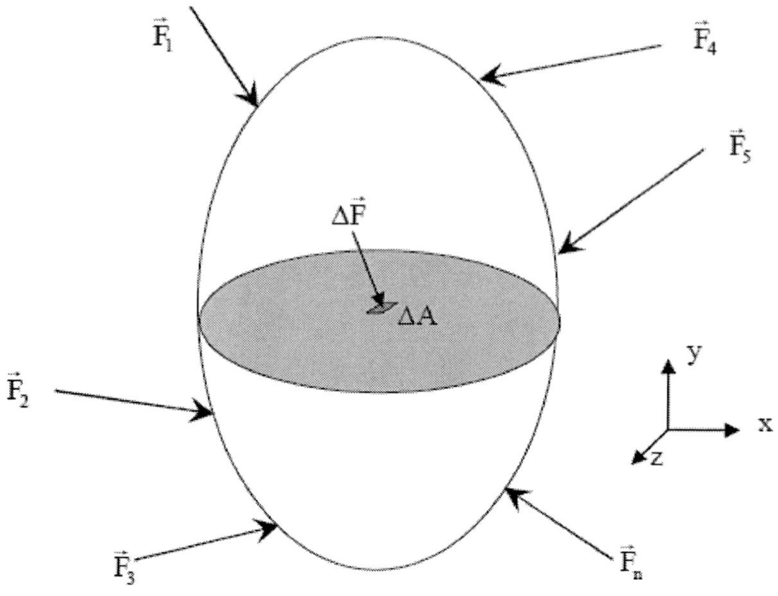

Figure 3. The general internal reaction force vector, $\Delta\vec{F}$, on a small cross sectional area ΔA in a general body.

Let ΔF_x, ΔF_y, and ΔF_z be the components of the internal reaction force vector, $\Delta\vec{F}$, acting on the small area, ΔA, along three mutually perpendicular coordinate axes, x, y, and, z respectively. In general, as the location of the small cross sectional area, ΔA, changes the magnitude and direction of the force vector, $\Delta\vec{F}$, therefore its components change (*i.e.* stress is not constant within the material but is a field). The cross sectional area, ΔA has a unit normal in the direction of y-axis. The force intensities (*i.e.* stresses) are defined as:

$$\tau_x = \lim_{\Delta A \to 0} \frac{\Delta F_x}{\Delta A} = \frac{dF_x}{dA}$$
$$\sigma_y = \lim_{\Delta A \to 0} \frac{\Delta F_y}{\Delta A} = \frac{dF_y}{dA} \qquad \text{(Eq.1)}$$
$$\tau_z = \lim_{\Delta A \to 0} \frac{\Delta F_z}{\Delta A} = \frac{dF_z}{dA}$$

Here, shrinking a finite area, ΔA, to zero and defining the stress at a point is possible only with continuum approximation of the real material. Stress is a field variable defined at every point in the continuum. In equation 1, the three components of the stress *state* is obtained by dividing the three components of the internal reaction force vector by the cross sectional area, keeping in mind the x and z components of the force are in the area therefore resulting shear stress and y component of the force perpendicular to the area therefore resulting normal stress. The cross sectional area in Figure 3 is a hypothetical surface and its normal being along y axis of the coordinate frame was selected arbitrarily. Stress at a material point is defined by assuming a cube of infinitesimal size around the point whose three surfaces have the outer unit normals in the direction of three mutually perpendicular coordinate axes as shown in Figure 4. Therefore stress components resolved at a Cartesian coordinate system, at a point may be represented by a square matrix.

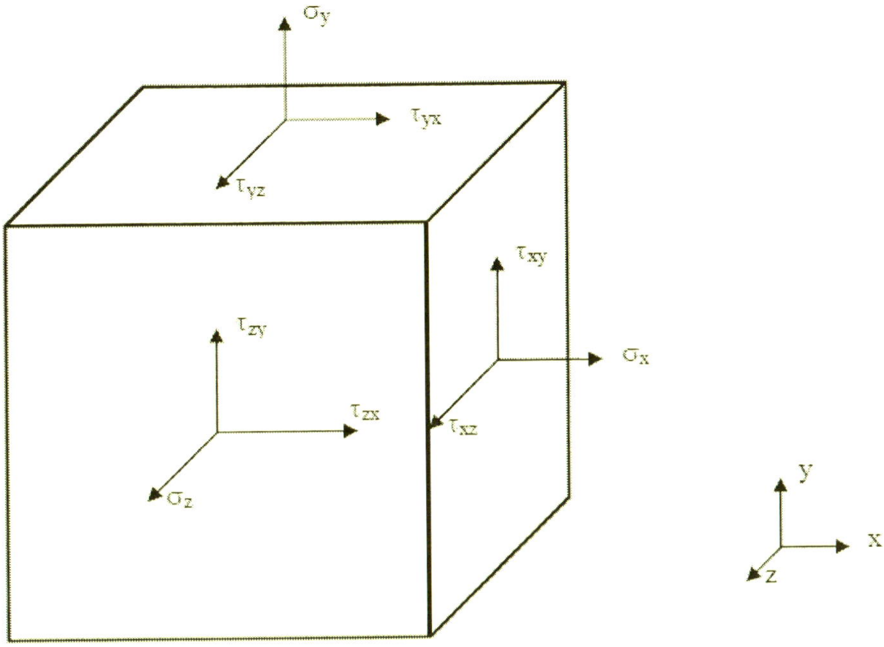

Figure 4. Geometric interpretation of stress tensor.

$$[\tau]=\begin{bmatrix}\sigma_x & \tau_{xy} & \tau_{xz}\\ \tau_{yx} & \sigma_y & \tau_{yz}\\ \tau_{zx} & \tau_{zy} & \sigma_z\end{bmatrix} \qquad \text{(Eq.2)}$$

In equation 2 sometimes the normal stresses may be denoted by τ_{xx}, τ_{yy} and τ_{zz} too to be compatible with the convention stating that the first subscript is the direction of the outer unit normal of the plane on which the stress is acting whereas second subscript is the direction of the stress itself[1]. The positive stress convention is, if the stress itself and the outer unit normal of the area stress is acting are both in the positive coordinate directions or both in the negative coordinate directions, the stress is positive. If one is in positive coordinate direction while other is in negative coordinate direction then the stress is negative. This convention implies tensile normal stresses are positive whereas compressive normal stresses are negative.

For materials where the mechanical action and reaction among particles are equipollent to only a force and there is no couple stress, the principle of angular momentum (or for equilibrium, the moment equilibrium) states that the components of a Cartesian stress tensor is symmetric (*i.e.* $\tau_{xy}=\tau_{yx}$, $\tau_{yz}=\tau_{zy}$ and $\tau_{xz}=\tau_{zx}$) and there exists only six independent

[1] Other conventions do exist in the literature, e.g. Truesdell and Noll (1965).

components. Details of the derivation may be found in most textbooks on mechanics of materials, elasticity or continuum mechanics (*e.g.* Malvern, 1969).

2.3. Strain

Strain (sometimes referred as engineering strain) is a measure of deformation under the action of forces. Length change per unit length is termed as normal strain (Figure 5) which is in response to normal stress whereas the angular deviation of two mutually perpendicular lines is termed as shear strain (Figure 6) which is in response to shear stress. Both normal strain (elongation per unit length) and shear strain (angle change from a right angle, in radians) are dimensionless quantities and units like micro[2] inches per inch or micro millimeters per millimeter are utilized in practice and με and % elongation are other designations for normal strain.

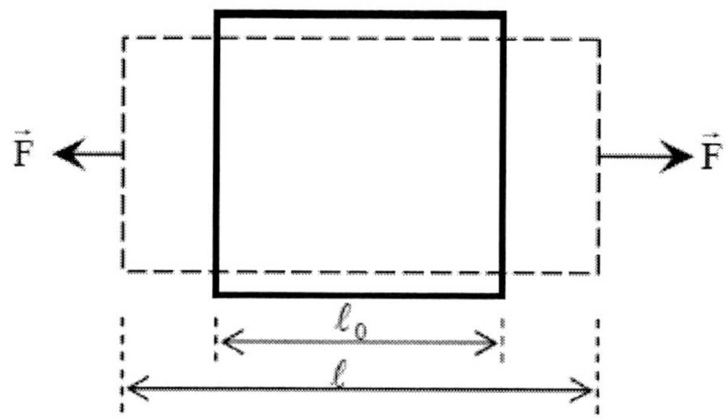

$$\varepsilon = \frac{\ell - \ell_0}{\ell_0}$$

Figure 5. Normal strain.

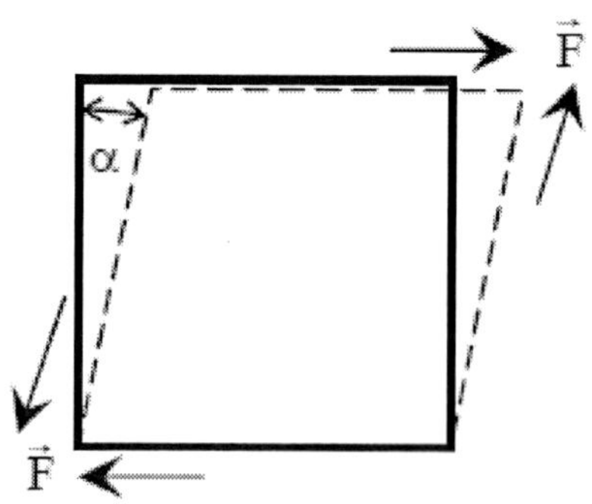

$\gamma = \tan \alpha$ □ α (for small angles in radians)

Figure 6. Shear strain.

[2] Micro (symbol μ) stands for 10^{-6}.

2.4. Constitutive Relation (Material Law)

Constitutive relation is a mathematical equation between stress and strain specific to certain material (but independent of geometric details like the size of the test specimen), mostly phenomenological and obtained by experiments. Here, only one constitutive relation, Hooke's law for isotropic, linear elastic solids will be presented.

For isotropic linear elastic solids under small strains the stress-strain relationship becomes

$$\begin{bmatrix}\sigma_x\\ \sigma_y\\ \sigma_z\\ \tau_{xy}\\ \tau_{yz}\\ \tau_{xz}\end{bmatrix}=\frac{E}{(1+\upsilon)(1-2\upsilon)}\begin{bmatrix}1-\upsilon & \upsilon & \upsilon & 0 & 0 & 0\\ \upsilon & 1-\upsilon & \upsilon & 0 & 0 & 0\\ \upsilon & \upsilon & 1-\upsilon & 0 & 0 & 0\\ 0 & 0 & 0 & \frac{1-2\upsilon}{2} & 0 & 0\\ 0 & 0 & 0 & 0 & \frac{1-2\upsilon}{2} & 0\\ 0 & 0 & 0 & 0 & 0 & \frac{1-2\upsilon}{2}\end{bmatrix}\begin{bmatrix}\varepsilon_x\\ \varepsilon_y\\ \varepsilon_z\\ \gamma_{xy}\\ \gamma_{yz}\\ \gamma_{xz}\end{bmatrix} \quad \text{(Eq.3)}$$

where E is the elastic (Young's) modulus and υ is the Poisson's ratio of the material. For plane stress state Equation 3 reduces to

$$\begin{bmatrix}\sigma_x\\ \sigma_y\\ \tau_{xy}\end{bmatrix}=\frac{E}{1-\upsilon^2}\begin{bmatrix}1 & \upsilon & 0\\ \upsilon & 1 & 0\\ 0 & 0 & \frac{1-\upsilon}{2}\end{bmatrix}\begin{bmatrix}\varepsilon_x\\ \varepsilon_y\\ \gamma_{xy}\end{bmatrix} \quad \text{(Eq. 4)}$$

It should be noted that for plane stress state although $\sigma_z = \tau_{xz} = \tau_{yz} = 0$, ε_z is nonzero.

For linear elastic (Hookean) materials the algebraic relations between stress and strain components are:

$$\begin{aligned}\sigma_x &= \frac{E}{1-\nu^2}\left(\varepsilon_x + \nu\varepsilon_y\right)\\ \sigma_y &= \frac{E}{1-\nu^2}\left(\nu\varepsilon_x + \varepsilon_y\right)\\ \tau_{xy} &= \frac{E}{2(1+\nu)}\gamma_{xy}\end{aligned} \quad \text{(Eq. 5)}$$

shear modulus (sometimes termed as modulus of rigidity), G, is not independent if elastic (Young's) modulus, E, and Poisson's ratio, ν, are known:

$$G = \frac{E}{2(1+\nu)} \qquad \text{(Eq. 6)}$$

so

$$\tau_{xy} = G\gamma_{xy} \qquad \text{(Eq. 7)}$$

3. Strain Measurement

For small strain approximation, u, v, and w being the components of displacement along x, y, and z directions, then the six components of strain can be evaluated as:

$$\begin{aligned}
\varepsilon_x &= \frac{\partial u}{\partial x} \\
\varepsilon_y &= \frac{\partial v}{\partial y} \\
\varepsilon_z &= \frac{\partial w}{\partial z} \\
\gamma_{xy} &= \frac{\partial v}{\partial x} + \frac{\partial u}{\partial y} \\
\gamma_{yz} &= \frac{\partial w}{\partial y} + \frac{\partial v}{\partial z} \\
\gamma_{xz} &= \frac{\partial w}{\partial x} + \frac{\partial u}{\partial z}
\end{aligned} \qquad \text{(Eq. 8)}$$

Because, strain measurements are mostly performed on the free surfaces of the bodies where there exits plane stress state, determination of only three in-plane Cartesian components of strain tensor, ε_x, ε_y, and γ_{xy} is sufficient for experimental stress analysis.

Stress and strain fields, in general, are not uniform and stress and strain are point functions. However, it is not practically possible to measure strain at a point. Instead, strain in a finite (but possibly small) area is averaged. Then the differential displacement-strain relationship is converted into an approximate difference equation as:

$$\begin{aligned}
\varepsilon_x &= \frac{\partial u}{\partial x} \cong \frac{\Delta u}{\Delta x} = \frac{\ell_x - \ell_{x_0}}{\ell_{x_0}} \\
\varepsilon_y &= \frac{\partial v}{\partial y} \cong \frac{\Delta v}{\Delta y} = \frac{\ell_y - \ell_{y_0}}{\ell_{y_0}}
\end{aligned} \qquad \text{(Eq. 9)}$$

where ℓ_{x_0} is the initial length without force and deformation on the body, and, ℓ_x is the final length of the specimen along x direction with force on it. Therefore, many strain measurement techniques are based on measuring length change of an initial length (sometimes termed as gauge length).

3.1. Strain Measuring Techniques

Mechanical: There are mechanical strain gauges, some of them are termed as extensometers, are capable of measuring elongation or shortening mechanically. They may have scales indicating elongation or strain (elongation divided by initial gauge length).

Optical: Using stereo photogrammetry the displacements upon forces on free surfaces of bodies may be determined using two or more pre-calibrated cameras from which, using strain-displacement relationships (Eq. 8), strain on the free surface of the body may be calculated. There are diffraction, interferometric and photoelastic strain measuring techniques as well which will not be detailed here and interested reader is referred to textbooks on these topics.

Electrical: Electric resistance of an electric conductor is a function of the elongation (therefore normal strain) on the conductor which was first recognized by Lord Kelvin (Thomson, 1856). Therefore electric resistance strain gauges are transducers converting elongation into electric resistance change which can be measured very accurately.

3.2. Basic Characteristics of Strain Measuring Devices

3.2.1. Gauge Length

For non-uniform stress state, strain varies at every point of the specimen. The current strain measuring devices available cannot measure strain at a point but averages the strain along a finite length or within a finite area. If the intention is to determine strain (therefore stress) at a point then gage length is important and small gauge lengths yield smaller averaging errors.

3.2.2. Gauge Sensitivity

Like all measuring devices, there is a finite lower limit for the strains that can be measured. Gage sensitivity is the lowest strain that can be measured.

3.2.3. Measuring Range

Measuring range is the maximum strain that can be measured by a strain gauge without resetting or replacing the gauge.

3.2.4. Accuracy

Accuracy is the closeness of the measurement to a set of accepted values of the quantity to be measured.

3.2.5. Precision (Repeatability)

During repeated measurement of the same quantity, precise devices yield very close values at every measurement but not necessarily close to the set of accepted values.

3.3. Electric Resistance Strain Gauges

3.3.1 Brief History

Lord Kelvin (W. Thomson) recognized the change in electrical resistance of copper and iron wires when in tension in 1856 (Thomson, 1856). He further recognized that iron wire showed a greater increase in resistance compared to copper wire under the same strain. He used a Wheatstone bridge to measure the small resistance change accurately. His experiment has all the basics of modern electric resistance strain gauges. It took about 80 years to have commercial strain gauges (Dally and Riley, 1991, Stein 2006).

3.3.2. Basic Theory

The governing equation of electric resistance of a conductor is

$$R = \rho \frac{L}{A} \qquad \text{(Eq. 10)}$$

where R is the electric resistance of the conductor (in Ohms), ρ is the resistivity of the material (in Ohm-meters), L is the length of the conductor (in meters) and A is the uniform cross sectional area of the conductor (in square meters). The effect of change of each parameter on the resistance is governed by

$$\frac{dR}{R} = \frac{dL}{L} - \frac{dA}{A} + \frac{d\rho}{\rho} \qquad \text{(Eq. 11)}$$

Here the differential quantities are the infinitesimal changes and ratios show the infinitesimal change of the quantity per original value. For practical applications this equation may be replaced by finite changes (differences) of the quantity per original value. From Eq. 9 the ratio, dL/L, is the length change per original length which is normal strain to be measured by measuring the resistance change per original resistance, dR/R. The change in cross sectional area of the conductor in the elastic range is related to the length change by Poisson's ratio, υ, of the gauge material therefore Eq. 11 becomes

$$\frac{dR}{R} = (1+2\upsilon)\frac{dL}{L} + \frac{d\rho}{\rho} \qquad \text{(Eq. 12)}$$

The sensitivity of gauge material to strain is obtained as

$$S_A = \frac{dR/R}{\varepsilon} = 1 + 2\upsilon + \frac{d\rho/\rho}{\varepsilon} \qquad \text{(Eq. 13)}$$

The commonly known term, gauge factor S_g, is the resistance change of the strain gauge per applied longitudinal strain. However in this case because of uniaxial stress state there is transverse strain, ε_T, as well in addition to longitudinal strain, ε_L, and the gauge is calibrated on steel specimen with the condition $\varepsilon_T = -0.285\varepsilon_L$. In other words, the strain gauge is calibrated on a uniaxial steel tensile specimen with Pisson's ratio, $\nu = 0.285$, and gauge factor differs from the sensitivity of gauge material to strain due to transverse sensitivity (detailed in the section Corrections Transverse Strain Effects). The gauge factor remains nearly constant (therefore strain versus change in resistance per original value is linear) over a wide range of strain and hysteresis is very small (however see the section Gauge Factor Variation with the Temperature). Gauge factor is supplied by the gauge manufacturer. It should be noted that there is generally a slight difference for the gauge factor in tension and in compression (generally less than 0.2%) and manufacturers mostly provide an average of the two values (Watson, 2008).

$$S_g = \frac{dR/R}{\varepsilon} \quad \text{(Eq. 14)}$$

3.3.3. Typical Strain Gauges

Theoretically it is possible to measure strain using a single electric wire. However resistance of such a wire would be quite low, causing considerable heat production under electric current. Increasing the length of the wire to increase its electric resistance would increase the gauge length therefore not very desirable for estimating strains and stresses at a point. The other alternative is to decrease the cross sectional area of the wire to increase the resistance, however, such very thin wires is very hard to handle. Foil strain gauges, produced by Saunders and Roe in England in 1952 (Juvinal, 1967) for the first time solved this problem. Like a printed circuit board, the very thin gauge wires were etched on a small backing film and the gauge was bonded on the specimen. A typical foil strain gauge is presented in Figure 7. Other types of electrical resistance strain gauges commercially available are unbonded and bonded wire gauges.

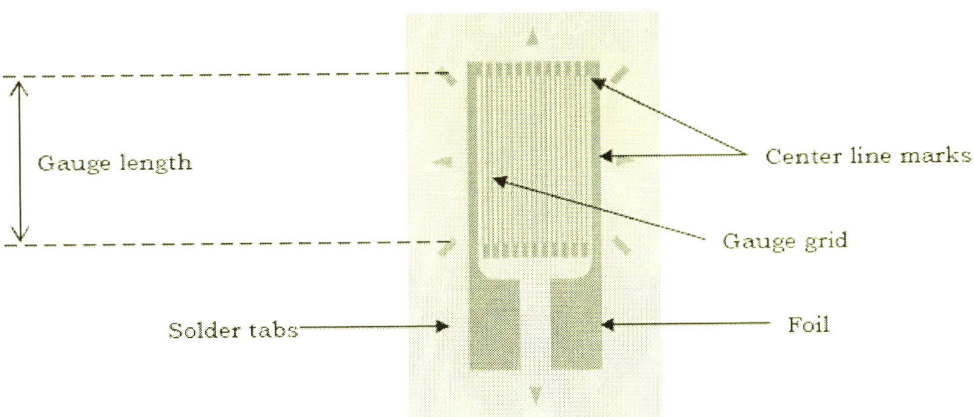

Figure 7. Single element foil strain gauge.

3.4. Measuring Circuits

Electric resistance of a strain gauge changes in proportion with the strain as

$$\frac{\Delta R}{R} = S_g \varepsilon \qquad \text{(Eq. 15)}$$

therefore it is essential to measure resistance change, ΔR, which is small (typically around 0.7 mΩ/με for a 350 Ω strain gauge) compared to the original resistance, R. Two common circuits used for this purpose are the potentiometer and the Wheatstone bridge. Both circuits convert resistance change into a potential difference (voltage), ΔE, which can be amplified, filtered, measured, logged to a computer rather easily. Many of the commercial strain measuring instruments use either of these circuits in addition to other electronics.

3.4.1. The Wheatstone Bridge

Wheatstone bridge may be used for static as well as dynamic strain measurement applications. It was S. H. Cristie in 1833 who invented the circuit called *diamond method* to compare wires of different electrical resistances however C. Wheatstone published a paper in 1843 explaining a *differential resistance measurer* therefore his name was associated with the bridge (Watson, 2008). The Wheatstone bridge circuit is presented in Figure 8.

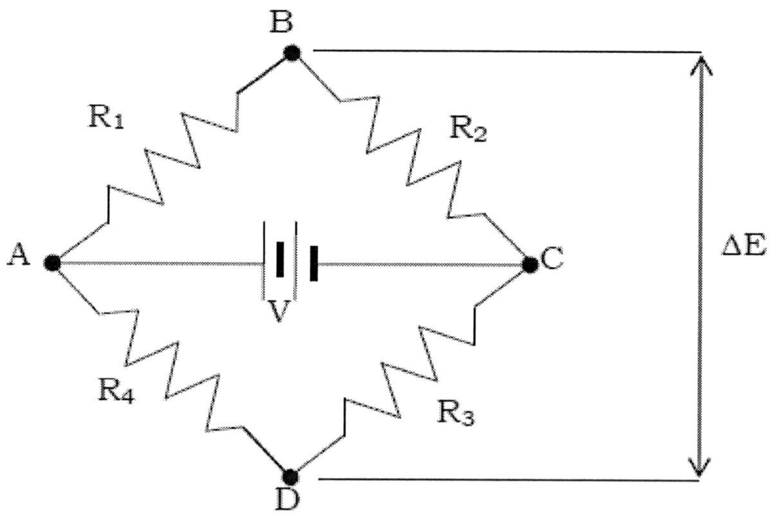

Figure 8. The Wheatstone bridge.

The potential difference, ΔE can be obtained as

$$\Delta E = V_{BD} = V_{AB} - V_{AD} \qquad \text{(Eq. 16)}$$

where the potential differences between nodes A and B, and, A and D may be calculated as

$$V_{AB} = \frac{R_1}{R_1 + R_2} V$$
$$V_{AD} = \frac{R_4}{R_3 + R_4} V \qquad \text{(Eq. 17)}$$

Substituting Eq. 17 into Eq. 16 and simplification yields

$$\Delta E = \frac{R_1 R_3 - R_2 R_4}{(R_1 + R_2)(R_3 + R_4)} V \qquad \text{(Eq. 18)}$$

The bridge is balanced (i.e. $\Delta E = 0$) when $R_1R_3 = R_2R_4$ and this condition is employed when the gauges are unstrained. For changes in these resistances (proportional to strain on gauge) the equation takes the form

$$\Delta E \simeq \frac{r}{(1+r)^2}\left(\frac{\Delta R_1}{R_1} - \frac{\Delta R_2}{R_2} + \frac{\Delta R_3}{R_3} - \frac{\Delta R_4}{R_4}\right) V \text{ where } r = \frac{R_2}{R_1} \qquad \text{(Eq. 19)}$$

This equation is approximate because a nonlinear term which is negligible when strains measured are below 50 000 $\mu\varepsilon$ is neglected.

For many practical applications, the initial values of all four resistances are chosen to be equal therefore Eq. 18 simplifies to

$$\Delta E = \frac{1}{4}\left(\frac{\Delta R_1}{R_1} - \frac{\Delta R_2}{R_2} + \frac{\Delta R_3}{R_3} - \frac{\Delta R_4}{R_4}\right) V \qquad \text{(Eq. 20)}$$

3.4.2. Quarter Bridge

One of the resistors, R_1, of the Wheatstone bridge is the strain gauge, others are fixed (ballast) resistors. In this case

$$\Delta E = \frac{1}{4} S_g \varepsilon V \qquad \text{(Eq. 21)}$$

The strain measured in a quarter bridge is the total strain. If temperature changes considerably and a gauge having the same thermal expansion coefficient with the specimen in that temperature range is not used (termed as self-temperature compensation, STC), measured strain will include some thermal strain in addition to mechanical strain. However thermal strain does not cause stress like the mechanical strain does. Therefore some sort of temperature compensation is needed to estimate stress.

3.4.3. Half Bridge with a Dummy Gauge

R_1 is an active strain gauge whereas R_2 or R_4 is so-called a dummy gauge. Other resistors are fixed resistors. Active gauge is on the specimen where the strain is to be measured.

Dummy gauge may be on the same specimen but at such a location that is not experiencing any mechanical strain (like on neutral axis in pure bending) or on a dummy specimen made of same material with the original specimen and held at the same temperature. Therefore the strains measured by these gauges will be

$$\frac{\Delta R_1}{R_1} = S_g \varepsilon_{mech} + S_g \varepsilon_{thermal}$$
$$\frac{\Delta R_{2\ or\ 4}}{R_{2\ or\ 4}} = S_g \varepsilon_{thermal} \qquad \text{(Eq. 22)}$$

In Eq. 19 the sign of $\Delta R_2/R_2$ (and $\Delta R_4/R_4$) is negative therefore the potential difference measured, ΔE will be proportional to only mechanical strain, ε_{mech}, because thermal strains, $\varepsilon_{thermal}$, cancel each other, which is responsible for stress as

$$\Delta E = \frac{1}{4} S_g \varepsilon_{mech} V \qquad \text{(Eq. 23)}$$

3.4.4. Half Bridge with Two Active Gauges

If one can find two such locations that strain in both locations are the same but one is tensile other is compressive then the two gauges of the half bridge may be utilized as two active bridges. The strains measured by these gauges will be

$$\frac{\Delta R_1}{R_1} = S_g \varepsilon_{mech} + S_g \varepsilon_{thermal}$$
$$\frac{\Delta R_{2\ or\ 4}}{R_{2\ or\ 4}} = -S_g \varepsilon_{mech} + S_g \varepsilon_{thermal} \qquad \text{(Eq. 24)}$$

In Eq. 19 the sign of $\Delta R_2/R_2$ (and $\Delta R_4/R_4$) is negative therefore the thermal strains will cancel and the potential difference will be

$$\Delta E = \frac{1}{2} S_g \varepsilon_{mech} V \qquad \text{(Eq. 25)}$$

which is twice the half bridge with dummy gauge. The same arrangement may be used to cancel out strain due to axial load and to measure strain only due to bending.

3.4.5. Full Bridge

All of the resistors are strain gauges and they are mounted such that gauges 1 and 3 are at a location such that they measure the negative strain of location 2 and 4. The resistance changes are

$$\frac{\Delta R_1}{R_1} = \frac{\Delta R_3}{R_3} = S_g \varepsilon_{mech} + S_g \varepsilon_{thermal}$$
$$\frac{\Delta R_2}{R_2} = \frac{\Delta R_4}{R_4} = -S_g \varepsilon_{mech} + S_g \varepsilon_{thermal} \quad \text{(Eq. 26)}$$

In Eq. 19 the sign of $\Delta R_2/R_2$ and $\Delta R_4/R_4$ are negative therefore the thermal strains will cancel and the potential difference will be

$$\Delta E = S_g \varepsilon_{mech} V \quad \text{(Eq. 27)}$$

which is twice the half bridge with two active gauges or quadruple the half bridge with dummy gauge or quarter bridge. Full bridge applications are mostly used in strain gauge based force sensors (sometimes called load cells too) to measure static or quasi-static forces (see section Force and Torque Measurements Using Strain Gauges).

3.4.5. The Potentiometer Circuit

The potentiometer circuit is often employed in dynamic strain gauge applications (Figure 9).

The potential difference, ΔE due to strain can be computed as

$$\Delta E = \left(\frac{R_1 + \Delta R_1}{R_1 + \Delta R_1 + R_2 + \Delta R_2} - \frac{R_1}{R_1 + R_2} \right) V \quad \text{(Eq. 28)}$$

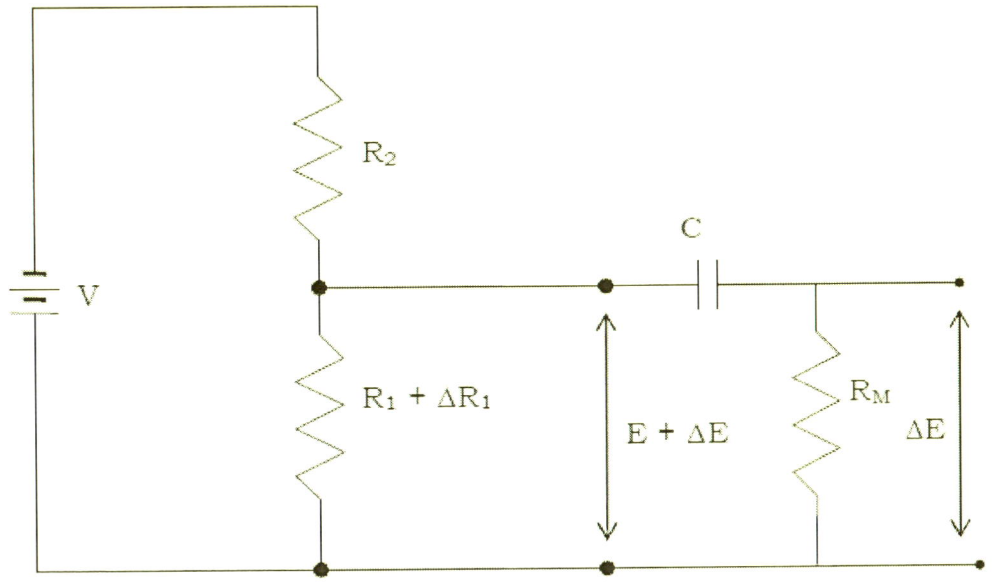

Figure 9. Potentiometer circuit.

For R_1 being the strain gauge and R_2 is a fixed resistor. Like the Wheatstone bridge, the potentiometer circuit has nonlinearity in relating ΔR to ΔE however this nonlinearity is negligible for strains below 2% elongation (*i.e.* 2000 με). It is also possible that R_2 is a dummy gauge (like in half Wheatstone bridge with a dummy gauge) for temperature compensation or an active gauge measuring negative of the strain on gauge 1 (like in half Wheatstone bridge with two active gauges). If the two gauges are identical then

$$\Delta E = \frac{1}{4}\left(\frac{\Delta R_1}{R_1} - \frac{\Delta R_2}{R_2}\right)(1-\eta)V$$
$$\eta = 1 - \frac{1}{1+\frac{1}{2}\left(\frac{\Delta R_1}{R_1} + \frac{\Delta R_2}{R_2}\right)} \quad \text{(Eq. 29)}$$

The output of potentiometer circuit is a small voltage fluctuation (ΔE) over a relatively large DC voltage obtained by the voltage divider circuit ($E = R_1 / (R_1 + R_2)\,V$). This signal may be monitored by filtering out the common mode by a suitable filter without any considerable distortion on the time varying part, ΔE provided that the strain, therefore ΔE are dynamic (*i.e.* changing in time). The simple filter circuit shown in Figure 9 is a high-pass filter with cutoff frequency, $f_c = \frac{1}{2\pi R_m C}$. Potentiometer circuit is not suitable for static strain measurements because the filter that blocks the constant potential, E, would block the stationary signal ΔE too.

3.4.6. Constant Current Potentiometer Circuit

Rather than using a constant voltage source, a constant current source may be used in a potentiometer circuit as shown in Figure 10.

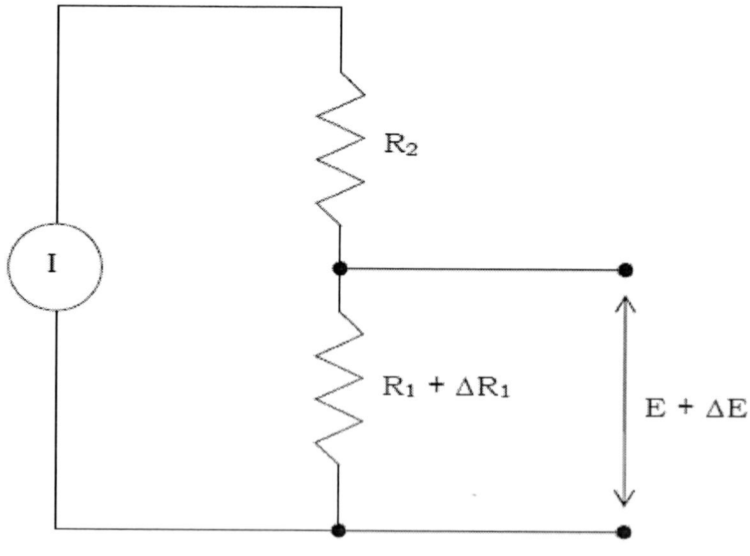

Figure 10. Constant current potentiometer circuit.

In constant current potentiometer circuit the potential difference, E, across the strain gauge, R_1 is

$$E = IR_1 \qquad \text{(Eq. 30)}$$

For a change, ΔR_1, in resistance of the strain gauge, R_1, the potential difference becomes

$$\Delta E = I\Delta R_1 \qquad \text{(Eq. 31)}$$

In Equation 30 or 31 the fixed resistor R_2 does not appear and it may be eliminated totally for a constant current circuit as shown in Figure 11.

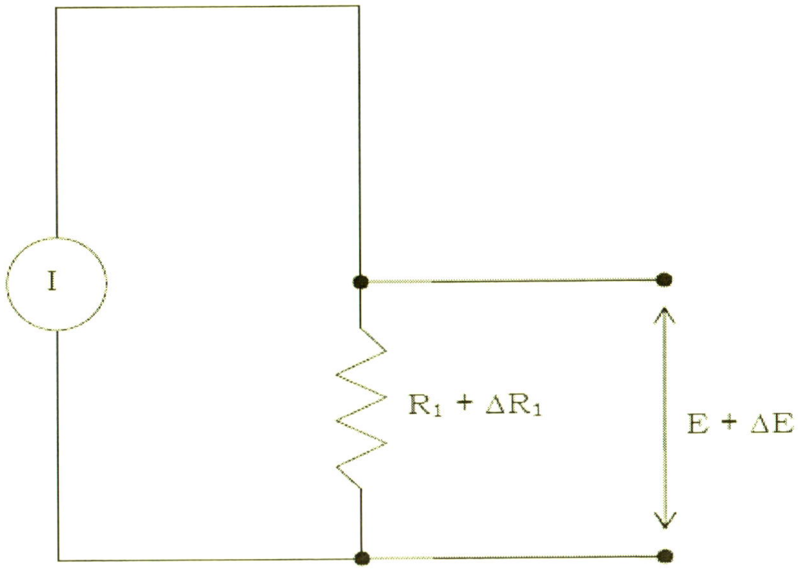

Figure 11. Constant current strain gauge circuit.

Equation 30 holds for this circuit as well. The strain for this circuit would be:

$$\varepsilon = \frac{\Delta E}{IR_1S_g} \qquad \text{(Eq. 32)}$$

For constant current circuits, in selecting the current source, the power dissipated at the gauge should be considered (which appears as heat, causing temperature increase and may damage the gauge if it is beyond the permissible limit of the gauge):

$$P_{dis} = I^2R_1 \qquad \text{(Eq. 33)}$$

3.4.7. Double Constant Current Circuit

Since in constant current potentiometer circuit R_2 does not show up in equations it is not possible to perform temperature compensation or signal cancelling by using R_2 as a dummy or active gauge. However, double constant current circuit presented in Figure 12 may be employed for these purposes.

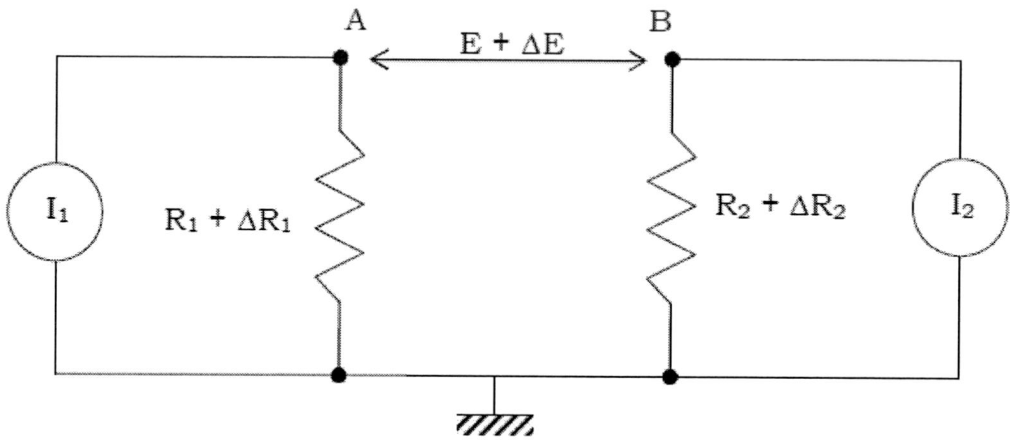

Figure 12. Constant current strain gauge circuit.

For the two strain gauges unstrained (i.e. $\Delta R_1 = \Delta R_2 = 0$) the potential difference measured between A and B, E, would be

$$E = I_1R_1 - I_2R_2 \qquad \text{(Eq. 34)}$$

This potential may be set to zero if

$$I_1R_1 = I_2R_2 \qquad \text{(Eq. 35)}$$

This condition may be achieved either by adjusting the current sources or by using identical strain gauges and identical constant current sources. For an initially balanced double constant current circuit the potential difference, ΔE, due to changes in the resistances of two strain gauges would be

$$\Delta E = I_1\Delta R_1 - I_2\Delta R_2 \qquad \text{(Eq. 36)}$$

Therefore, double constant current circuit may be used both for static and dynamic strain measurements.

If only R_1 is a strain gauge and R_2 is a fixed resistor the strain would be

$$\varepsilon = \frac{\Delta E}{I_1R_1S_g} \qquad \text{(Eq. 37)}$$

If R_1 is an active and R_2 is a dummy gauge (identical gauges, therefore for balance $I_2 = I_1$) then one would perform thermal compensation and cancel out thermal strain

$$\varepsilon_m = \frac{\Delta E}{I_1 R_1 S_g} \quad \text{(Eq. 38)}$$

If both R_1 and R_2 are active gauges (again identical gauges, therefore for balance $I_2 = I_1$) then like in Wheatstone bridge with two active gauges the strain measured by gauge 2 must be negative of that of gauge 1 then,

$$\varepsilon_m = \frac{2\Delta E}{I_1 R_1 S_g} \quad \text{(Eq. 39)}$$

Using two active gauges in double constant current circuit it is possible to cancel out signals other than temperature as well, provided that the signal appears the same in both gauges but not negative of each other.

3.5. Calibration

Like all measuring devices, strain measuring devices need periodic calibration to assure accuracy and precision of the equipment. More often, it is required to scale the sensitivity of the equipment so that the output of the device registers some predetermined input.

3.5.1. Direct Calibration

If it is possible to apply a precisely known mechanical input to a measuring system then the register of the measuring system may be calibrated directly. Direct calibration is practical for strain gauge based transducers because a number of precisely known values of inputs may be applied to transducer.

3.5.2. Indirect Calibration

Indirect calibration, commonly termed as shunt calibration applies a precisely known simulated strain gauge output to the terminals of the strain measuring equipment. This signal is generated by applying a shunt calibration resistor, R_s, in parallel with the strain gauge as shown in Figure 13.

The equivalent resistance of shunt calibration resistor and resistor 1 becomes

$$R_{eq} = \frac{1}{R_1} + \frac{1}{R_s} = \frac{R_s R_1}{R_1 + R_s} \quad \text{(Eq. 40)}$$

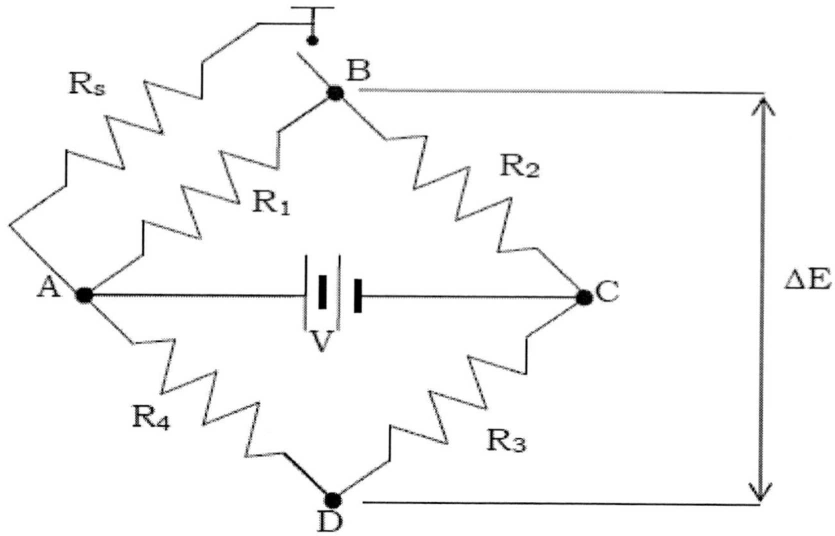

Figure 13. Shunt calibration for strain gauge1 on the Wheatstone bridge.

and the simulated change in this resistance would be

$$\Delta R = R_1 - R_{eq} \qquad \text{(Eq. 41)}$$

which is a negative number therefore simulated strain is compressive. The simulated strain would be

$$\varepsilon_s = \frac{-R_1}{S_g (R_1 + R_s)} \qquad \text{(Eq. 42)}$$

The resistances of some common shunt calibration resistors and their simulated strains are presented in Table 1 for a gauge factor of 2.

Table 1. Some Shunt Calibration Resistors

Gauge Resistance (Ω)	Shunt Resistance (Ω)	Simulated Strain (με)
120	14 880	4 000
350	11 880	5 000
1 000	5 880	10 000

3.6. Strain Gauge Arrangements for Different Stress States

3.6.1. Uniaxial Stress State

In uniaxial stress, normal stress along one axis is nonzero, normal stresses along the other two mutually perpendicular axes are zero. Shear stresses along all three mutually perpendicular axes are all zero. The strain gauge is aligned along the direction of the stress.

Knowing the elastic modulus, E, of the specimen and measuring the strain, stress can be obtained as

$$\sigma_x = E\varepsilon_x \qquad \text{(Eq. 43)}$$

3.6.2. Biaxial Stress State

In biaxial stress state the principal directions of stress (which coincides with the principal directions of strain for elastically isotropic materials) are known. Rectangular two-element strain gauge is aligned along the principal directions. Knowing the elastic modulus, E, and Poisson's ratio, ν, of the specimen, from the two strains measured by the two gauges the principal stresses can be obtained as

$$\sigma_1 = \frac{E}{1-\nu^2}(\varepsilon_1 + \nu\varepsilon_2) \qquad \text{(Eq. 44)}$$
$$\sigma_2 = \frac{E}{1-\nu^2}(\varepsilon_2 + \nu\varepsilon_1)$$

3.6.3. General Plane Stress State

If nothing is known about the plane stress state then three strains, two normal strains along two mutually perpendicular directions and one shear strain, the angular deviation from the right angle of the two mutually perpendicular directions in undeformed state has to be measured. However strain gauges can only measure elongations not angular changes directly. Using transformation equations (also known as Mohr circle equations), by measuring normal strains along three different directions it is possible to calculate two normal strains and one shear strain along any Cartesian coordinate system x-y as

$$\varepsilon_1 = \varepsilon_x \cos^2\theta_1 + \varepsilon_y \sin^2\theta_1 + \gamma_{xy}\cos\theta_1 \sin\theta_1$$
$$\varepsilon_2 = \varepsilon_x \cos^2\theta_2 + \varepsilon_y \sin^2\theta_2 + \gamma_{xy}\cos\theta_2 \sin\theta_2 \qquad \text{(Eq. 45)}$$
$$\varepsilon_3 = \varepsilon_x \cos^2\theta_3 + \varepsilon_y \sin^2\theta_3 + \gamma_{xy}\cos\theta_3 \sin\theta_3$$

where ε_1, ε_2, ε_3 are the three normal strains measured by three strain gauges and θ_1, θ_2, θ_3 are the angles between the axes of the three strain gauges with the x-axis.

For two very common strain gauges, the three element rectangular gauge (Figure 14) the strains are presented in Eq. 46 and for delta rosette (Figure 15) the strains are presented in Eq. 47.

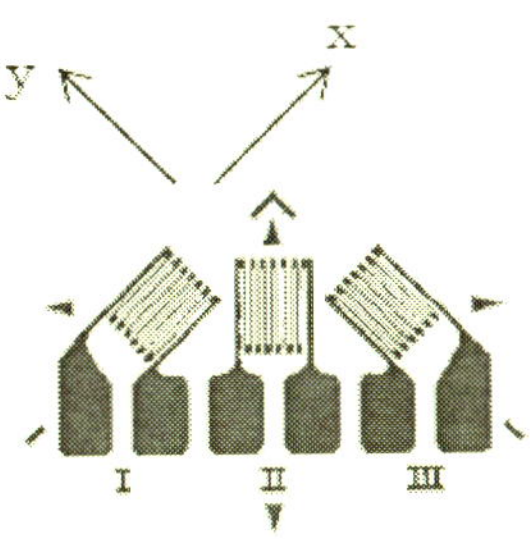

Figure 14. Three Element Rectangular Gauge.

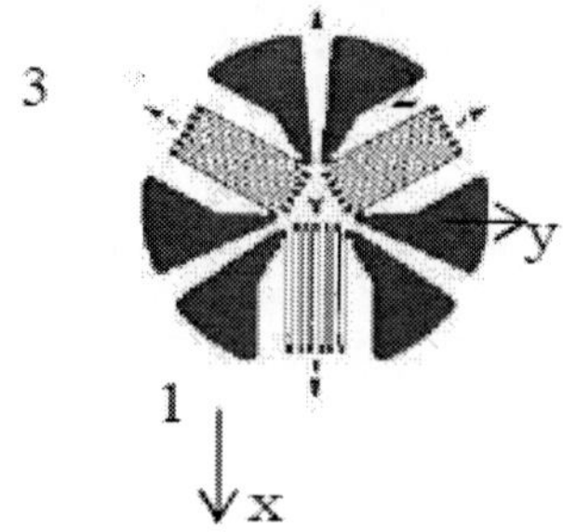

Figure 15. Delta Rosette.

$$\varepsilon_x = \varepsilon_1$$
$$\varepsilon_y = \varepsilon_3 \qquad \text{(Eq. 46)}$$
$$\gamma_{xy} = 2\varepsilon_2 - (\varepsilon_1 + \varepsilon_3)$$

$$\varepsilon_x = \varepsilon_1$$
$$\varepsilon_y = \frac{1}{3}\left[2(\varepsilon_2 + \varepsilon_3) - \varepsilon_1\right] \qquad \text{(Eq. 47)}$$
$$\gamma_{xy} = \frac{2}{\sqrt{3}}(\varepsilon_3 - \varepsilon_2)$$

3.7. Strain Gauge Installation Guidelines

The transfer of surface strains to the strain gauge is essential for strain measurement. Therefore proper surface preparation is essential. It is strongly recommended to follow the instructions of individual strain gauge manufacturer. In this section the general guidelines for strain gauge application to metal surfaces are presented.

3.7.1. Solvent Degreasing

Oil, grease, organic contaminants and chemical residues due to production, handling or storage are removed mostly using aerosol type degreasers although hot vapor degreaser or ultrasonically agitated liquid bath may be utilized too. For porous materials like cast metals and titanium, to remove dirt out of pores, heating is recommended. If possible, degrease the whole specimen, if the specimen is very large then an area of 10 to 15 cm around the specimen site is recommended to be degreased. While wiping, the strokes should be from center to sides to keep the center cleanest.

3.7.2. Surface Abrading

First, if any, scale, rust, galvanized coatings must be removed. The suitable surface texture is obtained by using the abrasive grit recommended by the manufacturer, together with a liquid chemical (if recommended by the manufacturer and applicable to that specific test material).

3.7.3. Gauge Location Layout Lines

Gauge location layout lines are drawn to locate and align the strain gauge correctly. The best tool to draw the lines is a pencil. Hard tip tools which scores the surface rather than burnishing, like a pen or a pointed metal are not recommended.

3.7.4. Surface Conditioning

The specimen surface should be conditioned using the proper conditioner, never letting the surface to dry. At the end of the process, the surface should be wiped from center to the sides with a clean sponge.

3.7.5. Neutralizing

Before bonding the gauge the surface pH level should be brought back to a value recommended by the adhesive manufacturer. The liquid conditioner should be never let to dry. After wiping the conditioner to sideways the specimen is ready for strain gauge bonding.

3.7.6. Handling the Gauge

Many strain gauges arrive chemically clean and manufacturers specially treat the surface to be bonded therefore the gauges should not be touched with bare hands and should not be contaminated prior to bonding.

3.8. Sources of Measurement Errors

3.8.1. Electrical Sources of Errors

Strain measurements are mostly done in the presence of electric and/or magnetic fields that may superimpose electrical noise on the measurement signals. If noise is not handled properly it may cause inaccurate measurements or in extreme cases may obscure the measurement signal. For accurate measurements, signal to noise ratio needs to be maximized. In order to achieve this goal first the types, characteristics and sources of such noise has to be analyzed.

Noise is an unwanted signal that interferes with the measurement signal. Any electrical device that produces, transmits, transforms or consumes electrical power is a potential source of electrical noise. Higher potentials and higher currents produce more noise and with decreasing distance to the source of the noise, strain gauges pick up more. There are two basic types of electrical noise, electrostatic and magnetic. Many of the electrical devices produce both types.

3.8.2. Electrostatic Noise

Electrostatic noise is generated by the presence of electric potential without need for a current flow. Alternating current (AC) causes electrical noise in any electrical conductor, including stain gauges and wiring by capacitive coupling. The simplest and most effective way to reduce electrostatic noise is an electrical conductor shielding (sometimes simply called shield or Faraday shielding) which is connected to ground or a constant potential with a low resistance connection.

3.8.3. Electromagnetic Noise

Magnetic fields are generated by either permanent magnets or by the electric current. In the presence of alternating current, noise will be induced in stationary electric conductors. The methods to reduce the electromagnetic noise during strain gauge measurements include using twisted pair cables, eliminating the auxiliary terminals and connecting the lead wires directly to strain gauge tabs, using non-inductive gauge types and using magnetic shielding materials on strain gauges. For high frequency fields (like electronic ballasts, eco-bulbs *etc.*) the shield material should be suitable for these frequencies.

One recommended method to determine amount of noise picked up by the circuit is to remove the excitation and determine the self generating noise or effects of various precautions on this noise (Watson, 2008).

Another source of noise is the leakage to the ground through the strain gauge or the cables. The electrical resistance between the specimen and the strain gauge and related cabling is recommended to be over 10 GΩ. It should be mentioned that the resistance may decrease in time, with applied strain or with elevated temperatures.

3.8.4. Electrical Resistance of Measuring Equipment

In calculations all potential measuring devices were idealized with an infinite internal resistance whereas all current measuring devices were idealized with zero resistance devices. The actual voltmeters have finite resistances. This causes some current to be drawn by the meter causing the potential to be measured to drop. The actual ammeters have nonzero internal resistances therefore would cause a rise in the equivalent resistance of the circuit and cause a reduction in the current. Most of the modern measuring devices utilizing electronic circuits and amplifiers have very high internal electrical resistances while measuring potential difference and a very low internal resistance while measuring current therefore the errors due to electrical resistance of measuring equipment is mostly negligible. The high internal resistance voltmeters are termed as voltage sensitive deflection bridge by Holman (1989).

3.8.5. Effect of Lead Wire Resistance

If lead wires are long, and/or strain gauge resistance is low, the electrical resistance of lead wires, R_L, may become significant when compared to the initial resistance of the strain gauge, R_g. Therefore the equivalent resistance of lead wires and strain gauge would be $R_{eq} = R_g + R_L$ which causes a de-sensitization, $D = R_g / (R_g + R_L)$ which can be compensated by using a corrected gauge factor on strain measuring device, $S_g' = D\, S_g$, where S_g is the gauge factor supplied by the manufacturer.

3.8.6. Thermal Effects on Lead Wire Resistances

The lead wires, commonly made of copper, have a high thermal coefficient of electrical resistance. These wires act as an integral part of the strain measuring system with their resistances.

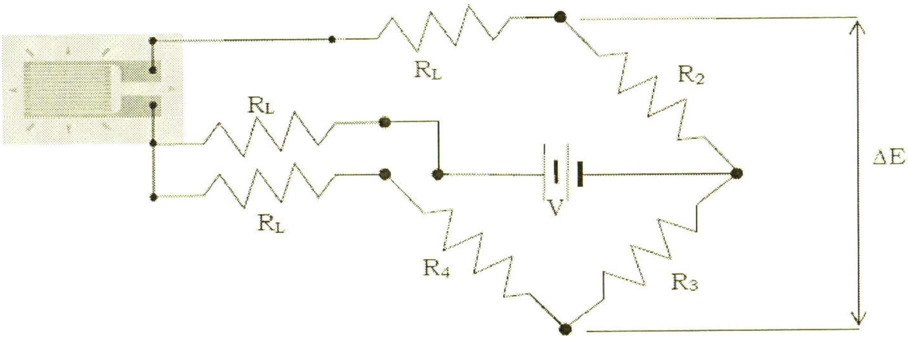

Figure 16. Three lead system for a single active gauge.

Any change in ambient temperature will influence the equivalent electrical resistance of the strain gauge and wire unit which will change the equipment reading. This error source may be eliminated by using a so called three lead system as shown in Figure 16. The three lead system is based on equal change in lead resistances therefore to keep all three wires at the same temperature it is recommended to keep the wires together by twisting them.

3.8.7. Strain Gauge Thermal Output

For a strain gauge bonded on a specimen, a temperature change would cause a change in the electrical resistance of the strain gauge which is called the thermal output of the strain gauge. An approximate correction for the thermal output may be performed by using the graph supplied by the gauge manufacturer as

$$\varepsilon_{corr} = \varepsilon - \varepsilon_T \qquad \text{(Eq. 48)}$$

where ε is the strain registered at the test temperature, ε_T is the thermal output of the strain gauge obtained from the graph supplied by the manufacturer provided that the system is balanced at the temperature specified by the manufacturer and ε_{corr} is the corrected strain which should be used to estimate stress.

This procedure is an approximate procedure because the gauge factor was assumed to be 2.00 and the approximation is good for gauges with gauge factors close to 2.00. For gauges having a gauge factor different from 2.00, ε_T is obtained by multiplying the value in manufacturer's sheet by a factor of $2/S_g$.

Hines (1960) showed that for gauges mounted on curved surfaces with radius of curvature less than 12 mm, the thermal output of the gauge would be different than gauges mounted on flat surfaces.

3.8.8. Gauge Factor Variation with Temperature

For any strain measurement, ε_1, performed at gauge factor S_{g1} needs to be corrected to, ε_2, for another gauge factor S_{g2} the procedure of correction is

$$\varepsilon_2 = \frac{S_{g1}}{S_{g2}} \varepsilon_1 \qquad \text{(Eq. 49)}$$

In case of considerable temperature changes a simultaneous correction of thermal output and gauge factor errors may be necessary.

3.8.9. Corrections for Transverse Strain Effects

Gauge factor of a strain gauge is calibrated for uniaxial stress on a material with Poisson's ratio of 0.285. The transverse sensitivity error may be due to one or more of the following factors:

- Strain gauge is on steel (Poisson's ratio, $\nu = 0.285$, the calibrated value) but specimen is subjected to a stress state other than uniaxial,
- Strain gauge is on steel with uniaxial stress state but the gauge axis is not aligned with the principal axis,
- Strain gauge is installed on a specimen with different Poisson's ratio.

Errors due to transverse sensitivity are generally small since transverse sensitivity of a strain gauge is small. However if there is a huge difference between principal strains, the percent error on the measured small strain may be considerable.

$$\frac{\Delta R}{R} = S_a \varepsilon_a + S_t \varepsilon_t \qquad \text{(Eq. 50)}$$

where S_a is the axial sensitivity of the strain gauge and S_t is the transverse sensitivity. The gauge factor, S_g (Eq. 14) is obtained for uniaxial stress with a Poisson's ratio of 0.285 therefore under the condition of $\varepsilon_a = -0.285\ \varepsilon_t$. If S_t were zero, the transverse strain would not affect the resistance of the strain gauge. For a strain gauge with gauge factor S_g, supplied by the manufacturer, S_a sensitivity of the gauge material to strain,

$$S_t = \frac{S_a - S_g}{0.285} \qquad \text{(Eq. 51)}$$

Defining the cross sensitivity factor

$$K = \frac{S_t}{S_a} \qquad \text{(Eq. 52)}$$

and

$$F = \left(1 - K^2\right) S_a \qquad \text{(Eq. 53)}$$

correction for uniaxial stress state can be performed as

$$\varepsilon_L^{corr} = \frac{1 - 0.285K}{1 - \nu K} \varepsilon_L \qquad \text{(Eq. 54)}$$

where υ is the Poisson's ratio of the test specimen.

Correction for biaxial stress state where the two gauges are aligned with the principal axes, x and y,

$$\varepsilon_x^{corr} = \frac{1-0.285K}{1-K^2}\left(\varepsilon_x - K\varepsilon_y\right)$$
$$\varepsilon_y^{corr} = \frac{1-0.285K}{1-K^2}\left(\varepsilon_y - K\varepsilon_x\right) \quad \text{(Eq. 55)}$$

Correction for three-element rectangular rosette where gauge A is along x-axis, gauge B makes 45° with positive x and y axes and gauge C is along y axis,

$$\varepsilon_x^{corr} = \frac{1-0.285K}{1-K^2}\left(\varepsilon_A - K\varepsilon_C\right)$$
$$\varepsilon_y^{corr} = \frac{1-0.285K}{1-K^2}\left(\varepsilon_C - K\varepsilon_A\right) \quad \text{(Eq. 56)}$$
$$\gamma_{xy}^{corr} = \frac{1-0.285K}{1-K^2}\left[2\varepsilon_B - \left(\varepsilon_A + \varepsilon_C\right)\right]$$

For a three element delta rosette where gauge A is along x-axis, gauge B making 120° and gauge C in 240° (in counter-clockwise direction) with x-axis,

$$\varepsilon_A^{corr} = \frac{1}{3}\frac{1-0.285K}{1-K^2}\left[\left(3+K\right)\varepsilon_A - 2K\left(\varepsilon_B + \varepsilon_C\right)\right]$$
$$\varepsilon_B^{corr} = \frac{1}{3}\frac{1-0.285K}{1-K^2}\left[\left(3+K\right)\varepsilon_B - 2K\left(\varepsilon_A + \varepsilon_C\right)\right] \quad \text{(Eq. 57)}$$
$$\varepsilon_C^{corr} = \frac{1}{3}\frac{1-0.285K}{1-K^2}\left[\left(3+K\right)\varepsilon_C - 2K\left(\varepsilon_B + \varepsilon_B\right)\right]$$

3.8.10. Reinforcement Effects of Strain Gauges

Strain gauge itself is supposed to have an equivalent elastic modulus in the range 5-30 GPa (Watson 2008) therefore for a metallic specimen of thickness a few millimeters, the reinforcement effect of the strain gauge is negligible. However for materials with lower elastic moduli, if the thickness is small the strain gauge reinforces the original material causing strain values read from the strain gauged specimen to be considerably lower than the ungauged case. Reinforcement effects are analyzed in detail in Perry (1985), Little et. al. (1990), and, Ajovalasit and Zuccarello (2005).

3.9. Strain Gauge Excitation Levels

Strain gauges, being electrical resistances, dissipate electrical energy into heat. There is a trade off between the excitation voltage (or alternatively current) and permissible heat generation. As the excitation voltage (or alternatively current) increases the signal output of the strain gauge increases however increasing voltage (or alternatively current) increases heat generation as well. The following factors affect the excitation level of a strain gauge:

- Strain gauge grid area, larger areas can dissipate heat easier.
- Strain gauge resistance, large resistance gauges generate less heat therefore may be excited with higher voltages.
- Conduction properties of the test specimen, high thermal conductivity specimens like cooper and aluminum conduct generated heat very well, on the other hand stainless steel, non-metallic materials, plastics, polymers and many biological materials are poor heat conductors.

Allowable gauge power dissipation per unit area is mostly specified by the gauge manufacturer depending on measurement type (static or dynamic), heat conduction capacity of the test specimen, and, required accuracy. Using proper value of power density in grid, P_g, the bridge excitation voltage may be obtained as:

$$V = 2\sqrt{R_g A_g P_g} \quad \text{(Eq. 50)}$$

where R_g is the resistance of the gauge (in Ohms), A_g is the grid area of the gauge (grid length times grid width, in mm) and P_g is the power density in grid recommended by the manufacturer (in Watts per square millimeters). The potential on the strain gauge would be half of this voltage provided that the ballast resistor has the same resistance as the strain gauge.

Typical power allowable power dissipation per unit area (sometimes termed as power density) recommendations are presented in Table 2.

Table 2. Typical recommended maximum power density (kW/m^2) for strain gauges [Micro-Measurements, 2010]

Measurement Type	Heat Conductivity of Specimen	Excellent Aluminum or Copper	Good Steel	Fair Thin stainless steel, titanium	Poor Reinforced plastics	Very Poor Plastics, polymers
	Accuracy					
Static	High	3.1-7.8	1.6-3.1	0.78-1.6	0.16-0.31	0.016-0.031
	Moderate	7.8-16	3.1-7.8	1.6-3.1	0.31-0.78	0.031-0.078
	Low	16-31	7.8-16	3.1-7.8	0.78-1.6	0.078-0.016
Dynamic	High	7.8-16	7.8-16	3.1-7.8	0.78-1.6	0.016-0.078
	Moderate	16-31	16-31	7.8-16	1.6-3.1	0.078-0.31
	Low	31-78	31-78	16-31	3.1-7.8	0.31-0.78

3.10. Gauge Selection

Gauge selection is dependent on the following factors:

Operating Temperature

The gauge alloy and the carrier (backing) of the foil strain gauge should be suitable for the intended operating temperature. Manufacturer's recommendations for each individual gauge should be the source of primary information. The general guidelines are as follows:

Paper backing gauges are suitable only to temperatures up to 50° C, epoxy backing gauges may be used in a temperature range of -200 to 115°C and phenol resin and glass fiber backing gauges have a temperature range of -240° to 230°C. Advance alloy (constantan) gauge grids tend to exhibit continuous drifts at temperatures above 65°C, karma alloy grids may be used in between -260 and 260°C and K-alloy encapsulated gauges may be exposed to temperatures up to 400°C for short periods of time.

Strain State

In addition to points mentioned in the section Strain Gauge Arrangements for Different Stress States, the following points need to be considered for selecting strain gauges:

- For sharp strain gradients gauge length should be kept at a minimum (0.2 or 0.4 mm),
- It is recommended to have the gauge length no larger than 0.1 times the hole radius, fillet, notch or any other stress concentration geometries,
- For strains larger than 5% annealed constantan (P-alloy) grid material is recommended,
- For strains around 10% to 15% copper-nickel grid post-yield gauges are recommended,
- For strain measurement on non-homogeneous materials, it is recommended that the gauge would have sufficient size to average the strain on all constituents of the material,
- For static measurements advance (constantan) or karma alloy is recommended due to temperature stability, for low levels of dynamic strain measurement isoelastic alloy is recommended,
- In using rosette strain gauges single plane rosette gauges are better in terms of heat dissipation but the gauge size is larger so the measurement may be inaccurate for high strain gradients whereas stacked rosettes have smaller gauge size but poor heat dissipation.

Other Factors

- If the time period over which strain is measured would be long, the gauge size should be selected as large as possible to minimize the relaxation in the gauge adhesive.
- For shock measurements dynamic gauges should be preferred. If the number of shocks is large, fatigue resistant gauges and adhesives should be preferred.
- For vibration measurements if only frequency and amplitude of stress are desired dynamic gauges may be preferred, for low frequencies static gauges may be used.
- For strain gauges running in magnetic environments constantan and platinum-tungsten grids are preferred, for gauges to be used under radiation constantan and platinum-tungsten 1200 alloys are relatively stable.

3.11. Force and Torque Measurements Using Strain Gauges

It may be necessary to construct custom made force or torque transducers in dentistry. One candidate element for such transducers is the strain gauge. In force and torque measurements these quantities are converted into a mechanical deformation causing strain which is converted into a potential difference by using a strain gauge circuit. Such a device is termed as a force/torque transducer or sometimes a load cell.

3.11.1 Link Type Load Cell

The axial load on a link would cause longitudinal and transverse strains in proportion with the load. The two strain gauges (1 and 3) are bonded in the axial direction on the link whereas the other two (2 and 4) are bonded in the transverse direction.

The force to be measured would be

$$F = \frac{2AE}{S_g(1+\nu)V}\Delta E \qquad \text{(Eq. 51)}$$

where A is the cross sectional area of the link, E is the elastic modulus, υ is the Poisson's ratio of the link material, S_g is the gauge factor, V is the bridge excitation voltage and ΔE is the potential difference of the bridge under load.

3.11.2. Beam Type Load Cell

Beam type load cells may measure lower loads compared to link type load cells. The beam type load cell utilizes a full Wheatstone bridge as presented in Figure 18. Gauges 1 and 3 would be on the top surface of the beam, 2 and 4 on the bottom surface, close to the fixed end so that the bending moment due to the force F is larger. The strain gauges are aligned in the longitudinal axis of the beam.

The force to be measured would be

$$F = \frac{Ebh^2}{6S_g\ell V}\Delta E \qquad \text{(Eq. 52)}$$

where E is the elastic modulus of the beam material, b is the width, h is the height of the beam, S_g is the gauge factor, ℓ is the distance between the force and strain gauge center, V is the bridge excitation voltage and ΔE is the potential difference of the bridge under load.

3.11.3. Torque Cells

Torque cells are used to measure the twisting effect of forces. They mostly contain a circular (sometimes hollow for small torques) shafts as the elastic element. However, torque causes shear strain along the shaft axis on the shaft surface therefore the normal strain along the principal directions are measured for torque measurement as presented in Figure 19. by 45 degrees aligned gauges.

The torque to be measured would be

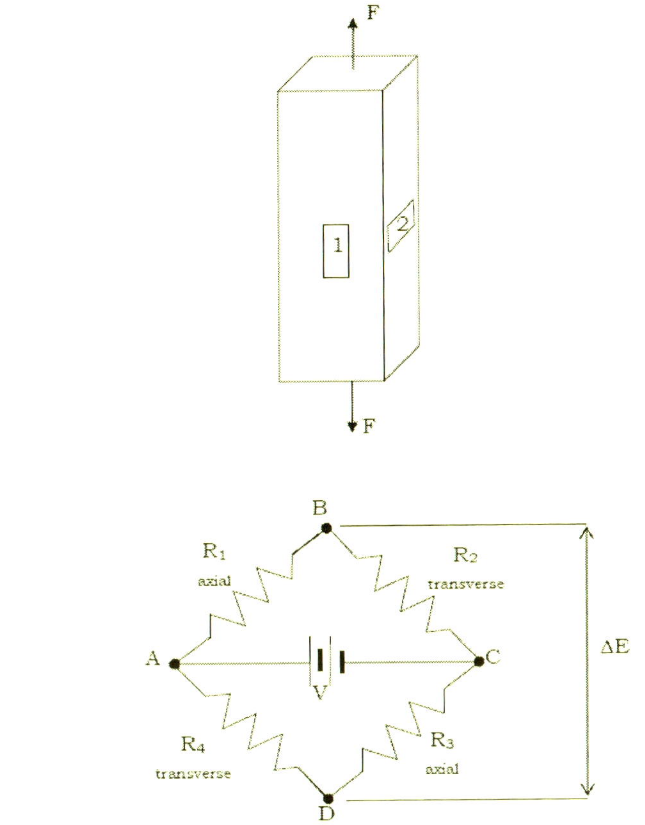

Figure 17. Link type load cell and its wiring.

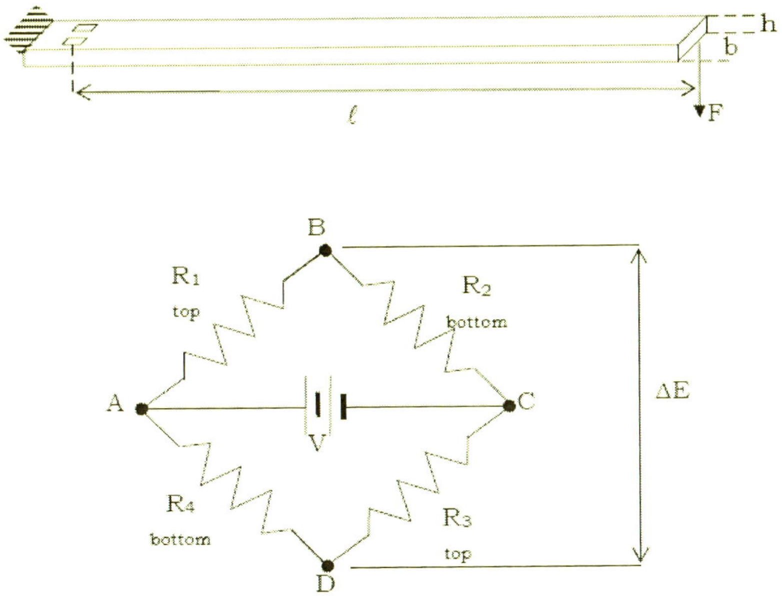

Figure 18. Beam type load cell.

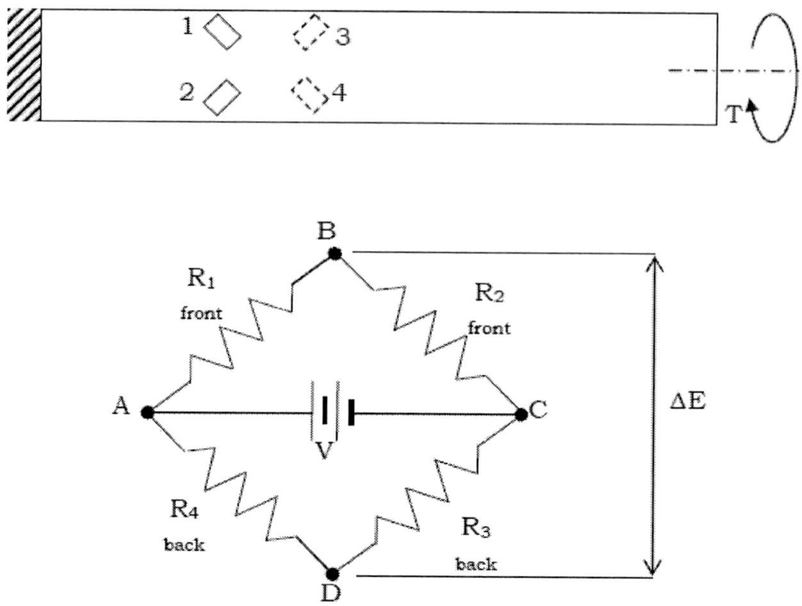

Figure 19. Torque cell and its wiring.

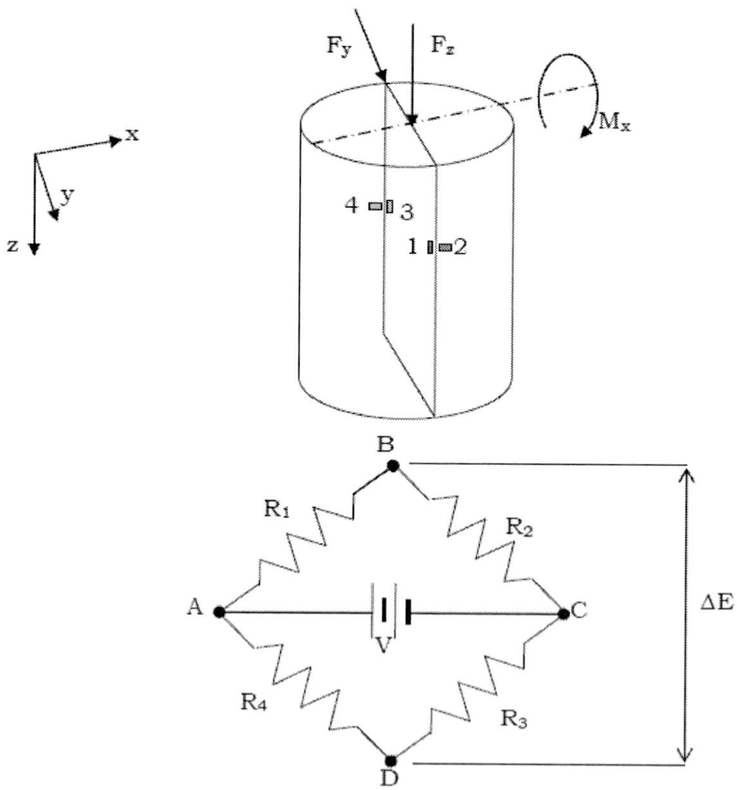

Figure 20. Measurement of axial component, F_z.

$$T = \frac{\pi D^3 E}{16(1+\nu)S_g V}\Delta E \qquad \text{(Eq. 53)}$$

where D is the diameter of the shaft, E is the elastic modulus, υ is the Poisson's ratio of the shaft material, S_g is the gauge factor, V is the bridge excitation voltage and ΔE is the potential difference of the bridge under load.

3.11.4. Simultaneous Force and Moment Measurement

In some cases it may be necessary to measure more than one component of a force or force and moment components simultaneously whereas in other cases it may be necessary to measure only one component of a force. An example application might be as presented in Figures 20, and 21 (Berme, 1990). In these force transducers, rather than calculating the force and moment components from strains, the transducer is calibrated by applying known force/moment and measuring the corresponding potential difference, ΔE, obtained from the Wheatstone bridge to account for all effects like transverse sensitivity, cross-talk etc.

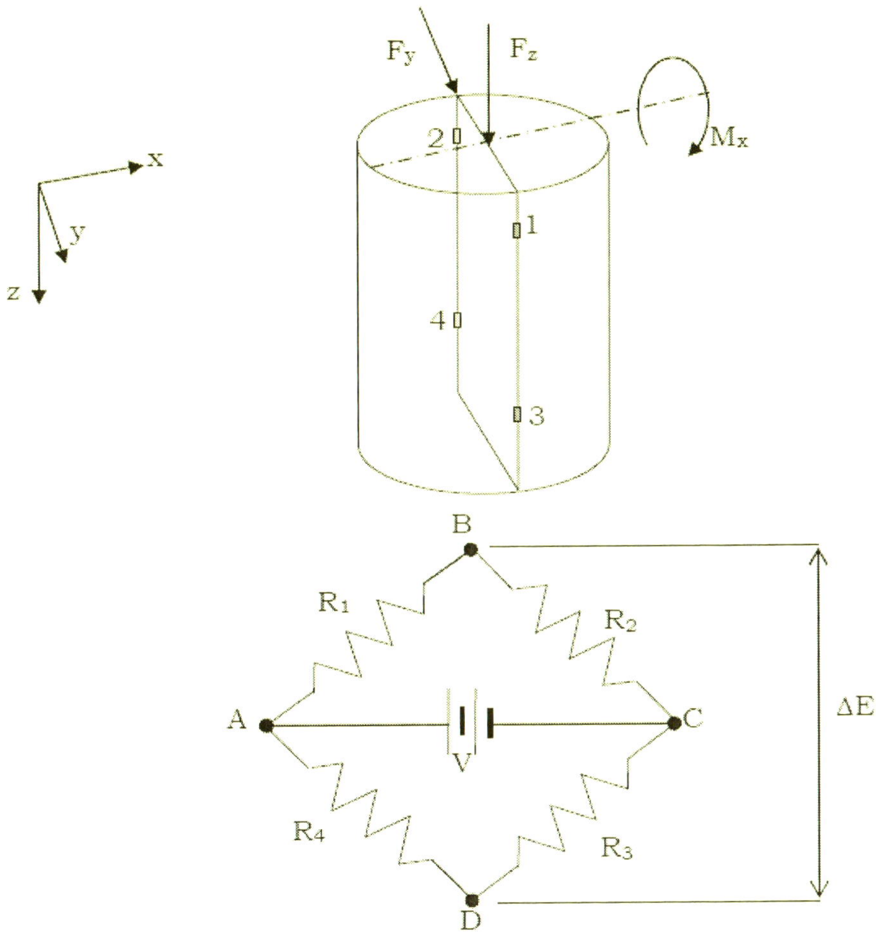

Figure 21. Measurement of shear component, F_y by bending moment difference method.

ACKNOWLEDGEMENT

Unpublished lecture notes of Ömer Gündüz Bilir, Professor Emeritus of Mechanical Engineering Department of Middle East Technical University provided a very neat and clear starting point for the author.

REFERENCES

Ajovalasit, A., Zuccarello, B., 2005 Local reinforcement effect of a strain gauge installation on low modulus materials, J. Strain Anal. 40(7), 643–653.

Berme, N, Load Transducers, *in Biomechanics of Human Movement, Applications in Rehabilitation, Sports and Ergonomics,* Berme and Capozzo (eds), Bertec Corp, Worthington, Ohio, USA, 1990.

Daly, James W., Riley, William F. *Experimental Stress Analysis*, McGraw-Hill, 1991.

Fung, Y. C., A *First Course in Continuum Mechanics*, Third Edition, Prentice Hall, 1994.

Hines, F. F., 1960, Effect of mounting surface curvature on the temperature coefficient of bonded resistance strain gages, Proceedings, Western Regional Strain Gage Committee, *The Society for Experimental Mechanics*, Bethel, pp 39–44.

Holman, J. P., *Experimental Methods for Engineers*, McGraw-Hill, 1989.

Juvinall, R.C., *Engineering considerations of stress, strain, and strength*, McGraw-Hill, 1967.

Khan, A. S., Wang, X., *Strain Measurements and Stress Analysis*, Prentice Hall, 2001.

Little, E.G., Tocher, D., O'Donnell, P., 1990, *Strain gauge reinforcement of plastics, Strain* 26(3), 91–97.

Malvern, L. E., *Introduction to the Mechanics of a Continuous Medium*, Prentice-Hall, 1969.

Micro-Measurements, 2010 Technical Note TN-502, Vishay Precision Group, Document Number 11052, Revision Date 1. November. 2010.

Perry, C. C., 1985, Strain gage reinforcement effects on low-modulus materials, *Exp. Tech.* 9(5), 25–27.

Peter K. Stein "1936 A banner year for strain gages and experimental stress analysis- a historical perspective", *Experimental Techniques*, v. 30, pp. 23-41, 2006.

Thomson, W. (Lord Kelvin), 1856, On electrodynamic qualities of metals, Phil Trans. R. Soc., v. 146, pp. 649-751.

Truesdell C., Noll, W, 1965, The nonlinear field theories of mechanics in *Encyclopedia of Physics,* ed. S. Flügge v. 3/3, Berlin: Springer-Verlag.

Watson, R. B., Bonded Electrical Resistance Strain Gauges in *Springer Handbook of Experimental Mechanics*, Sharpe W. N., Jr, (*ed.*) Springer Science + Business Media, New York, 2008.

Further Reading on Electric Resistance Strain Gauges

Holister, G.S., *Experimental Stress Analysis, Principles and Methods*, Cambridge University Press, 1967.

Hetéyi, M, *Experimental Stress Analysis*, Wiley, 1960.

Murray, W. M., Miller, W. R., *The bonded electrical resistance strain gage: an introduction*, Oxford University Press, 1992.

Window, A. L., *Strain gauge technology*, Springer, 1992.

ASTM E 251 Standard Test Methods for Performance Characteristics of Metallic Bonded Resistance Strain Gages.

VDI/VDE/GESA 2635 Part 1: Experimental structure analysis - Metallic bonded resistance strain gages - Characteristics and testing conditions.

VDI/VDE/GESA 2635 Part 2: Experimental structural analysis - Recommended practice for high-temperature strain measurements.

In: Biomechanics of Dental Implants
Editor: Murat Çehreli
ISBN 978-1-62100-780-7

Chapter 9

PHOTOELASTIC STRESS ANALYSIS

Murat Çehreli

1. INTRODUCTION

Developed at the turn of the 20th century, photoelasticity is a non-destructive, practical, and cost-effective method used in many industrial applications for qualitative and quantitative stress analysis. It is a full-field technique based on an optomechanical property called birefringence, inherent in many transparent polymers, and allows measuring the magnitude and direction of principal stresses. The intuitive insight from these full-field displays can be very self-explanatory that it may be unnecessary to convert them to numerical values, although the conversion can be done if desired. A loaded photoelastic specimen or a specimen coated with a photoelastic material exhibits fringe patterns that are directly related to the difference between principal stresses in the plane normal to the direction of ordinary light propagation. The fringes appear, because the chosen materials become optically anisotropic when loaded.

The method is traditionally used for analyzing two-dimensional plane problems, but has been also adapted to solve the three-dimensional problems in dentistry. The so-called quasi-three dimensional photoelastic stress analysis technique, developed by the work of Caputo and Standlee [1] is basically the combination of the two- and three-dimensional photoelastic techniques. A passage of polarized light through a model of an arbitrary or anatomic geometric configuration and the generation of colored patterns, isochromatic fringes, are observed in the circular polariscope. The model, i.e., implant-supported fixed prostheses in a resin mandibular model, is three-dimensional, but the fringes are observed and analyzed in two dimensions. Some details of this technique will be discussed later in this section. Another application of the photoelastic technique called "stress freezing", which will not be the basic emphasis in this chapter, is also an extension of the two-dimensional technique. A strain-sensitive photoelastic coating is applied on the body of a complex three-dimensional structure, i.e., vehicle water pump casting, for analyzing surface stresses. The coatings can be applied virtually on any test object regardless of its size, shape or material composition. In this technique, the liquid plastic is cast on a flat mould and while partially polymerized and in a pliable state, manually applied to the surface of the object and left for complete polymerization. Then, the coating is bonded using special reflective cement and tested.

Owing to the advent of superior computer processing, finite element modeling has become the dominant technique in stress analysis. As for finite element analysis technique, the idealization of the physical problem to a mathematical model requires certain assumptions that lead to differential equations governing the mathematical model, and since the procedure is numerical, it is imperative to verify its predictive accuracy. For example, a threaded joint experiences non-uniform contact and friction, and therefore, contact definition should be implemented at the interface for correct interpretation of stresses (Figure 1) [2].

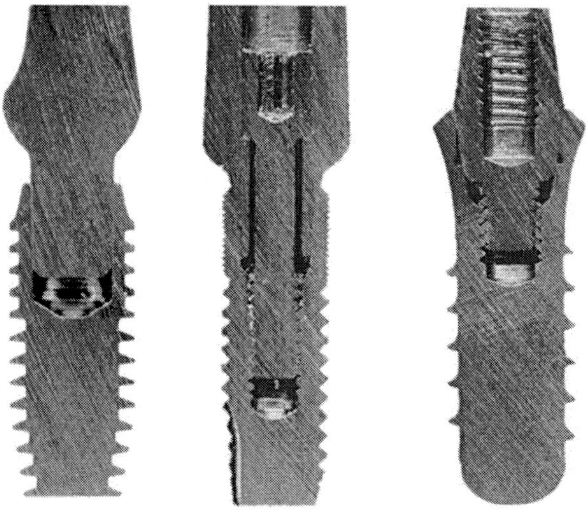

Figure 1. An example of screw joint could be seen in 2-piece conventional oral implants. The evaluation of the mechanical behavior of the components by means of finite element analysis requires implementation of contact definition at the joint interface.

This is certainly a time-consuming procedure. In such cases, photoelasticity still remains as a rapid and effective tool in modern stress analysis, although its assumptions may not always well-fit into biologic simulations (Table 1). The technique has been applied to bone and implants from late 1930s, when polymeric materials such as phenol formaldehyde and epoxy resins were synthesized.

Table 1. Assumptions for standard photoelastic stress analysis

Lineer elastic loading conditions
Photoelastic resin is homogeneous isotropic
Adiabatic conditions
Material properties of photoelastic resins are insensitive to small temperature changes
Constant ambient temperature

This development was parallel to those in engineering, especially in the fields of orthopedics and dentistry. Photoelastic stress analysis is being used for solving many engineering problems. Owing to increasing use of finite element stress analysis and strain-gauge analysis in dentistry, the use of photoelastic stress analysis technique is restricted to rather "descriptive" assessment of the biomechanical problems. As the details of the technique could be found in experimental stress analysis text books, some of the basics will be included in this chapter.

2. Polarization of Light

Light or luminous may be considered as electromagnetic vibrations similar to radio waves and each wave intersects its axis of propagation. The vibrations of each ray have different orientations with no favored direction and measurable intensities therefore refer to a superposition of many millions of waves. An incandescent source emits radient energy, which propagates in all directions and contains a whole spectrum of vibrations of different frequencies or wavelengths. A portion of this spectrum, wavelengths between 400-800 nm [15 and 30 × 10^{-6} in], is useful within the limits of human perception. Although the light emitting from ordinary sources like a bulb or Sun is unpolarized, it can easily become polarized as it interacts with matter (Figure 2). As the light waves containing vibrations in all perpendicular planes pass through a polarizing filter, namely the polarizer in the photoelastic technique, only one component of these vibrations, parallel to the privileges axis of the filter will be transmitted (Figure 3). The photoelastic effect (alternatively called the piezo-optical effect) is the change of refractive index caused by stress. When subjected to stress, some transparent materials such as glass, many plastics, some elastomers, semiconductors, various fluids, and certain biological materials such as collagen tissue exhibit the property of double refraction (birefringence). Birefringence can be induced in such materials by the presence of an applied load or residual stress. This changes the polarization state of the transmitted light. The extent of the change relies on the material properties and the thickness of the specimen. Birefringence is interpreted as transmitting light, as transverse electromagnetic waves, preferentially along two mutually perpendicular planes that coincide with the principal stresses. According to the principal stress magnitudes, the light is transmitted at different velocities, causing interference by their superposition on exiting the specimen. The resulting stresses cause the refractive index to vary with orientation.

Figure 2. The human eye usually cannot distinguish polarized light from natural light unless a polarizing filter is used. If the sky is viewed carefully on a very clear cloudless day, a broad dark blue band reaching from one horizon to the other 90° from the Sun could be noticed. The blue sky is polarized, because the air molecules scatter sunlight.

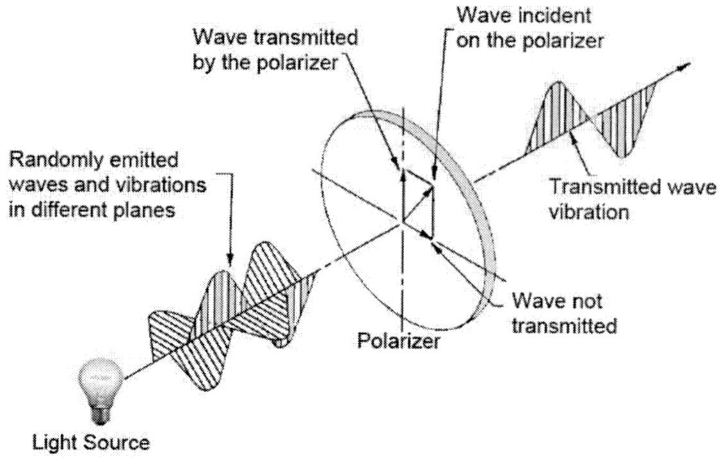

Figure 3. Filtering of plane polarized light from ordinary incident light.

To explain this phenomenon, recall the definition of refractive index, η, which is the ratio of the speed of light V in the medium to that in vacuum C:

$$\eta = V/C \qquad \text{(Eq. 1)}$$

The electric field vector, E, of the light beam travelling in space oscillates up and down at an angular frequency, ω, in a fixed plane, called the "plane of polarization" of the beam. (The wavelength of the light is $\lambda = 2\pi C/\omega$). In a birefringent material, the refractive index relies on the orientation of plane of polarization and magnitude of the birefringence is the difference in indices expressed as:

$$\Delta\eta = \eta_{\perp} - \eta_{II} \qquad \text{(Eq. 2)}$$

where $\eta_{\perp}$ and η_{II} are the refractive indices on the two planes.

The resulting interference fringes are a function of both principal stress direction and magnitude. In a two-dimensional model, loaded in its plane, light waves may be considered to be transmitted along the directions of principal stresses with the following relationship (stress-optical coefficient) describing the refractive indices [3]:

$$\eta_1 - \eta_2 = C(\sigma_1 - \sigma_2) \qquad \text{(Eq. 3)}$$

where η1 and η2 are the refractive indices corresponding to principal stresses σ1 and σ2, respectively, and C is known as the stress optic constant. The stress-optic coefficient is determined for the photoelastic material through an experiment on a specimen in which the stress state is known and must be measured and used in a way that accounts for the time-dependent behavior of most photoelastic materials. The stress-optic coefficient makes possible the measurement of stress in an unknown stress field by observation of the relative retardation. Relative retardation is the difference between the numbers of wave cycles experienced by the two rays traveling inside the body. This phenomenon is called double refraction or birefringence. In a photoelastic material, the birefringence depends on the induced stress, and many such materials can be described to a good approximation by the stress-optical law:

$$\Delta\eta = C(\lambda)(\sigma_1 - \sigma_2) \quad \text{(Eq.4)}$$

where C is the stress-optic coefficient, and the quantity in the second parentheses is the difference between the two principal stresses in the plane normal to the light propagation direction. The difference between principal stresses is twice the maximum shear stress in that plane. The stress-optic coefficient is generally a function of the wavelength λ.

The stress distribution in an irregularly shaped body can be inspected by replicating the actual structure (sometimes in a lower or upper scale for convenience) using a birefringent material like epoxy resins. As stated, polarizers are essentially just birefringent materials that pass only light polarized in the polarizer's principal optical plane. When a polarized beam propagates through a transparent plastic of thickness *t*, where X and Y are the directions of principal strains at the point under consideration, the light vector splits and two polarized beams are propagated in planes X and Y (Figure 4).

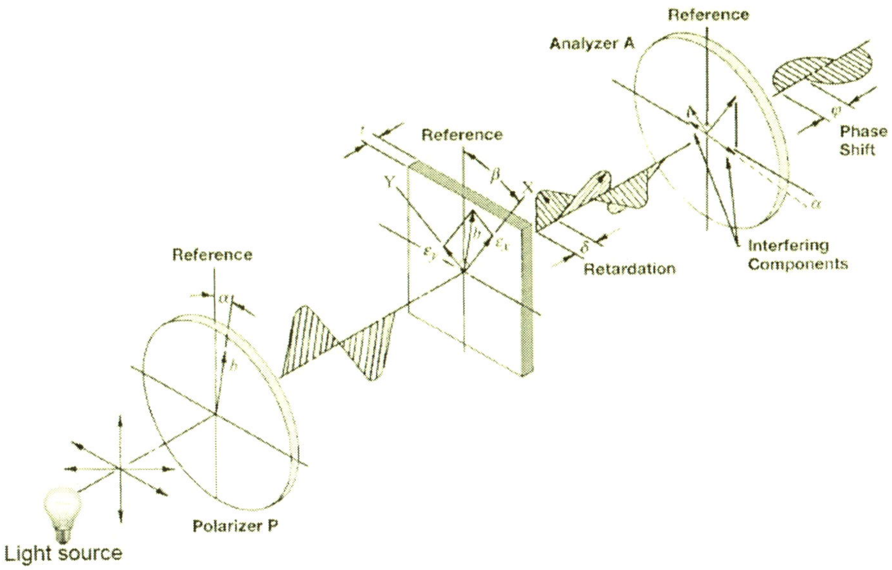

Figure 4. The plane polariscope. The ordinary light entering the polarizer becomes polarized. The polarized light then enters the birefringent material, i.e., crystal, where it is refracted and divided into two separate components vibrating perpendicular to each other and parallel to the axes of anisotropy. The polarized light waves then pass through the analyzer which passes only those components of the light waves that are parallel to the polarization direction of the analyzer (Adapted from Tech Note 702-2, Vishay, Micromeasurements).

If the strain intensity along X and Y is ε_x and ε_y, and the speed of the light vibrating in these directions is Vx and Vy, respectively, the time necessary to cross the plate for each of them will be t/V, and the relative retardation between these two beams is [4]:

$$\delta = C\left(\frac{t}{V_x} - \frac{t}{V_y}\right) = t(\eta_x - \eta_y)$$

(Eq. 5)

where δ is the retardation, C is the speed of light in a vacuum or in air. According to Brewster's law "The relative change in index of refraction is proportional to the difference of principal strains", or (4):

$$\eta_x - \eta_y = K(\varepsilon_x - \varepsilon_y)$$

(Eq. 6)

where K is the strain-optical coefficient. It is a dimensionless constant, similar to gauge factor of electrical resistance strain gauges. Adding optical filters known as quarter-wave plates in the path of light propagation produces circularly polarized light (Figure 5), and the image observed is not influenced by the direction of principal strains. The intensive of emerging light thus becomes:

$$I = b^2 \sin^2 (\pi\delta / \lambda)$$

(Eq. 7)

In a circular polariscope, the light intensity becomes zero when $\delta = 0$, $\delta = 1\lambda$, $\delta = 2\lambda$ . . . , or in general [4]:

$$\delta = N\lambda$$

where N is 1, 2, 3, etc.

Once $\delta = N\lambda$ is known, the principal strain difference is

$$\varepsilon_x - \varepsilon_y = \frac{\delta}{2tK} = \frac{N\lambda}{2tK} = Nf$$

(Eq. 8)

where fringe *value*, f, contains all constants, and N is the result of measurements.

Each ray that enters the model is divided into two components and after emerging from the model they combine to yield either constructive or destructive interference. However, that light waves cannot combine to yield optical interference unless they are coherent and are polarized in the same plane. The photoelastic polariscope serves to bring these waves into a common plane, so that optical interference can ensue. The polariscope does not rotate the planes of polarization of these rays, but instead, it transmits only those components of the two interfering rays that lie in a common plane (plane of polarization of the analyzer) (Figure 6). The apparatus used to reveal isochromatic fringe patterns is called a circular polariscope, shown schematically in Figure 5. Several other versions of the photoelastic polariscope have been advanced, notably the diffused-light polariscope, the reflection or doubling polariscope, and the lens-system polariscope, generally accepted as the most versatile instrument for photoelastic investigations [5,6].

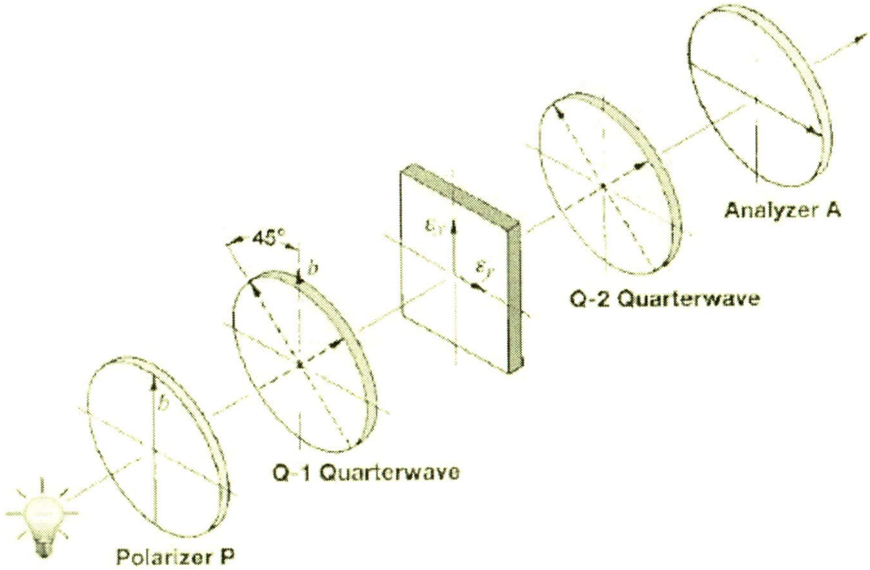

Figure 5. The circular polariscope. The radiation source can produce either conventional white (polychromatic) or filtered (monochromatic) light. The electric field vector of light striking the first polarizer with an arbitrary orientation can be resolved into two components, one in the polarization direction and the other perpendicular to it. The polarizer will block the transverse component, allowing the parallel component to pass through to the specimen.

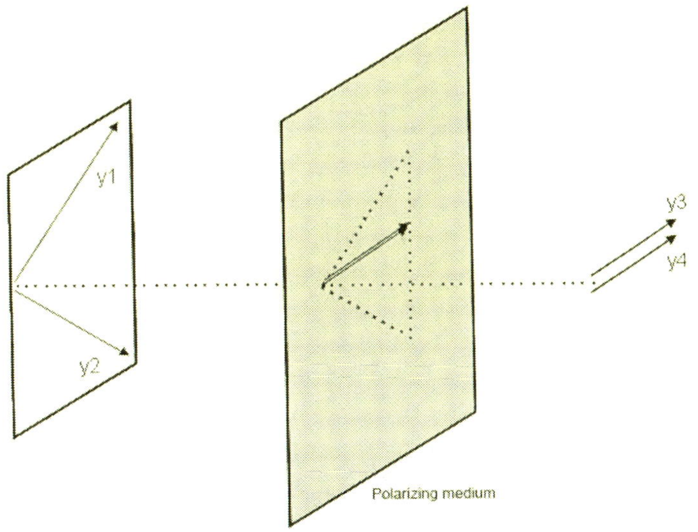

Figure 6. If the two light rays, y1 and y2, (arrows represent the planes of polarization of the two wave components) pass through a polarizing medium oriented with its transmission axis horizontal, the vertical component of these waves will be stopped (absorbed) and the horizontal components, y3, and y4, transmitted. y3, and y4 are waves polarized in a common plane, and they can combine to produce optical interference.

The components of the circular polariscope can be divided into two categories. The first category includes elements that direct and focus the light rays. These are the two field lenses and the camera in the lens-system polariscope. The second category consists of elements that control the wavelength and polarization of the waves. The functions of the polarizing elements, polarizer, quarter-wave plates and analyzer are shown in Figure 5. The polarizer divides the incident light waves into two components; one in the polarization direction and the other perpendicular to it. The polarizer will block (absorb) the transverse component and allow the parallel component to pass through to the specimen. A quarter-wave plate is a member that behaves exactly like a photoelastic model having uniform birefringence of 1/4.6. It is oriented with its principal planes (or principal axes in this case) at an angle of 45 degrees □to the axis of the polarizer and used to provide equal quantities of light along each of the two planes of polarization at every point in the experimental model. The quarter-wave plates are crossed, i.e., the plane of polarization of the higher-velocity waves in the first plate coincides with the plane of the slower waves in the second quarter-wave plate. The analyzer, actually a second polarizer, oriented with its polarizing axis crossed to the axis of the polarizer, a dark-field isochromatic pattern is formed. The centers of dark fringes are points of integral values of N (N = 0, 1, 2, 3,...). If the polarizer and analyzer axes are parallel, a light-field isochromatic pattern is formed, and the centers of light fringes are points of integral values of N [5,6]. The light-field pattern is generally preferred in analysis concerning biomechanical experiments in dentistry.

3. Remarks for Preparation of Models and Photography

Let us explain some important details in making of the test models and photography. Looking at the picture of a loaded photoelastic model, the apprentice may ask how on earth the background of an experimental model looks pretty "white" with the use of a rather "yellowish" low-intensity light source. Unfortunately, most scientists do not share their "secrets" or "experiences" regarding such analysis. To tell the truth, there are no secrets, but it rather seems that one is condemned to discover by himself the myth of taking the pictures, if he does not have experience and knowledge in the art of photography. For this reason, I have decided to share my humble experience without boring the reader. I also believe that it would be beneficial to the reader to provide some tips on how to cast such experimental models with high accuracy and some technical details in photography.

As previously stated, with regard to applications in oral implants, the so-called quasi three-dimensional technique is used. As for this technique, the implant or implants, embedded in the photoelastic resin is actually three-dimensional, but its image (projection) is analyzed in two-dimensions. In the quasi-three dimensional photoelastic technique, it is not worth making anatomic models for the analysis, because complex problems such as simulation of bone and soft tissue with their actual physical properties, distortion of soft tissue under load during assessment of removable dentures or implant-supported overdentures, and boundary conditions of the jaws can never be solved with this relatively simple descriptive approach.

First and foremost, it should always be kept in mind that photoelastic resins are highly sensitive to force and changes in temperature:

1. This implies that even at the stage of mixing of the resin according to the manufacturers recommendations, the procedure must be carried out with gentle movements and in constant temperature. Too rapid/vigorous mixing will create permanent stress lines as well as incorporate hundreds of air bubbles in the mix and the setting resin. Eventually, the cast experimental model will not be suitable for use. These permanent stress lines appear as translucent waves (strands) in the model that never disappear and interfere with generation of isochromatic fringes.

Another cause for the stress lines and air bubbles is rapid cooling of the resin components during mixing. To avoid this, place the mold (usually made from an impression material) that is going to be cast, the resins (according to the instructions), two glass bowls, one glass probe (for mixing), and some water in the heater. When the temperature in the resin rises to the recommended level for mixing, start mixing the resin following the Bain Marie technique (water bath) using the two glass bowls; one for the resin components, one for the water, and the glass probe for the mixing. The Bain Marie technique is traditionally used in science to heat materials gently and gradually to fixed temperatures, or to keep materials warm over a period of time. During this procedure, keep the mold in the heater. Photoelastic resins used for applications in dentistry are heated to a temperature approaching to or slightly exceeding 50°C, and this temperature will not ruin the mold. Quite the contrary, an increased temperature of the material will decrease surface tension, and upon gentle pouring of the resin from one edge into the mold, air bubbles will not be present on the outer surface of the experimental model upon setting.

2. Even a small amount of external constant force can produce stresses in a photoelastic block/model. Upon removal of the applied force, it might take hours for complete stress-relief and recovery to the previous stress-free condition. Despite the early use of these resins (mostly PL-2, Vishay Company) as bone simulant, photoelastic resins can never mimic the complex nature of living skeletal tissue. Nevertheless, recent developments has shown that cancellous bone structure could be replicated on the trabecular scale in the form of photoelastic models [7,8].

In some circumstances, it may be required to trim, carve or drill in the model. First, the excess resins are removed to finalize the shape to the model. This should be done slowly and under profuse water cooling. Assume one intends to study load distribution around an immediately-placed orthodontic implant. In such a scenario, the initial contact with the bone as well as stresses created by torque tightening of the implant needs to be simulated. If a self-tapping implant is to be placed, the pilot hole drilling should be done under profuse water cooling and at very low speeds (i.e., < 250 rpm). The implant socket is finalized by inserting the implant. The implant, however, should be inserted into its respective socket progressively/incrementally with utmost care, removing the debris at the interface in each increment, because only stress-free models (maximum 0.45 fringe order) are used for the analysis.

Another issue when fabricating photoelastic models for assessment of implant designs or implant-supported prostheses is whether the model should be made from one resin or two. As stated previously photoelastic resins cannot simulate the complex mechanical nature of living bone tissue including applications in the microscale [7,8]. Photoelastic resins are homogeneous in nature, which does not apply to the bone tissue. According to this author, making composite models for better assessment of peri-implant stresses does not make much sense. A good example for this may be the fabrication of composite models by means of different photoelastic resins to simulate both the cortical and the cancellous bone [9]. As the Young's modulus of a photoelastic resin increases, its strain/stress optical coefficient does. Therefore, in comparison with the use of a resin having lower Young's modulus, one has to apply more loads to observe the same amount of fringes, when a resin with relatively higher Young's modulus is used. Because the photoelastic stress analysis technique has been mostly used for "descriptive" purposes rather than quantitative assessment of stresses around implants so far, the rationale for fabrication of composite models is unclear. Another reason may be creation of permanent stresses at the interface between two resins during polymerization. Two make a composite model one has to cast the second layer for cancellous bone after polymerization of the first layer. As photoelastic resins are very responsive to force, forces arising from the polymerization shrinkage of the second layer will result in low-magnitude stresses at the interface, which will not disappear. This will eventually result in a discontinuity between the cortical bone and cancellous bone layers [9], which is far from what is observed in finite element analysis of such cases (Figure 7).

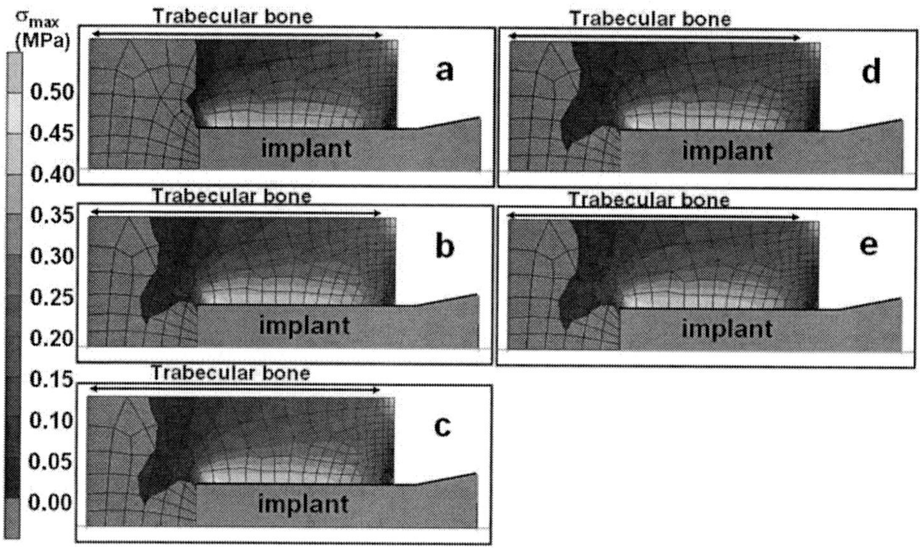

Figure 7. Time-dependent changes in maximum principal stress distributions in a finite element analysis of an implant according to Stanford's Theory. (a. Loading after osseointegration; b. At 3 months of loading; c. At 6 months of loading; d. At 9 months of loading; e. At 12 months of loading.) Note that there is not any discontinuity between the stress distribution between the cortical and trabecular bone layers at any time interval.

Photography in this analysis is combination of science and art. Camera settings are of utmost importance for good images. The aim is to capture the isochromatic fringes with colours "saturated" as much as possible, because pale fringes will not allow evaluation of stresses. Increasing saturation makes colors in photos look richer. Exposure is not just a function of aperture. It is a complex relationship between aperture, shutter speed, and medium sensitivity (ISO). The shutter speed is the length of time that the light capture medium is open to the light. 1/30 is 1/30 of a second. In order to manipulate the process of exposure, filters are being used in conventional photography. However, such filters are not used in photoelastic stress analysis. As a rule of thumb, the longer the exposure, the light gets in the camera, and therefore, over-exposure (improper exposure causing an image to look too light. There is a loss of detail in bright) is likely. In over-exposed images, details are lost (along the dynamic range) and true-to-life colors are blowing out (lost of saturation). Nevertheless, if the aperture setting compensates to the lower shutter speed, over-exposure does not occur. Now, let us show an example on these factors. Assume that in a "moving" object only the shutter speed is reduced i.e., from 1/1000 to 1/15 to reveal its movement, and aperture settings avoid over-exposure, the image will express the movement of the object with a degree of blurriness, while the image of the objects that are not moving will still appear sharp (Figure 8).

Figure 8. The picture of a water fall. Because the shutter speed is reduced to 1/15, the image of the flowing water gains some blurriness, while the rocks on the right side appear sharp. The aperture was set to f22 so that the reduced shutter speed is compensated in the sunny daylight to avoid over-exposure.

Although dynamic analysis is possible in the photoelastic technique, in quasi-three photoelastic stress analysis concerning oral implants, dynamic forces are not applied and therefore, there is not any fluctuation of fringes over time. There is no moving object or fringes. The images are captured under static loading, presuming that it would coincide a "moment" of the actual dynamic loading (mastication) scenario. In normal photography, one should try to keep the shutter speed between 1/90-1/125 or faster to avoid camera blur. Anything less than 1/60 indicates use of a tripod or one should be aware of the camera movement. Also, the longer the focal length, telephoto versus wide-angle lens, the more camera movement will affect image sharpness.

The aperture of the camera is the opening formed by the blades of the iris or diaphragm in the lens, through which light passes to expose the film. Aperture size is usually given in f-numbers, the larger the number, the smaller the opening. Aperture size together with shutter speed determines the amount of light falling on the film (exposure). The aperture is sometimes called the "stop". Numbers on the lens aperture ring and the camera's LCD (where applicable) indicate the size of lens aperture. The lower the number, the larger the aperture. As the scale rises, each number is multiplied by a factor of 1.4. Standard numbers are 1.0, 1.4, 2, 2.8, 4, 5.6, 8, 11, 16, 22, 32, etc., each change resulting in a doubling or halving of the amount of light transmitted by the lens to the film.

There is a general linear relation between light intensity, shutter speed and diaphragm (f-stop). If one of them is increased, the other must be decreased by the same amount, calculated in f-stops, in order to keep the light level on the film plane constant. However, there are limits to the linearity of this relation, because the latent picture is formed in a three-step process, where the first two photons excite the silver-chloride molecule and only the last one fixes the state. If a considerable amount of time elapses between the three steps, the excited molecule may spontaneously fall down to a lower energy level again, and then more than the minimum number of three photons will be needed, requiring a longer exposure. The same effect may also apply to high photon densities, as two photons may hit one molecule at the same time, but only excite it by one energy level. The time range (shutter speed) in which this reciprocity law failure effect (or the Schwarzschild effect) does not play a role depends on the type of film and it is indicated on the technical data (Reciprocity Law failure/reciprocity information) sheets from the manufacturer. As a general guideline, for daylight films it is between 1/1000 and 1 s, for tungsten films it is between 1/100 and 10 s. Color films, however, are composed of three light sensitive layers, each of which react slightly differently on changes in light intensities, therefore, also changes in the color balance will occur. These can be equalized with color correction filters in conventional photography. Although the reciprocity law failure effect can take place with daylight photography, the long shutter exposures of night photography definitely suffer from this. Digital images do not suffer from reciprocity law failure, although some state that it could be observed in the histograms. Because of such low resolution compared with film, highlight areas will burn out much more prominently in the image. For lightning, 100 ISO film is also well-suited, but one might also consider using 200 for more distant lightning, and closing the aperture, to make the photo sharper. For astronomy and other night scenes, one will need 400 or 800 speed film.

In photoelastic stress analysis, images obtained with digital cameras could work well and therefore, these details may not be always necessary. In the photographic stage of the photoelastic technique, a tripod should always be used, because shutter speeds longer than 20-25 seconds are generally used. A macro lens should be used and the camera should allow adjustment of shutter speeds more than 20 seconds and preferably have a "bulb" option. The room should be dark, when the photographs are taken. If autofocus is a problem in the dark, either manually focus to the fringes or aim a flash light to the object for autofocusing. The intensity of the ordinary light does not need to be high, a 10W soft light bulb will be sufficient. As evident from Figure 5, one has to locate the camera at the opposite end of the set-up with regard to the light source. As the light source is located behind the experimental model, implants and superstructures are typically observed as dark silhouettes [9,10]. This does not impair the assessment of isochromatic fringes, but reduces the quality of the picture. To overcome the problem of dark silhouettes, the author has introduced a simple and effective method [11,12]. A low-intensity light source, i.e. a flash-light, exposed to the model from the "camera side" for about only one second completely eliminates the dark image- and to some extent- reveals the isochromatic fringes on the surface of the implant (Figure 9).

The fringes on the surface of the implant are not used in this technique for two reasons; first there is not any full-field image that would allow visualization of the stress field, and more importantly, attempts to determine the fringe order does not make sense, as it can be determined using the conventional way, which will be explained in this section. If the analyzer is arranged so that a pale yellow background is obtained, one should start practicing the most appropriate shutter speed with aperture closed as much as possible. The purpose is to obtain a "white" background from a light source that is originally "yellowish" in color and in the meantime, capture clearly the full path of isochromatic fringe orders. This might depend not only on the intensity of the light source, but could vary depending on the thickness of the photoelastic resin model. Photographs with long shutter speed, i.e., 20-30 seconds, will result in images having clearly visible and colorful fringes (Figure 10a-c).

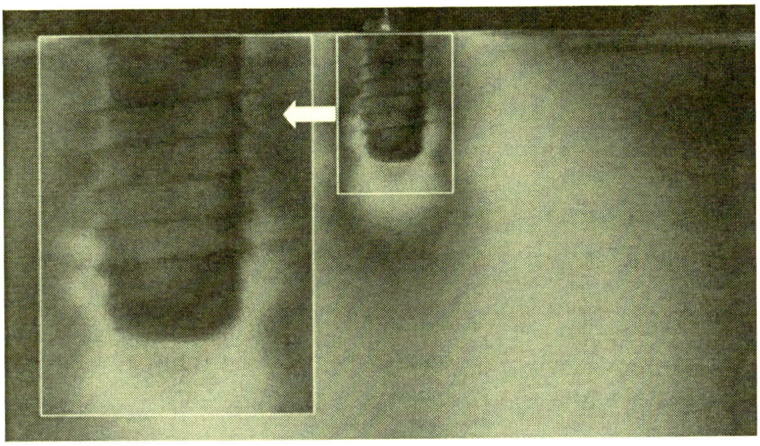

Figure 9. The implant is not viewed as a dark silhouette. The magnified view of the implant reveals the actual sandblasted surface and the spiral design of the threads. Note also the generation of fringes on the surface of the implant.

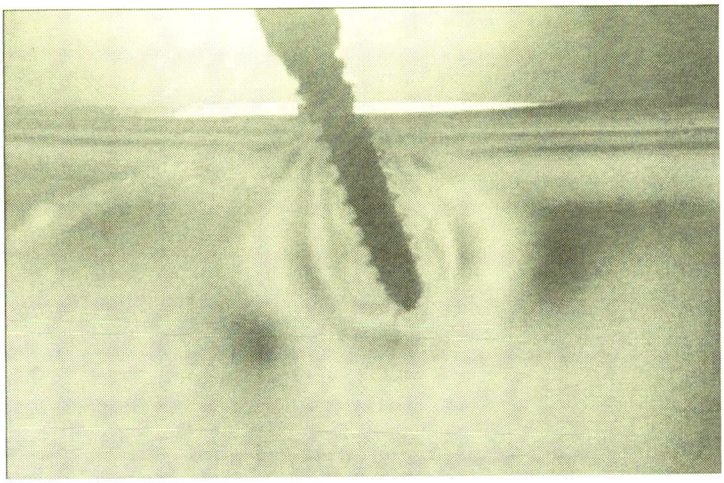

Figure 10a. The image of an orthodontic implant screwed into the resin incrementally. The fringes are a consequence of insertion torque. The image fails to meet the requirement for use in the analysis. This picture was taken at a shutter speed less than 15 sec. Note that glowing light source could be detected from the background. The shutter speed is still too high, because the full path of isochromatic fringes is not clearly visible. This is a case for over-exposure.

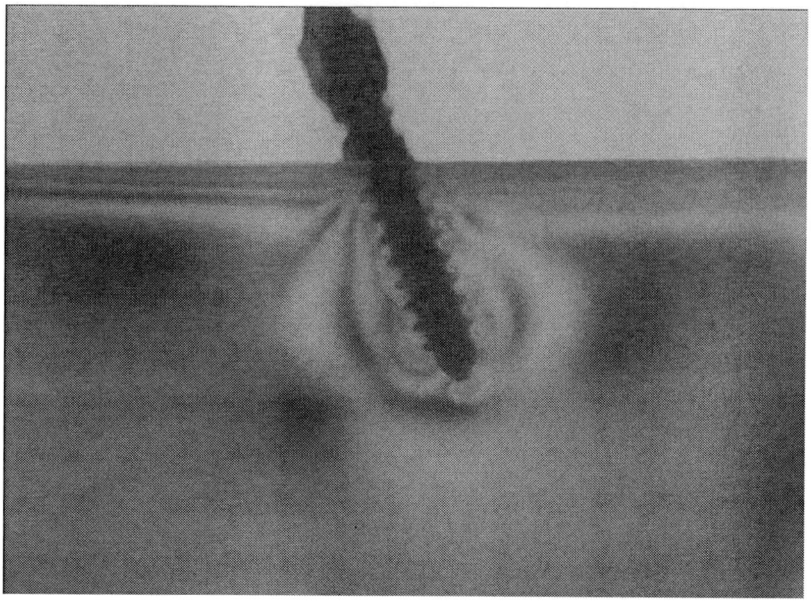

Figure 10b. The picture of the same model with 30 seconds shutter speed and f29. Note that the isochromatic fringes are more visible. The background is now more homogenous, but its color is still not "white". The main problem with this image is that it is slightly dark. This is a case for under-exposure.

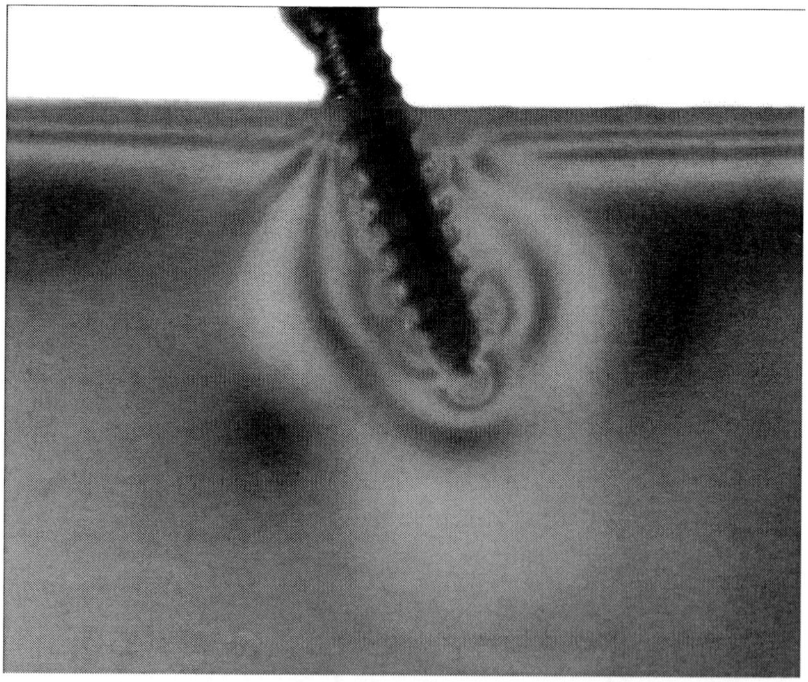

Figure 10c. After a series of adjustments for shutter speed, the intensity of the ordinary light, and the aperture, a white background is could be obtained and at the same time, the isochromatic fringe orders could be clearly observed with sufficient saturation. This image could be used for the analysis. Note that spiral thread structure is also slightly visible; the top of the implant is slightly illuminated by 1 sec light application from the camera side.

4. PHOTOELASTIC FRINGES

The foregoing is a rather general description of the formation of photoelastic patterns. It applies equally well to two-dimensional and three-dimensional photoelasticity and to the method of photoelastic coatings. Photoelastic fringes comprise two groups, isoclinics and isochromatics, which represent the direction and magnitude of principal stresses, respectively. Isoclinic fringes occur whenever either principal stress direction coincides with the axis of polarization of the polarizer. Isoclinic fringes, therefore, provide information about the directions of the principal stresses in the test model. Isoclinic fringes are often traced on a screen or photographed at increments of rotation of the polarized light field, i.e. 10°, and lines of principal stress directions (isostatics) plotted by geometrical construction. Complete solution a two-dimensional stress problem could be undertaken with combination of values of (σ_1- σ_2) from the photoelectric stress pattern and isoclinic fringes. Through a standard plane polariscope, both isochromatic and isoclinic fringes could be observed, but this makes quantitative stress analysis difficult. Isoclinic fringes can be removed by using a circular polariscope.

Isochromatic fringes are lines of constant principal stress difference, (σ_1- σ_2). If the source light is monochromatic these appear as dark and light fringes, whereas with white light illumination of colored fringes are observed. The difference in principal stresses is related to the birefringence and hence the fringe color through the Stress-Optic Law. Isochromatic fringe orders are a function of the principal stress magnitudes according to the following relationship [6]:

$$\sigma_1 - \sigma_2 = Nf\sigma/t \qquad \text{(Eq. 9)}$$

where $\sigma_{1,2}$ =principal stress magnitudes, N =fringe order, $f\sigma$ =material fringe constant, and t=model thickness.

Isochromatic fringe orders are assigned integer numbers from zero upwards, in order of appearance with increasing stress, and being proportional to principal stress difference (see Eq. 9) represent loci of constant shear stress since:

$$\tau = \sigma_1 - \sigma_2/2 \qquad \text{(Eq. 10)}$$

where τ is shear stress.

In the polariscope, the isochromatic fringes in a test material appear as a series of successive and contiguous different-colored bands in which each band represents a different degree of birefringence corresponding to the underlying stress in the tested part. The contour of an isochromatic fringe is determined by the flow of stresses in that particular region and represents equal differences in principal stresses. Hence, the color of each band uniquely identifies the birefringence, or fringe order (and stress level), everywhere along that band. The tint of passage is a dividing zone between red and blue in the first-order fringe and red and green in the second-order fringe, indicating first and second fringe orders, respectively. Beyond this point, a repetition of pink and green colors is observed and each transition indicates a new fringe order (Table 1) [4]. The number of fringes indicates the stress or strain magnitudes, and the fringes being close to each other demonstrate higher stress concentrations at that region. This technique has been also used in photoelastic stress analysis related to the force transfer characteristics and load partitioning of oral implants supporting fixed and removable prosthesis.

When observing an unloaded experimental model through a polariscope, the model will appear uniformly black. This implies a stress-free model. As load is gradually applied, the most highly stressed region begins to take on color - first gray, then white, and, when the violet is extinguished, yellow. Simultaneously with the increase in the load and generation of new fringes, the earlier fringes are pushed toward the areas of lower stress. With further loading, additional fringes are generated in the highly stressed regions and move toward regions of zero or low stress until the maximum load is reached (Figure 11). The fringes can be assigned ordinal numbers, i.e., first, second, as they appear, and they will retain their individual identifies "orders" throughout the loading sequence. The fringes are ordered not only in terms of serial numbering, but also never superimpose one another and always keep their respective positions in the ordered sequence.

Let us come back again to the generation of fringes under increasing load. The blue is extinguished to produce orange; and then green, to give red. The next color to disappear with increasing load is yellow, leaving a purple color followed by the extinction of orange, generating a deep blue fringe. The purple fringe, easily distinguished from the red and blue on either side, and very sensitive to a small change in strain level, is referred to as the "tint of passage". Owing to its distinctiveness and resolution, the purple tint of passage is selected to mark the increment in relative retardation equal to a fringe order of unity (Table 1). Subsequent recurrence of the tint of passage with greater relative retardation signifies the presence of higher integral fringe orders.

Table 1. Isochromatic fringe characteristics [4]

	Color	Approximate Relative Retardation		Fringe order N
		nm	in$\times 10^{-6}$	
	Black	0	0	0
	Pale Yellow	345	14	0.60
	Dull Red	520	20	0.90
	Red-Blue Transition	575	22.7	1.00
	Blue-Green	700	28	1.22
	Yellow	800	32	1.39
	Rose Red	1050	42	1.82
	Red-Green Transition	1150	45.4	2.00
	Green	1350	53	2.35
	Yellow	1440	57	2.50
	Red	1520	60	2.65
	Red-Green Transition	1730	68	3.00
	Green	1800	71	3.15

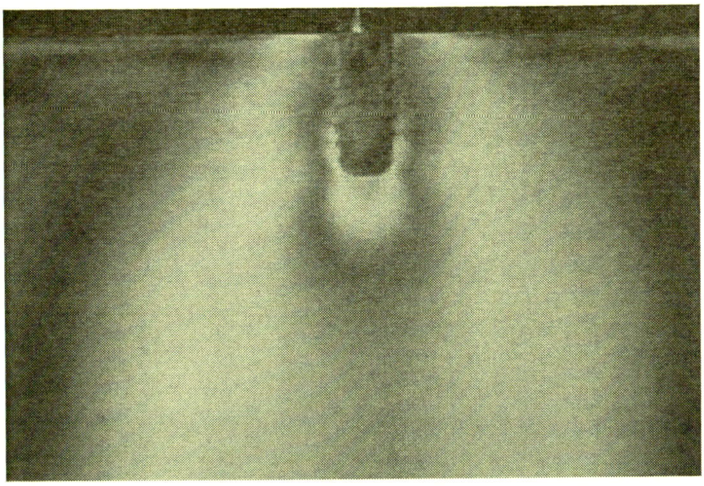

Figure 11. Isochromatic fringe orders around a loaded oral implant under static load in the quasi-three dimensional method. Prior to application of load, a black zone (zero fringe order) was surrounding the implant. As the load was applied, this was quickly replaced with successive fringes in accordance with the direction and magnitude of stresses and therefore, moved to a distance from the initial point of load application, where the stress-free zone exists.

As the load increases on the experimental model and produces additional relative retardation, the red is extinguished from the white light spectrum, and the fringe color is blue-green. Under increasing load, the relative retardation approaches the level where it corresponds to twice the wavelength of violet, extinguishing this color for the second time and starting the fringe cycle over again. However, the deep red color at the far end of the white light spectrum also has twice the wavelength of violet, and thus undergoes its first extinction simultaneously with the second extinction of violet. The result is that the fringe color is the combination of two complementary colors, yellow and green (Figure 12).

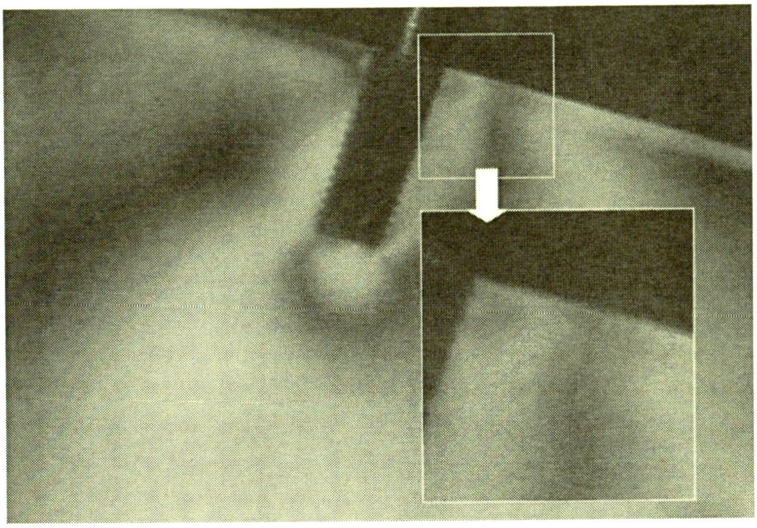

Figure 12. The isochromatic fringe orders around an obliquely loaded implant. At the tension side (left), low magnitude stresses are observed. At the compression side (right), note the yellow-green transition (magnified view) and generation of more fringes at close proximity to the implant neck.

As the load and relative retardation continue to increase, the fringe color cycle is repeated, but the colors are not exactly the same as in the first cycle because of simultaneous extinction of two or more colors. With each successive complete color cycle, the effect of increasingly complex simultaneous extinctions is to cause the fringe colors to become paler and less distinctive. Because of this effect, fringe orders above 4 or 5 are not distinguishable by color in white light. Because of simultaneous multiple extinction of colors, the second-order tint of passage is fainter than the first, and falls in the transition area between red and green fringes. At fringe orders of 3 and 4 the tint of passage is not distinctly visible as a purple band, but the well-defined transition between red and green in each case serves the same function and represents the integral fringe order (Figure 13).

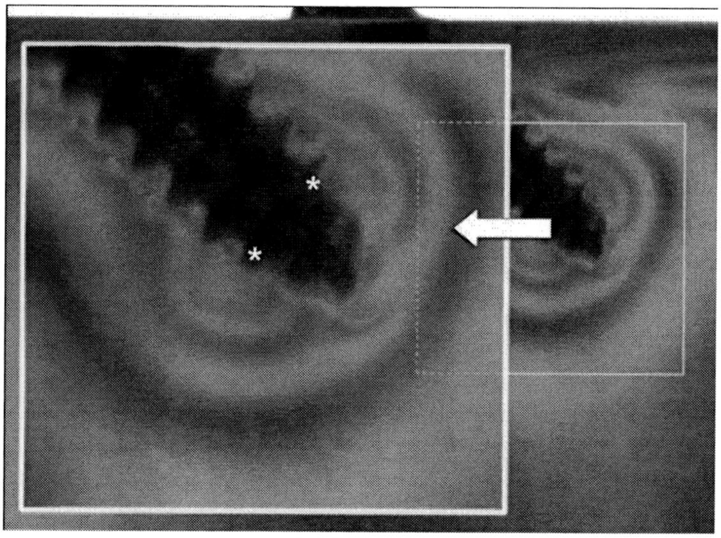

Figure 13. Close view of isochromatic fringe orders produced around a self-drilling orthodontic micro-implant 1.4 mm in diameter. Note that the fringe orders 3-4 (next to the threads with *) is not distinctly visible as a purple band and after the 4th fringe is not clearly distinguishable.

Using PL-2 resin and loads more than 10-15 kg would not make sense for oral implants, because there will probably be more than 3-4 fringes at some parts, i.e., the collar of the implants. The photographs of the fringes should be captured after ensuring that the number of fringes does not increase in time. This may take usually 1-2 minutes at most. If the same model is to be used in a different loading condition, i.e., oblique loading, the model should be inspected to verify the pre-loading stress-free state prior to each loading experiment. It may take a few hours for the isochromatic fringes to disappear completely in an ordinary model.

References

[1] Caputo, A.A., Standlee, J.P., eds. (1987) *Biomechanics in clinical dentistry,* Chicago:Quintessence Publishing. 13–26.

[2] Akca, K., Cehreli, M.C., Iplikcioglu, H. (2003) Evaluation of the mechanical characteristics of a reduced-diameter morse-taper implant: A non-linear finite element stress analysis. *Clinical Oral Implants Research*,14,444-455.

[3] Orr, J.F., Finlay, J.B. (1997) Photoelastic stress analysis. In Orr, J.F., Shelton, J.C. *Optical measurement methods in biomechanics*. Chapman and Hall, London. 1-196.

[4] Tech Note 702-2, Vishay, Micromeasurements.

[5] Post, D. Photoelasticity. In: Doyle JF, Phillips JW (Eds). *Manual on experimental stress analysis*. Fifth Edition. Society for Experimental Mechanics 1989;6:1-32.

[6] Orr, J.F.. (2003) Images from waves-photoelastic modelling of bones. *Irish Journal of Medical Science*,172,209-231.

[7] Tancred, D.C., Carr, A.J. and McCormack, B.A.O. (1998) Development of new synthetic bone graft. *Journal of Materials Science: Materials in Medicine*,9,819-23.

[8] Mushipe, M.T., Orr, J.F. (2001) Fabrication of life-size photoelastic cancellous bone models. *Strain*,37,123-6.

[9] Fanuscu, M.I., Iida, K., Caputo, A.A. and Nishimura, R.D. (2003) Load transfer by an implant in a sinus-grafted maxillary model. *International Journal of Oral and Maxillofacial Implants*,18,667-74.

[10] Sadowsky, S.J. and Caputo, A.A. (2004) Stress transfer of four mandibular implant overdenture cantilever designs. *Journal of Prosthetic Dentistry*,92,328-36.

[11] Cehreli, M., Duyck, J., De Cooman, M., Puers, R. and Naert, I. (2004) Implant design and interface force transfer: A photoelastic and strain-gauge analysis. *Clinical Oral Implants Research*,15,249-257.

[12] Akça, K. and Cehreli, M.C. (2008) A photoelastic and strain-gauge analysis of interface force transmission of internal-cone implants. *International Journal of Periodontics and Restorative Dentistry*,28,391-9.

In: Biomechanics of Dental Implants
Editor: Murat Çehreli

ISBN 978-1-62100-780-7

Chapter 10

Reliability of Experimental Stress/Strain Data

Murat Çehreli

1. Introduction

Validity and reproducibility of scientific data is of utmost importance. When the experimental conditions or the stress/strain analysis techniques differ from one another, the qualitative and quantitative information obtained by those techniques may not always be the same. Eventually, this may undermine the reliability of the data and its relevance to the actual biologic condition.

There have been two basic reasons for biomechanical experiments on oral implants. The assessment or prediction of the biologic response of an osseointegrated implant under various loading conditions has long been a topic of research interest. After introduction of immediate and early loading of implants, skeletal tissue differentiation around implants under various loading conditions has also been studied to a certain extent, as peri-implant tissue formation and mineralization are dependent on several factors including the mechanical status of the bone-implant interface. It is frequently pronounced that certain thresholds of strains induced in bone around implants rule its response [1-3]. Although the magnitude of species- and site-specific tissue strains and the mechanisms that guide bone modeling/remodeling are still a topic of controversy and argument, there is a consensus that mechanical factors and loading history play a substantial role as a feedback mechanism in bone remodeling around implants. It is, therefore, crucial to fully understand how bone responds to implants under loading in terms of intraosseous strains, particularly at the marginal bone level. In this context, interpretation of the data of in vitro experimental stress/strain analysis and animal studies and extrapolation of these findings into the clinical situation must be tempered with caution and skepticism due to discrepancies between the outcomes of different engineering techniques employed for biomechanical experiments on implants.

2. Limits of Experimental Stress Analysis Methods on Bone/Implant Biomechanics

When dealing with a complex stress/strain analysis problem in which a complete theoretical solution may prove impractical with respect to time, cost or degree of difficulty, experimental stress/strain analysis methods are used. Most frequently, photoelastic stress analysis, two- or three-dimensional finite element (FE) stress analysis, and strain-gauge analysis are used for biomechanical assessments.

2.1. Photoelastic Stress Analysis

Studies using two- or three-dimensional photoelastic stress analysis [4,5] and holographic interferometry [6,7] have been performed in dentistry. Through observation of isochromatic fringe patterns in the photoelastic coating under load, reflective photoelasticity allows identification of stress fields on the surface of a studied object accessible to normally incident light. Photoelastic coatings allow point determinations of shear stresses and directions of measurements of principal stress axes. While the method is generally used to test industrial prototypes, its applications in dentistry are limited to evaluation of removable dentures and misfit in implant suprastructures [8-10]. Although a high correlation (r=0.98) could be obtained between strain-gauge readings and strains calculated from reflective photoelasticity [11], it cannot be used to study peri-implant tissues. Photoelastic models are used to simulate implants placed in bone and maximal shearing stresses (isochromatics) within the loaded model, the "*bone*" are observed through the circular polariscope. In dental applications, this technique has several limitations undermining its scientific value. Unlike conventional FE analysis of the same scenario, the anatomic form of the bony structure and the way it is attached to surrounding tissues is not considered in this analysis.

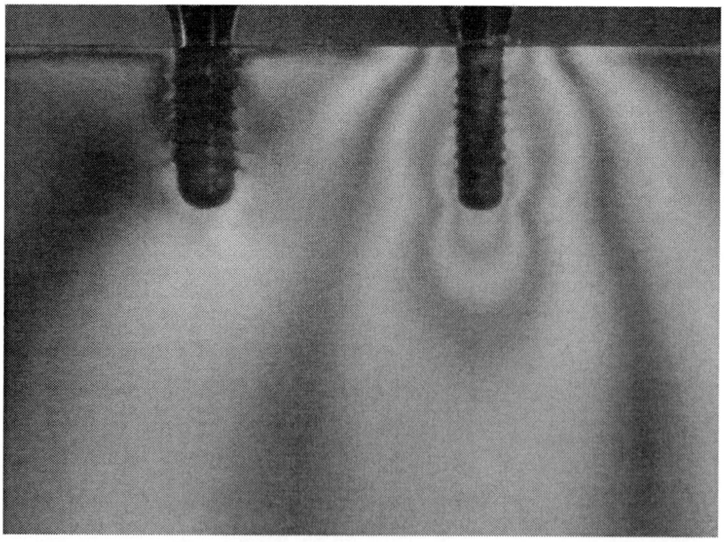

Figure 1. A typical photoelastic image of an axially- loaded implant supporting a fixed prosthesis. This technique is rather descriptive.

In addition, local cortical bone thickness and density of trabecular cannot be simulated [12]. Moreover, the isotropic photoelastic material cannot simulate an orthotropic material, namely, the bone tissue. The Young's modulus of photoelastic resins are also lower than bone tissue and fabrication of composite models to simulate cortical and trabecular parts of bone has limitations [4].

An implant embedded in a photoelastic resin block has 100% contact with the resin, which certainly does not mimic the actual biologic relationship at the bone-implant interface (Figure 1). The three-dimensional structures (implants, prosthesis (if included) and a "block of bone") are viewed in two-dimensions. The resolution of isochromatic orders decreases after the third order. Therefore, the experiments cannot be performed using high loads. Normal stresses within the material simulating the bone cannot be detected by the photoelastic technique.

Quantification of peri-implant stresses in the photoelastic resin will result in data that will certainly not mimic the outcome of in vivo biologic tissues. Owing to its inherent limitations, this technique, therefore, could rather be used only for "*descriptive*" evaluation of different implants/prosthetic designs.

2.2. Strain-Gauge Analysis

In 1917, Thompson [13] stated in his classic treatise, *On Growth and Form*, that the general bone adaptation is dependent on load-induced mechanical deformation, namely, strain. It was probably Evans [14] who first measured in vivo strains in canine tibia during gait by means of strain-gauges. Following the introduction of the strain-gauge technique after two decades [15,16], it became possible to measure in vivo strains under physiologic loading conditions. There is no doubt that, the technique made a great contribution to current knowledge on mechanobiology of bone, as it would be impossible to analyze strain-related bone response without direct quantification of strains on bones of different species.

2.2.1. Apparent Strain and Fatigue Life

Bonded electrical resistance strain-gauges have a good reputation both in applications in engineering and dentistry. They are relatively inexpensive, can achieve overall accuracy of better than +/-0.10%, are available in a short gauge length, have small physical size and low mass for most applications in industry and engineering, and are highly sensitive. The bonded resistance strain-gauge is suitable for a wide variety of environmental conditions and temperatures, such as measurement of strain in jet engine turbines operating at very high temperatures and in cryogenic fluid applications at temperatures as low as -452°F. Bonded resistance strain-gauges can be used to measure both static and dynamic strain. Strain-gauges are moderately responsive to changes in temperature. Their use in fluctuating temperatures will affect their reading (Figure2), because fluctuations induce changes in the resistance of the gauge and gauge factor. Alterations in environmental temperature upon bonding and balancing of a strain-gauge will lead to strains even in the absence of load. This reading is called thermal output or apparent strain. Apparent strain is any change in gauge resistance that is not caused by the strain on the force element. Apparent strain is relatively small (approximately 0.5 $\mu\varepsilon$/°F) when gauges are used between room temperature and body temperature. If in vitro experiments are performed under constant temperature, this may not have any effect on the measurements. Likewise, body fluids will also allow constant temperature for in vivo experiments. However, one must be aware that friction between overlying soft tissue and gauges might increase local temperature. Since high-resistance gauges (1000 or 5000 Ω) aid in heat dissipation, provide lead wire densensitization and improve signal to-noise ratio, their use on bone tissue, which has poor thermal conductivity would be the most appropriate procedure. As in vivo temperature could be considered constant, the use of these relatively expensive gauges is not essential. 120 Ω or 350 Ω gauges will be sufficient for most in vivo applications [17].

Moreover, strain-gauges cannot withstand infinite number of loading cycles. When strain is repeatedly applied to a strain-gauge, it causes increased resistance under zero strain, peeling-off of the gauge, or disconnection, resulting in failure. The number of cycles a gauge can endure is called its fatigue life, indicated by repetition number under the specified conditions of strain amount repetition speed as apparent strain drifts to $100X10^{-6}$ strain from the beginning. The fatigue life of the gauge relies on the backing material, backing size and the adhesive used to bond the strain-gauge. Normally, the larger the backing the longer is the fatigue life. Considering that miniature strain-gauges are often used for dental purposes, frequent problems and maintenance requirements are unavoidable. In addition, aging and instability of the metal and the bonding agent can change apparent strain.

Compensation for apparent strain is necessary, if the temperature varies while the strain is being measured. In most applications, the amount of error depends on the alloy used, the accuracy required, and the amount of the temperature variation. If the operating temperature of the gauge and the apparent strain characteristics are known, compensation is possible.

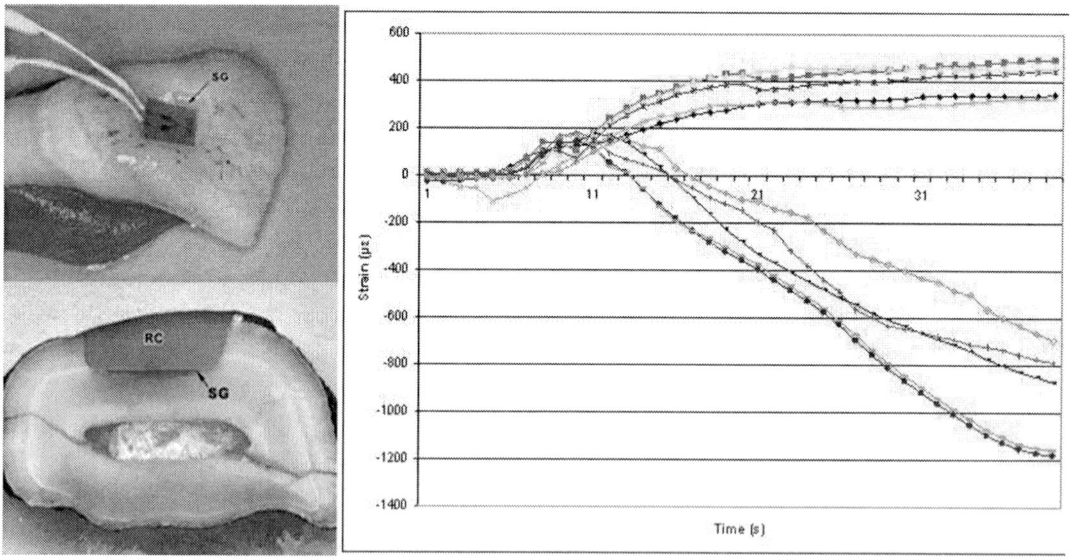

Figure 2. Strain vs time graphics of two light-polymerized resin composite (RC) samples by strain-gauge (SG) analysis in wet conditions. The increase in strain amplitudes during first 15 sec is not a consequence of forces arising from shrinkage of the resin, but is solely a consequence of heat induced by the light source. In these experiments 120 Ω gauges were used. When low excitation voltages and the least thermal effects are desired, higher-resistance gauges are a better choice.

2.2.2. Limits of Application

Strain-gauges have been used for bone strain measurements for more than 5 decades. Some of the factors that complicate the use of strain-gauges with bone include:

1. Presence of body fluids, including saliva, blood, bone fluid, and fatty tissue.
2. The body's defense mechanism to attack foreign bodies including strain-gauge wires, waterproofing materials, gauge backings, which can play a role in long-term animal experiments.
3. Mechanical forces and consequent temperature increase that act on the gauge due to movement of muscle and overlying tissue.
4. Limited access to the area of interest, i.e., posterior buccal alveolar zone.

This technique has also been implemented to measure bone strains around load-bearing implants for static loading experiments [18-20]. This method could only be used for measurements on cortical bone. Even miniature gauges are not appropriate for direct strain measurements on the trabecular bone or in the vicinity of implants. The method of bonding strain-gauges to moist cortical bone of tibia and fibula and waterproofing the gauges and lead wires has been demonstrated by Lambert [21]. Although, this technique is also applicable in the oral cavity (Figures 3 and 4), bonding of the gauges to the periosteum is sometimes a hard task. In addition, the direct measurement of strains from the periodontium may be difficult owing to the changing nature of the tissue response to load and the alteration in blood supply [22]. When gauges are bonded to prolonged experiments, traditional waterproofing materials such as expoxy and silicone may induce adverse tissue response. In addition, shearing forces arising from muscles may potentially debond gauges. There seems to be more risk of debonding and incidence of short circuit leading to ruining of the gauges during the in vivo experiments, particularly on moist surfaces in comparison to gauges bonded on dried bone surface. Indeed, some animal experiments included gauges bonded on dried bone surfaces. However, removal of the periosteum by scraping, cleaning, dehydrating and degreasing of the bone surfaces may create macroscopic trauma to cortical bone structure and decrease its local thickness [23]. When the objective of the experiment is to quantify functional strains in a time-dependent manner at the in vivo level, this surface damage may cause a remodeling or healing response that alter bone properties and recruit cells that degrade and remove the adhesive more quickly.

Owing to ethical issues, in vivo strain-gauge applications to bone tissue in the oral cavity of humans has been very limited. In vivo implanted/bonded gauges pose a risk of infection due to transcutaneous leads and transdermal connectors. Asundi and Kishen [24], reported direct bonding of rosette strain-gauges on cortical bone adjacent to maxillary central incisor of a patient, who underwent endodontic surgery. They bonded gauges by a conventional technique, in which a fast-setting adhesive was used to bond the gauges, although the method of waterproofing was not explained. In addition, the gauges were removed immediately after measurements. When, time-dependent strain-gauge measurements on bone are intended at the in vivo level, it should be taken into account that cyanoacrylate adhesives could degrade in 2-3 weeks leading to measurement inaccuracies [17]. Alternative methods are utilization of dental adhesives originally used on enamel surfaces [25], polymethylmethacrylate on abraded or drilled and roughened bone surfaces [26], and most promisingly, calcium phosphate ceramic coated strain-gauges [27] that allows a durable bone-gauge interface up to 9 months [17].

Strain-gauges are available with gauge lengths in range of 0.2-100 mm. The configurations of strain-gauges often used for implant biomechanics are uniaxial and/or rosette and they can be bonded on implants, abutments, prostheses or bone to measure an "average" strain value, as determined by the area of their measuring grid (Figures 2 and 3). This implies that electrical resistance strain-gauges cannot measure strain at a predefined point, but in an area. This is one of the major limitations of the technique. It is, therefore, imperative to select the best grid area for the anticipated strain region; high strain regions require gauges with small grid sizes, because errors in excess of 20% may arise when large gauges are bonded to high-strain areas [17].

It has long been recognized that the implant-bone complex should be stressed within a certain range for physiologic homeostasis and optimum functioning, where strains within the skeletal tissue would range between 200-1500 $\mu\varepsilon$ [1-3]. It should, however, been taken into account that these data have been obtained from strain-gauge analyses that provided global strains occurring in an area determined by the measuring grid of the gauge. Recent evidence using digital image correlation show that experimentally determined bone matrix strains around osteocyte lacuna resulting from macroscopic strains of approximately 2000 $\mu\varepsilon$ (0.2%) can reach levels of over 30.000 $\mu\varepsilon$ (3%) over fifteen times greater than the applied macroscopic strain, possibly due to local inhomogeneities in the bone matrix and strain concentrating effects of the local microstructure [28,29].

The digital image correlation technique, a proven flexible and useful tool for deformation analysis, enables strain measurement by correlating the position of pixels in images collected by cameras from undeformed and deformed objects . The technique has been applied to measure local cortical bone images and to a certain limit homogenized strains in trabecular bone images [30,31]. The critical regions in the test specimen should be determined prior to gauge attachment and testing. Strain measurement must be used in conjunction with material properties derived from separate in vitro tests to determine the quality and quantity of stresses.

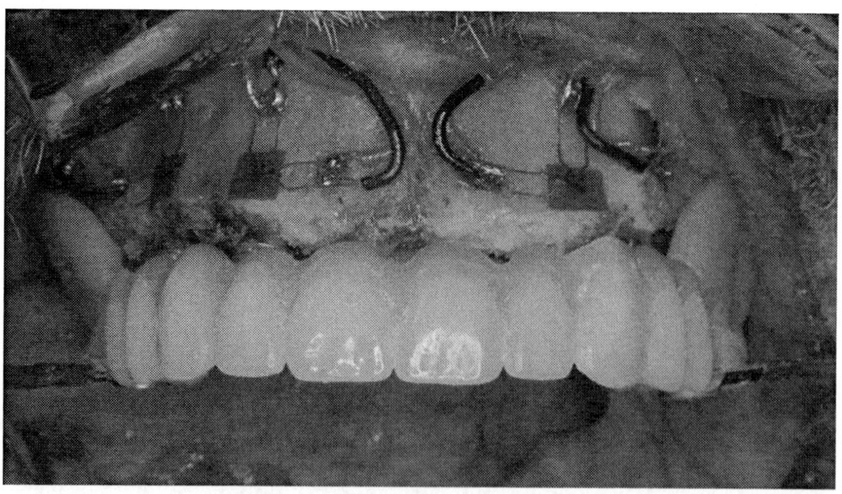

Figure 3. Resistance strain-gauges bonded on the surface of maxillary alveolar cortical bone of a cadaver. In this ex vivo experiment, strain-gauges with attached lead wires that are completely

encapsulated were used to prevent fluid infiltration into the gauge-wire junction. The gauges and wires including solders are covered by a transparent water-proof material. The wires soldered to the connecting terminals rise toward the labial sulcus, pass through both nostrils and connect to the data acquisition unit. Alignment of gauges is of most importance when measurement of principal strains is planned. If alignment along a specific loading axis is a problem or only peak strains are of interest, rosette gauges may be used, because they are relatively insensitive to misalignment errors.

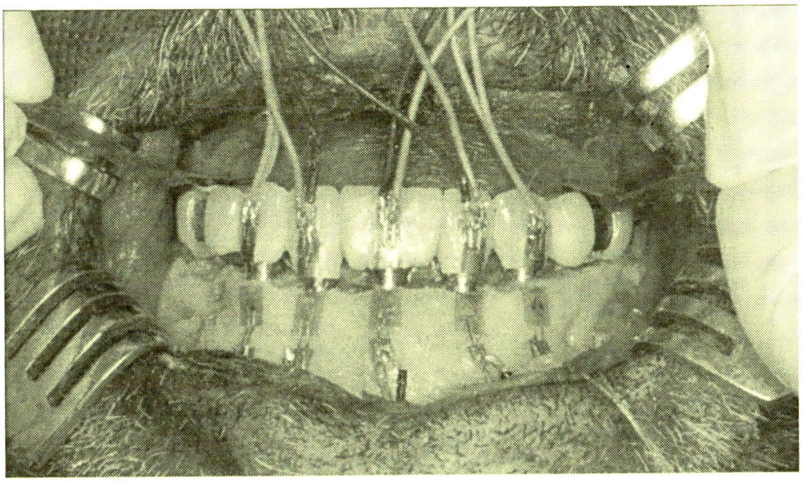

Figure 4. Linear strain-gauges bonded on the superstructure and cortical bone adjacent to the neck of the supporting implants in a cadaver model. In such experiments, it is essential to control the applied load. The amplitude of load was controlled by the load cells (LC).

Therefore, strain-gauges are often used to measure strains in a few critical regions of the test specimen, and are used in association with material properties and analytical models to draw inferences about the stress state of the bone. The strain-gauge analysis technique also has some limitations in applications for implants. They can be used in the following areas including:

1. Cortical bone adjacent to the neck of an implant
2. On a non-threaded smooth surface of an implant body
3. On implant abutments
4. On the outer surface of a fixed or removable dentures
5. Embedded in a resin such as photoelastic resin.

Owing to their dimensions and delicate nature electrical resistance strain-gauges cannot be used in critical areas such as:

1. Trabecular bone adjacent to an implant
2. The bone-implant interface unless the surface is coated
3. The implant-abutment interface
4. The thread of an implant
5. Soft tissue under removable dentures

The maxilla and mandible often have curved surfaces, and use of narrow gauges may minimize measurement inaccuracies. These devices have been fabricated to measure deformations on smooth and flat surfaces and application on a curved surface could potentially damage the gauge (backing) and its reading. Even the smallest miniature gauge's backing will not allow placement in the vicinity of an implant thread and measurement of peri-implant bone strains. Strain-gauges less than 3 mm in length are difficult to handle, orient, bond, and wire/solder in the mouth, and have relatively poor heat dissipation properties. Moreover, their maximum elongation, measurement stability during static loading, and endurance during fatigue loading are poorer than those of larger gauges [17]. When it is desired to measure the in vivo strain in a flowable material such as resin composites, resin cements and even coagulum, one must ensure that the material has sufficient tackiness. Otherwise, the measurements might have errors (Figure 5).

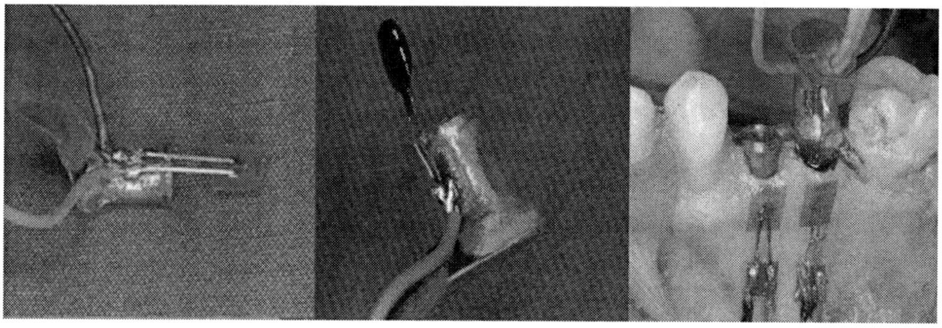

Figure 5. A linear strain-gauge soldered to a coping to measure peri-implant strains in coagulum. The length of the wires is adjusted to coincide the cervical level of the implan. Note the fresh blood can adhere to the polyimide backing of the gauge's both surfaces. It is very important to ensure tackiness of the material tested, when strain-measurements are made in such materials The strain-gauge covered with coagulum is inserted into the "jumping distance" of an immediately placed implant in a cadaver model. Controlled loading is undertaken by the miniature load cell connected to the top of the coping (Adapted from Cehreli et al.[19]).

2.3. FE Stress Analysis

In 1972, Brekelmans et al. [32] first used a FE model to assess stresses in a human bone under the action of physiological loads. Since then, the finite-element method has been increasingly used as a superior theoretical tool to analyze the mechanical behavior of bone segments with or without load-bearing implants. One of the main purposes of computational analysis technique is to solve physical problems or to determine the effectiveness or behavior of an existing structure or structural component subjected to certain loads. The idealization of the physical problem to a mathematical model requires certain assumptions that lead to differential equations governing the mathematical model and since the procedure is numerical, it is imperative to assess the solution accuracy. Additionally, the production of an appropriate and effective mathematical model is crucial to elucidate the physical phenomena, which requires the inclusion of comprehensive structural simulation of tissues and load-bearing implants for accurate quantification of induced stress or strain. Two major bone model types are available in the literature. The most common is the use continuum-level models, which suffer from a simplistic modeling of the cortical shell and density-based trabecular bone modeling. Sometimes the shell is not modeled or a constant shell thickness is assumed throughout the model. The continuum model itself could be of digitized type or smooth type. The second approach is computationally intensive high-resolution FE models, which contain the cortical bone layer automatically. Automated two-dimensional techniques are developed to digitally identify the geometrically complex cortical bone layer using high-resolution computed tomography (CT) scans [33].

2.3.1. FE Modeling

An almost actual representation of stress behaviors can precisely be provided by the three-dimensional FE stress analysis and with recent advances, time-dependent analyses are also possible with implementation of bone adaptation theories [34,35]. Using FE models, the system geometry can be individualized by computerized tomography or MRI and mechanical properties of biologic tissues can be defined. As local mechanical bone properties can be estimated from radiographic bone density measurements, CT seems as a valuable tool to develop individualized FE models to predict microscopic bone properties and patient-specific FE models. Specific commercial software such as ScanFE (SimplewareTM), Mimics (MaterialiseTM), and 3D Doctor has also been developed to automatically generate FE meshes from digital sectional data in a more efficient way. Using advanced digitalized imaging techniques, the efficient creation of an FE model for each specific patient will become a routine process in the future.

The FE offers the advantage of modeling structures, with intricate shapes, and indirectly measures their complex mechanical behavior at any theoretical point. As the numeric method uses the theories of elasticity and static equilibrium, the effects of external forces applied on an object can be evaluated as physical events in terms of stress and strain. Because the elastic properties of bone are dependent upon a number of the factors including anatomical site, tissue heterogeneity, architecture, loading direction, and age, it is imperative to implement the site-specific information into the numeric model. The heterogeneity of bone complicates attempts to model bone FE and photoelastic stress studies. The mechanical anisotropic nature of bone further perplexes attempts to model bone structure. In contrast to age-related deterioration of the skeleton, the density of the trabecular bone in the mandible does not decrease [36]. Quite the contrary, the trabecular bone density is reported to increase in the basal bone of the mandible upon tooth loss [37]. Studies dedicated to explore the mechanical properties of bone have been mainly focused on load-bearing long bones, such as the femur and the tibia. Information regarding site-specific mechanical properties of human cortical and trabecular bone in the mandible and maxilla in dentate and edentulous state are relatively scarse [38-40]. Moreover, the site- and species-specific information of bone is also lacking for most animals that are used for assessment of peri-implant tissue differentiation by conventional histomorphometry and numeric analysis. In such cases, an "available" bone information is used and this estimation might decrease numerical accuracy [41]. Owing to the scarcity of elastic modulus properties of human cortical and trabecular bone in the jaws, trabecular bone properties of long bones are frequently used in numeric models of the mandible and the maxilla. Extrapolating values from other parts of the skeleton interfere not only the interpretation, but also accurate prediction of the clinical scenario. Another factor that could influence the outcome of FE analysis is whether the three-dimensional trabecular structure of the bone surrounding the implant together with their mutual sliding contact interactions, strains, and micromotions are considered in the analysis [42]. Unfortunately, several studies have disregarded the implementation of this information into numeric analysis, which makes the prediction relatively unreliable.

2.3.2. Individualized FE Modeling

When the compatibility/agreement of two measurements techniques are explored, FE analysis could serve as a reliable tool. Keyak et al [43]. evaluated FE analysis as a tool for estimating proximal femoral fracture load and assessing hip fracture risk. In this regard implementation of individualized information is of most importance (Figure 6 and 7). FE models constructed from arbitrary computer-aided design (CAD) models result in data that should be considered descriptive only. In essence, the information regarding the implant with respect to bone should be incorporated in the individualized FE model, where the computed tomography images are first used to create the CAD model consisting of analytical surfaces and solids, which is then meshed. The effect of mechanical stimulation on bone tissue alone or bone surrounding an implant is traditionally investigated by histomorphometric measurements of histological sections at different time-intervals in animal studies. This attempt requires inclusion of a number of animals into the study, which are sacrificed at predetermined stages of the experiment. Despite efforts to keep experimental conditions standard, there are inter-individual variations in such an approach. Moreover, some animal models like the goat are more prone to implant loss. While this approach provides an insight into the factors studied, adaptive response within the same animal could not be monitored unless an approach that allows survival of the same animal, i.e., bone chamber, is used. Dynamic histomorphometry could provide more information regarding tissue differentiation, although such procedures are relatively time consuming. Alternatively, microfocus computed tomography (μCT)-based FE modeling could be used for assessment of individual-specific conditions especially in small animal research [41]. In μCT, the resolution of a system can be in the order of 10 μm x10 μm x10 μm depending on the object size, which is much better compared to the typical 60 μm x 60 μm x 1 mm resolution of the medical computed tomography (CT). Its resolution seems comparable to FE models obtained by the quantitative serial sectioning and serial grinding technique. μCT systems usually allow scanning of small specimens (4 cm^3) of long bones and small laboratory animals (for example SkyScan), while it is also possible to scan objects up to 5 kg (Philips HOMX 161 microfocus system). The radiation dose is always an important issue in animal research and low radiation dose regime could create a general noise, which does not dominate the entire image. μCT data show good agreement with histomorphometric data of bone [46,47] and therefore, μCT seems as a non-destructive tool that could be used to record individual bone images for FE modeling. However, quantitative value of μCT-based FE modeling needs to be improved by experimental studies conducted to provide sufficient information regarding species- and site-specific mechanical properties of bone, because many computational studies still implement an "estimated" bone mechanical property.

In μCT, bone, soft tissue and, implants are represented by voxels with different grey values. These voxels are not recognized as belonging to bone tissue during the segmentation process, because metal artefacts in μCT images alter grey values of voxels located in vicinity of implant. Owing to their higher atomic number, metallic implants attenuate x-rays more than surrounding bone and soft tissues. Fewer photons can reach the detectors, giving rise to the effect of the so called “missing” or “hollow” projections. Metal artefacts, typically seen as starburst streaking, result primarily from image reconstruction techniques, such as conventional filtered backprojection, which do not consider properly the problem of missing projections [48]. The artefact-affected region depends on the implant diameter and the radiation parameters, and could show a wide range of entension (24-200 μm) from the bone-implant interface [49,50]. Jaecques and collaborates [51] proposed the use of standard triangulated language (STL)-based approach to reduce the effects metal artefacts. A STL surface, generated from the segmented geometry, is converted into a closed volume that can then be meshed. The process, however, requires several software packages, is time-consuming, and not easy to be automatized. The final STL is an assembly of triangular faces and the final FE mesh is limited to tetrahedrons, limiting meshing options. Jaecques and collaborates [51] also claimed that there is need for more experimental data between μCT grey values and bone mechanical properties and that the reliability of such a relation strongly depends on the reproducibility and on the quality of the μCT images, which remains an important issue for the future [41]. The grey values also need to be calibrated (e.g. by including a phantom during scanning) so that the relation is independent of the scanning protocol.

As there is a growing need to adapt/amend implant designs due to inter-subject variations particularly in the field of orthopedics, individualized FE modeling is to be used for treatment planning and follow-up of patients. Indeed, recently developed modeling techniques allow generation of finite-element models that implement the subject-specific morphology of the bone as well as the distribution of the bone tissue mechanical properties [46,47]. At present, conventional CT, MRI, and peripheral quantitative CT (pQCT) can be used to generate individualized FE models for humans. pQCT was developed as a small-field, high-resolution extension of the existing CT to measure the peripheral skeleton such as the whole human head or neck, to tiny excised bones such as mouse femurs or vertebrae. The technique allows determination of cross-sectional images (slices) by scanning after which the images are integrated by a computer system through a complex procedure known as backprojection with filtration. The scanned field is divided into pixels, which actually represent the bases of volume units referred to as "voxels" used to determine bone density. The voxel intensities in the CT scan could be used to retrieve inhomogeneous properties of bone, which can be assigned to the mesh as patient-specific mechanical properties. The number of voxels per field is a fixed characteristic of each type of machine (usually 1024X1024), which does not change by increasing or decreasing the area of the selected field. When properly calibrated and processed, CT images can provide accurate information about the cross-sectional geometry of bone, which can be used to derive its actual three dimensional shape.

In the context of micro FE models, creation of the highly complex meshes in such a numeric analysis seems rather challenging. Today, the voxel-based method is probably the fastest and easiest way to build such a mesh. The method consists of associating a voxel representing bone with a cubic element in the mesh. Generation of such models is time-consuming and automatic, but the surfaces generated are composed of jagged edges. This may complicate numeric simulations, as stress concentrations may appear at the sharp corners. Inevitably, the accuracy and validity of these models can be low as compared to other conventional methods. To overcome this problem, the external surface of the model can be smoothed by extracting the outer surface of the mesh. Indeed, smoothing of voxel-based FE meshes by geometric signal processing approach has been shown to improve accuracy in the simple case of a sphere and trabecular bone structure [52,53].

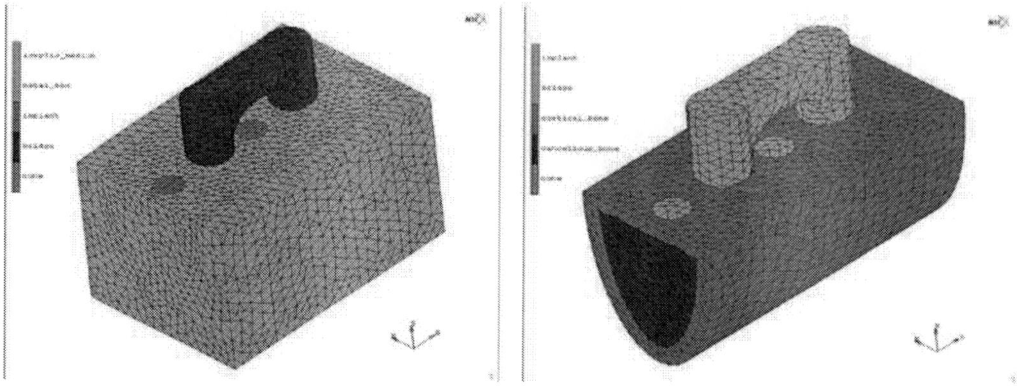

Figure 6. Two-FE models incorporating dental implants supporting a three-unit bridge. The actual in vitro model, in which strain-gauges are bonded to the neck of the implants, is simulated on the left side. Supporting as well as sleeping implants are embedded into acrylic resin. On the right side, the same scenario is simulated in a rather simple representation of non-anatomic bone constructed from an arbitrary CAD model. The strains induced even on the supporting implants (solely metal-not acrylic or bone) are different in both models (Adapted from Akca et al. [44]).

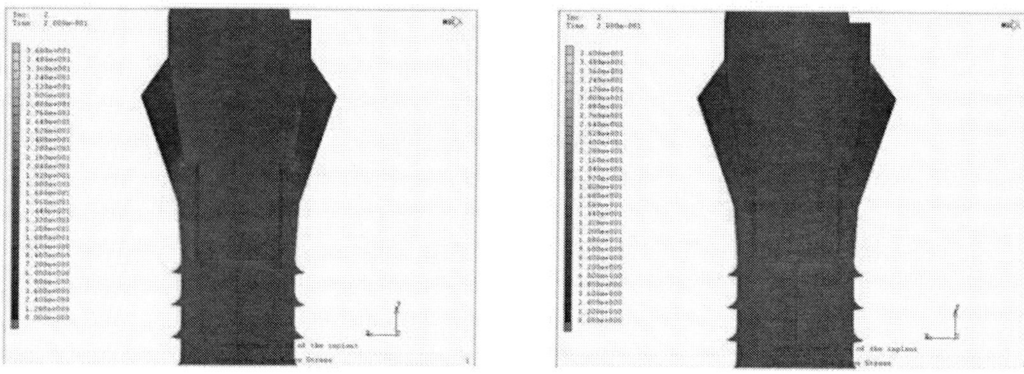

Figure 7. The von mises stresses induced in a loaded internal-cone implant. The FE model on the left side incorporates contact definition at the abutment-implant interface and the model on the left side considers the interface as "glued" as many numeric models. The distribution of von mises stresses are different in both models, suggesting that one-piece modeling of conventional dental implants might alter numerical accuracy (adapted from Cehreli et al.[45]).

In the context of patient-specific FE analysis, a number of problems exits. Currently, data processing techniques used to extract information on the geometry and the local tissue properties of the bone segment from the CT data may be distorted by non-negligible errors that propagate in a concealed way through various steps of the FE modeling, and undermine the accuracy of predictions unpredictably [54]. The reason for the distortion is the resolution of the dataset that relies on the scan parameters setting. The segmentation process of the site, undertaken by complete manual contours extraction algorithms, complex fully automatic algorithms, and frequently threshold-based algorithms for most cases always result in an uncertainty on the boundary definition and create threshold-specific geometric error that could propagate up to the mesh generation step.

When patient-specific FE modeling is performed, the bone density information implemented into the numeric model derived from CT scans in Hounsfield Units is not error-free. As the density of bone tissue is related to its elastic modulus, which is derived from a wide-range experimental data, using uncertain information impair the predictive accuracy of the numeric model [54].

According to Viceconti and co-workers [55] the accuracy of an individualized FE model may be hampered by various potential sources:

1. The boundary conditions applied to the numeric model may not accurately simulate the actual biologic condition.
2. The constitutive laws used to model the mechanical behavior of the bone may not be adequate or their parameters accurately identified.
3. The FE mesh may be topologically inaccurate in the sense that geometry of the bone is inaccurately derived from the computed tomography data.
4. The FE mesh may be topologically ill-conditioned, i.e. the shape of the elements is so distorted that the numerical accuracy of the model is reduced.

In subject-specific FE analysis, the creation of three-dimensional FE meshes of the bone tissue should be performed from the computerized data. This procedure should be automatic (Creation of a FE mesh with the smallest possible human effort), accurate (Solution of the mathematical problem with the smallest numerical error), robust (Low sensitivity of the numerical accuracy to the inter-subject variability) and general (Low sensitivity of the numerical accuracy to the anatomical variability) [55].

The existing algorithms proposed for the automatic generation of FE meshes can be divided in three categories:

1. Algorithms based on a surface mesh built from the segmentation. The level of automation is rather low and the outcome is a closed polygonal simplex surface.
2. Algorithms that is exclusively image-based, such as the "voxel mesh method". They are fully automated, general, and robust. Nevertheless, the boundaries of the mesh have jagged edges, decreasing the level of accuracy.
3. Algorithms that rely on segmentation contours i.e., template methods or grid-based methods. The level of accuracy, automation and robustness of these algorithms fall between the first two algorithms.

Subject-specific FE analysis in the field of orthopedics has focused on predicting the failure loads and patterns of load-bearing bones such as the femur. The main problem with current models is their lack of generality, because more general models are required to predict and localize the fracture risk and pattern for a bone segment under a generic loading condition. This requires implementation of a bone tissue failure criterion as well as a structural collapse criterion and validation experiments to improve the reliability of the results. Another important factor is the strength criterion. Regardless of the modeling strategies adopted, it is apparent that studies follow different strength criterion for the bone tissue, although many focus on stress parameters. These criterions include:

1. Distortional energy [43,56,57]
2. von Mises stress yield criterion for cortical bone in conjunction with a Hoffmann stress criterion, or a more complicated crushing-cracking stress criterion for trabecular bone [58,59].
3. Principal stress [60]
4. Anisotropic stress criterion including fabric tensor information [61]

There are numerous factors that affect force magnitudes in peri-implant bone (Table 1). Because these factors are always subject-specific, their simulation in the patient-specific numeric model is important. The most apparent uncertainty is related to the fact that in vivo load amplitudes as well as directions are not well-defined, as such load characteristics including on oral implants and gait patterns are usually roughly estimated. In an individualized FE analysis, the patients actual occlusal forces should be recorded, i.e., by miniature load cells and implemented into the analysis. In addition to inter-individual variations, occlusal forces and lateral component of these forces in dentate humans rely on many factors such the age, gender, and region of the dental arch. Patients with implant-supported fixed prosthesis have a masticatory muscle function equal to or approaching to that of patients with natural teeth, or with tooth-supported fixed partial dentures [62]. Placement of a mandibular fixed implant-supported prosthesis in complete denture wearers is reported to improve masticatory function and the magnitude of bite force [63-65]. Haraldson and Carlsson [65] measured 15.7 N for gentle biting, 50.1 N for biting as when chewing, and 144.4 N for maximal biting for 19 patients who had been treated with implants for 3.5 years. In another study, Carr and Laney [66] reported maximum bite forces between 4.5 and 25.3 N before and 10.2–57.5 N after three months of treatment with implant-supported prosthesis, and emphasized that, the amount of increase was dependent on the duration of being edentulous. Forces on implants are also dependent on the location of the implant in the dental arch. Mericske-Stern and Zarb [77] investigated occlusal forces in a group of partially edentulous patients restored with ITI® implants supporting fixed partial prostheses and measured an average value of maximum occlusal force lower than 200 N for first premolars and molars and 300 N in second premolars.

Table 1. Subject-specific factors influencing load distribution on implants

Site-specific morphologic and mechanical properties of basal, alveolar, and peri-implant bone
Location, direction, and amplitude of applied occlusal forces on the prosthesis
Type and geometry of the prosthesis
Prosthesis material ?
Superstructure fit
Condition of the opposing arch (prosthesis versus natural dentition)
Geometry, number, length, diameter, and angulation of supporting implants
Percentage of bone-implant contact
Location of the implant(s) in the arch
Thickness and resiliency of supporting soft tissues for overdentures
Mandibular deformation

Beyond these issues, presence or lack of mesh convergence test to determine the adequacy of mesh density in FE modeling, the actual definition of the bone-implant and/or implant-abutment interface (bonded versus contact definition), and uncertainties related to material property assignments and surface geometry affect the numeric accuracy (Figure 8) [54,78,79]. Another important factor in animal as well as human research is the validity of computational models that simulate mechanobiological response of tissues. Relatively simple algorithms have been shown to simulate tissue response to loading [80,81]. A critical issue with such predictive models is to corroborate them against animal experiments, because the inter-individual variations is compounded by problems relating to differences in loading and skeletal geometry. Mechanobiological computer simulations to predict the deposition and resorption of tissue during fracture healing and bone chamber experiments with titanium implants [82,83] have showed satisfactory correspondence between simulation and the experiment, thus suggesting a level of collaboration/agreement at the in vivo level. It is stated that, for many researchers such a collaboration or "validation" is inadequate and micromechanical FEs are required [84], in which the detailed modeling of osteocyte lacunae, Howships lacunae, Haversian canals have to be constructed to provide modeling accuracy. Creation of such models are not easy and they are too large.

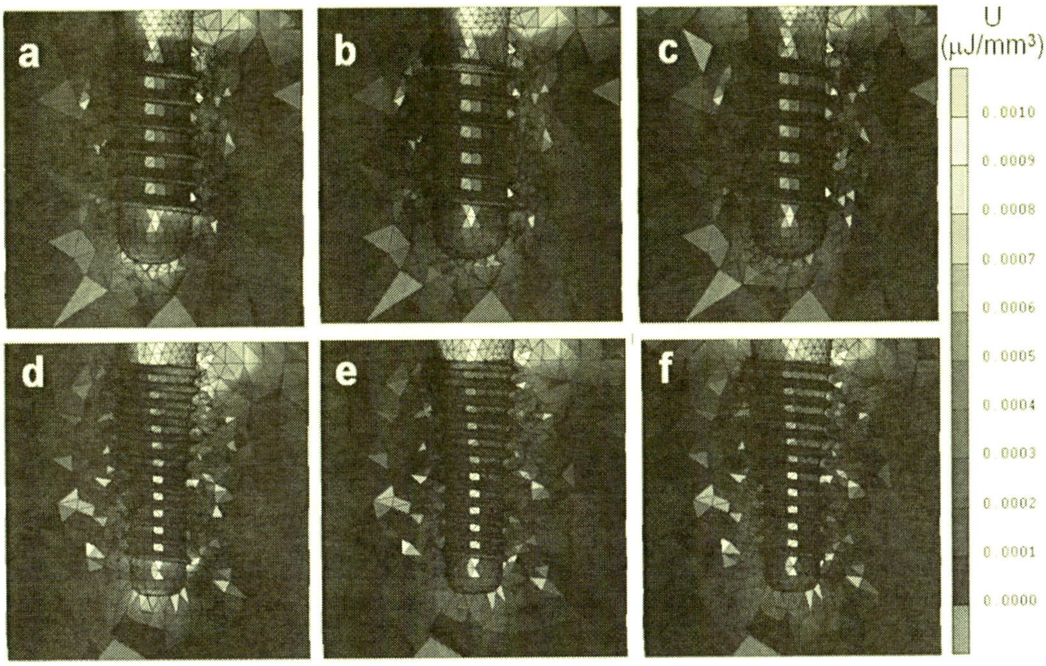

Figure 8. An example of how interface definition could affect strain Energy Density (μJ/mm^3) distribution for parallel-sided (upper row) and tapered (lower row) implant designs. (a, d. Non-linear Coulomb friction, b, e: Linear contact, c, f: Fully integrated implant with bone) (Adapted from Eser et al.[85]).

2.3.3. Implementation of Bone Remodeling Theories

In the field of dental implants, the prediction of peri-implant events by alterations in the material properties of bone and failure patterns of implants have been recently focused on time-dependent assessment by implementation of bone adaptation theories [34,35,85,86]. Bone including in the periphery of load-bearing implants responds by changing its mass density, when its mechanical loading conditions deviate from homeostatic levels, by a series of bone re/modeling processes [1-3], ruled by a physiological control system [87]. Bone remodeling theories distinguish between external modeling, where bone is added or removed at the periosteal and endosteal surfaces, and internal remodeling, characterized by changes in apparent bone density. A detailed overview of bone modeling and remodeling theories can be found in Hart's work [87]. Since 1980's, numerical solutions have verified mathematical theories of bone adaptation to predict changes in bone shape and density based on mechanical stimuli like strain, stress or strain energy density (the elastic potential energy per unit volume, stored in the elastic element due to deformation). Two isotropic bone adaptation hypotheses are the most employed in the literature. The first approach under the direction of Huiskes [88-90] assumes that, the variation of the elastic modulus is proportional to the difference between the actual strain energy density and a referential strain energy density. This stimulus is used for internal and external remodeling. The second hypothesis, proposed by Carter and Beaupre [91-93] establishes as stimulus a certain effective stress (related to the strain energy density), being the apparent density related to this effective stress. This also allows the combined formulation of external and internal adaptation. In both models, the results obtained are very similar, being almost indistinguishable in internal remodeling. These models are isotropic, but it is clear that actual bone is both quite anisotropic and heterogeneous. As an example, let us briefly explain implementation of the Stanford Theory into a numeric analysis related to oral implants [85,86]. The so called Stanford theory [91-93] is based on a daily stress stimulus which is defined by

$$\psi_b = \left(\sum n_i \sigma_i^{\,m}\right)^{1/m} \quad \text{(Eq.1)}$$

where n_i is the number of cycles of load type i, and σ_i is a continuum level effective stress defined as

$$\sigma = (2EU)^{1/2} \quad \text{(Eq.2)}$$

where E is the continuum elastic modulus and U is the continuum strain energy density. The strain energy density has been widely used as the mechanical stimulus in bone remodeling algorithm. Because strain energy is a scalar quantity, the directional properties of the strain/stress tensor can be eliminated when dealing with the remodeling control variables. According to Mellal et al. [94], the remodeling process in the alveolar bone may be calculated by analogy with the remodeling algorithm developed for the long bone, as long as some minor modifications to the reference values and constants in the equation are made. As the strain energy can be easily obtained directly from FE analysis programs, the strain energy bone remodeling algorithm is a feasible and practical method for predicting alterations in bone density.

The level of the daily stress stimulus determines the bone remodeling process. Under normal loading conditions daily stress stimulus remains in a determined range and bone remodeling does not occur. This range is called “lazy” or “dead” zone by Carter [91]. Just as in the Mechanostat remodeling algorithm, a "lazy zone" may be included in the strain energy density-based bone remodeling, in which remodeling does not occur in the equilibrium range between bone resorption and apposition (Figure 9).

The mathematical expression of the apposition, resorption rate of the bone by taking account the "dead" zone is defined as

$$r = \begin{cases} c.(\psi_b - \psi_{bas}) + c.w & (\psi_b - \psi_{bas}) < -w \\ 0 & -w \le (\psi_b - \psi_{bas}) \le +w \\ c.(\psi_b - \psi_{bas}) - c.w & (\psi_b - \psi_{bas}) > +w \end{cases} \quad \text{(Eq.3)}$$

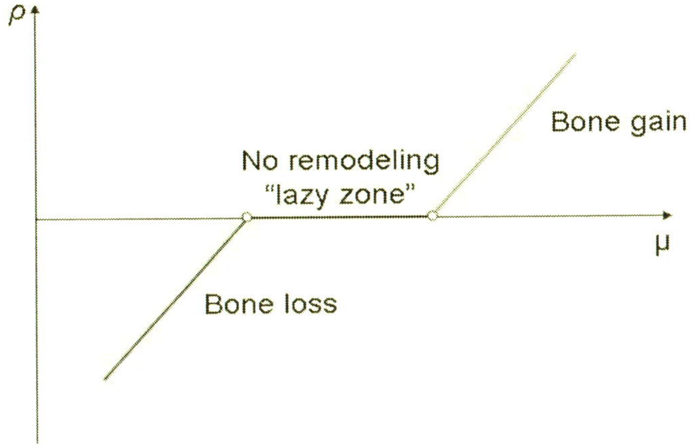

Figure 9. Bone density versus mechanical stimulus graph.

where $\dot{r}$ is the apposition-resorption rate (μm/day), ψ_b is the daily tissue level stress stimulus (Equation 1), ψ_{bas} is the attractor state stress stimulus (the level of daily stress stimulus in which bone mass change does not occur), c is an empirical rate constant and w is the half width of the central, normal activity region (dead zone). During implementation, the attractor stress stimulus ψ_{bas} is determined from the graph of cyclic energy stress magnitude and equivalent cyclic normal strain magnitude versus the number of cycles per day. The number of daily load cycles and amplitude of applied load should be determined according to the experiment and the implemented equivalent cyclic normal strain should preferably be species- and site-specific [85,86]. After assuming the empirical constant, m, to be equal to a certain level according to the experiment, attractor state stress stimulus is selected. The value of $m = 2$ has been proposed by Carter et al. [95] value has been applied in various simulation studies including the Stanford Theory. However, some of the concerns with daily stress stimulus in dental study consist in: (a) the number of loading cycles can vary across the different population groups; (b) the magnitude and of chewing pattern may vary depending on different eating habits. This will heavily jeopardize the outcome of daily stress algorithm, i.e., people who tend to eat more or harder food or will obviously, chew more and apply more occlusal force. Eventually, prediction of dental bone remodeling using this algorithm will require numerous statistical data gathering of the variation of human chewing pattern across different age, cultural and gender groups, which can be time consuming [96]. Then, the numerical value of c (μm/day)/(MPa/day) and value of w (MPa) are assumed and the initial elastic modulus and Poisson's ratio for the materials are implemented into the computational model. After determining apposition-resorption rate (μm/day), the new density of the bone can be calculated as follows:

$$\rho = rS_v\rho_t\Delta_t + \rho_o \qquad \text{(Eq.4)}$$

where ρ_o is the density of the bone in previous stage, S_v is the bone surface area per unit tissue volume, and ρ_t is the true density of the bone tissue (which is assumed to be equal to the density of fully mineralized tissue). The constants S_v and ρ_t are selected are selected for the analysis. The density (apparent density) of the bone can be related to the elastic modulus of the bone through the equation:

$$E = 3790\rho^3 \qquad \text{(Eq.5)}$$

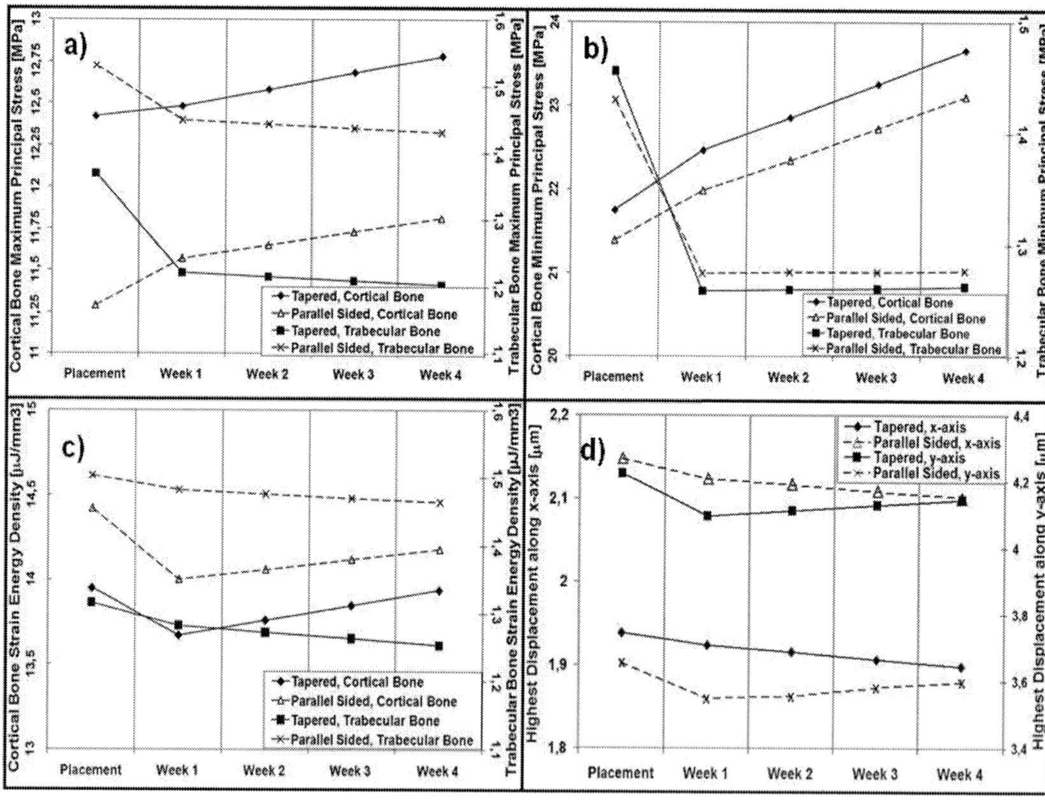

Figure 10. a. Time-dependent highest maximum principal stress values (MPa) in cortical (CB) and trabecular (TB) bone around parallel-sided and tapered implants. b. Time-dependent highest minimum principal stress values (MPa) in cortical (CB) and trabecular (TB) bone around parallel-sided and tapered implants. c. Time-dependent highest strain energy density values (μJ/mm^3) in cortical (CB) and trabecular (TB) bone around parallel-sided and tapered implants. d. Time-dependent highest displacement values (μm) in x-(perpendicular to implant) and y-(parallel to implant) axes of bone around parallel-sided and tapered implants (Adapted from Eser et al. [85]).

The equation above could be used for both cortical and trabecular bone, as the bone remodeling takes place on bone surfaces of marrow spaces/voids in trabecular bone and haversian canals in cortical bone [92]. The equation relating the elastic modulus with density is also used in a similar manner in the work of Mellal et al.[94]. Using the above relation, the new elastic modulus can be calculated for each step. The new elastic modulus is used for the next calculation. The number of iterations for the time periods are estimated using the convergence criteria of the change in apparent density for every element to be less than for example 0.02 g/cm^3 similar to Beaupre and co-workers [92,93] In this sense, the material properties of bone could be updated (i.e., remodelled) by a computer program (Figure10). An example [85,85] of such a program is presented in Figure 11.

Neither the Stanford Model nor any other theoretical model so far has been able to present time-dependent changes in the skeletal tissue at the microstructural level around implants. The trabecular bone in the present numerical approach is assumed to be a continuum in these theories, not simulating the actual microstructure of trabeculation. The reason is that the application of these theories does not comprise changing the number of elements in time, depending on the magnitude of load applied for woven or trabecular bone having a porous nature. This approach shows i.e., how bone density around a load-carrying implant changes in time, which could explain the increase or decrease in stiffness and apposition/resorption of the tissue. Beyond these issues, current bone remodeling theories fail to incorporate bone resorption induced by overloading. Van Oosterwyck et al. [97] developed a numerical remodeling algorithm to simulate consequences of overloading around implants, where an overload threshold of stress (31 MPa) was defined. In addition, Crupi et al. [98] used a linear function to establish a negative rate in the alteration of bone density when the stress exceeds a critical threshold. Li et al. [99] introduced a quadratic method assess overload-induced bone remodeling around dental implants [96].

3. Comparison of Experimental Stress Analysis Techniques for Compatibility and Validity

Coherence of predictions made by stress analyses with experimental data is a matter of debate. In accordance with the limits of experimental stress analysis, the level of agreement between stress/strain analysis techniques and correspondence of experimental stress/strain data to the actual in vivo scenario have been also a topic of research interest. In the context of bone as well as load-bearing bone implants, this is of utmost importance, because the amplitude of stress/strain measured by a technique that leads to bone loss around an implant will be interpreted as "correct" or not. In this section, studies related to dentistry and oral implants will be included.

The literature concerning such comparative analysis could be divided into three categories:

1. Comparison of solely experimental stress analysis techniques. This approach does not include in vivo experiments for validation whatsoever. It consists of comparing the quality and amplitude of stress/strain obtained by one technique with the outcome of another, i.e., photoelastic stress analysis versus strain-gauge analysis.

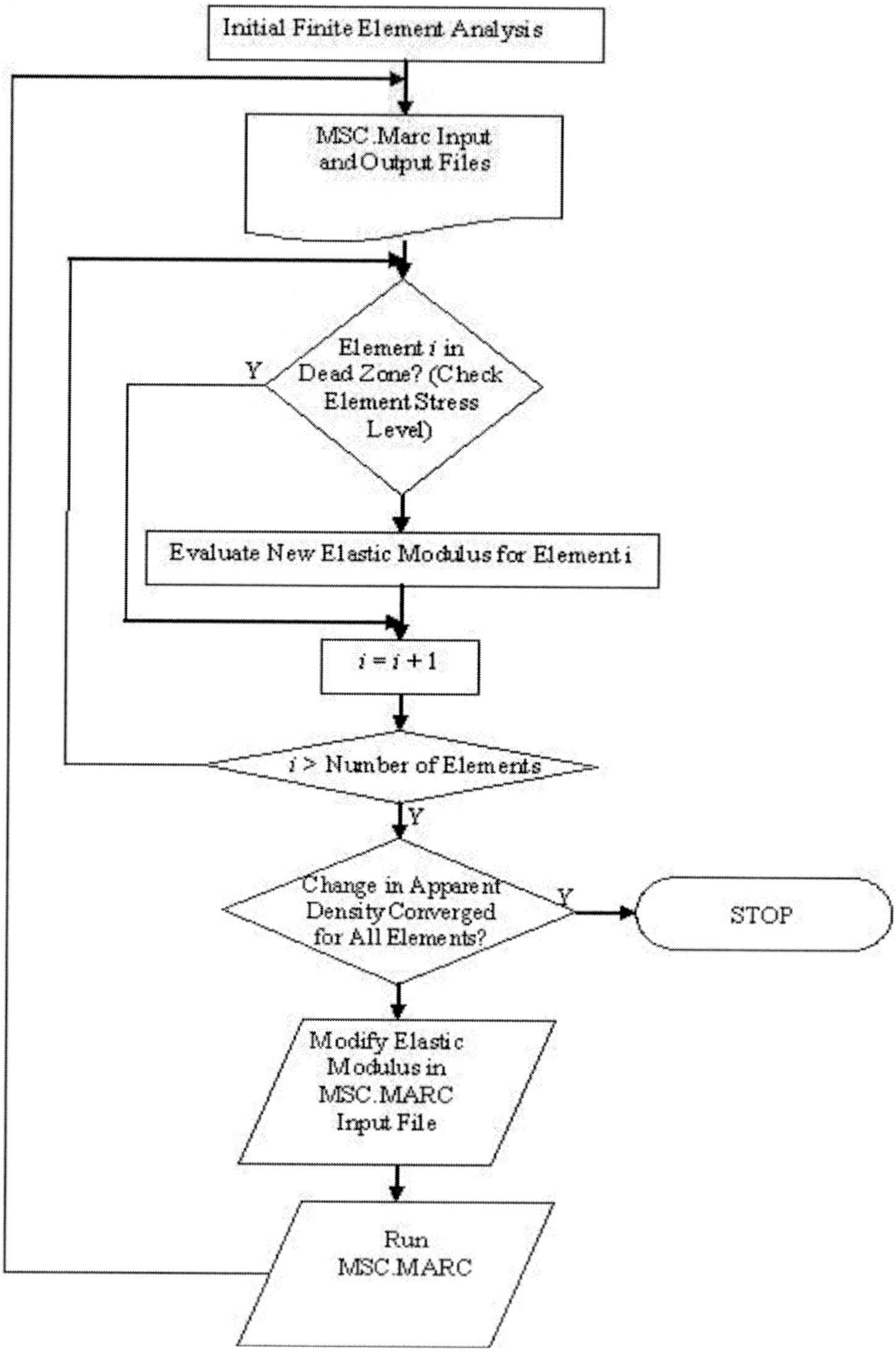

Figure 11. Flow chart of the code developed by Eser et al. [85,86]. The program Remodel reads the input file of the FE analysis, which contained information about the geometry, boundary conditions, the initial material properties and output file generated by the FE analysis code MSC. Marc that contained the calculated stresses and strains. Using the calculated continuum strain energy density U, which can be read from the MSC. Marc output file and the corresponding elastic modulus E, the continuum level effective stress σ_i can be calculated by using Equation 2. After achieving effective stress σ_i where

i=1 for one load type, the Daily Stress Stimulus ψ_b is calculated by the program by using the Equation 1 with the referenced values of m and n. In the next step, the program calculates the $\dot{r}$ (apposition-resorption rate) using the values of daily stress stimulus ψ_b from the previous step and by the referenced values of c and w from Equation 3. The program uses the value of $\dot{r}$ (apposition-resorption rate) in Equation 4 to calculate the new bone density of the bone elements by the referenced values of S_v and ρ_t and this new bone density ρ is used in Equation 5 to achieve the new elastic density for the bone elements. Using these information and Stanford theory for bone remodeling, the code Remodel generates a new input file for new FE analysis, which contains the information in the previous input file, but the updated material properties due to remodeling. This newly generated input file is analyzed by MSC. Marc again and the results are used by Remodel to update the next input file.

2. Comparison of the experimental stress/strain data of engineering methods with in vivo "biological" findings for validation. In such an approach, there is not any in vivo stress/strain measurement, but only biological findings. Studies usually comprise only one stress/strain analysis method. Lately, if FE approach is used, it is usually somewhat the individualized version of the actual in vivo animal or patient.
3. Theoretical models with tissue differentiation or bone remodeling theories, which compare their outcome with the in vivo experimental findings of independent studies. This approach attempts to "individualize" the experimental model of that study to eliminate the need for validation experiments.

3.1. Comparison of Solely Experimental Stress Analysis Techniques

3.1.1. Photoelastic Stress Analysis Versus in Vitro Strain-Gauge Analysis

There is only a small pool of inconclusive evidence in the dental literature. Brosh et al [100]. bonded linear miniature strain-gauges on the body of a non-threaded dental implant connected to straight and angled abutments. A photoelastic model including the implant was also constructed and the outcome of both methods as a sequel of static loading was compared. While the strain-gauge method showed 3-4 fold increase in strains in the cervical region with angled abutments, photoelastic method showed an increase of only 11% in fringe order, suggesting lack of correspondence between the two techniques. Their results are in contrast to the findings of Clelland et al [101]. who found correspondence between these methods using a similar experimental approach. It is tempting to speculate that, owing to the smaller measuring grid of miniature strain-gauges, it is likely that the level of measurement accuracy on a solid metal surface would be high. Because the resolution of the quasi-three dimensional photoelastic stress analysis technique relies on the resolution of the image ruled by the camera and isochromatic fringe orders above 3-4 have relatively low resolution, the level of accuracy in the photoelastic technique seems lower than the strain-gauge technique.

3.1.2. FE Stress Analysis Versus in Vitro/Ex Vivo Strain-Gauge Analysis

One of the main issues in such comparison is that the size of miniature gauges do not allow application many critical areas for biomechanical research, such as the implant thread or the implant neck without surface preparation. Trimming the backing of the strain-gauge may potentially alter its sensitivity and accuracy. Therefore, such comparative studies are relatively focused on large objects like the femur and smooth and frequently flat surfaces for proper bonding of strain-gauges.

The application of 3-dimensional FE and strain-gauge analyses on in vitro animal and human experiments have provided mutual compatibility and agreement of obtained results [102,103]. Keyak et al. [102] bonded strain-gauges to the surface of a human femur, where no detailed finite modeling was applied, and Baiamonte and co-workers [103] have used strain-gauges bonded on two cylindrical implant abutments, which resemble a solid structure more than a two piece implant collar. In the context of oral implants, however, a study on the neck of two-piece implants suggested that the data recorded from in vitro strain-gauge analysis are relatively higher than the results of a linear solution FE analysis [44]. The authors argued that conventional linear solution FE analysis cannot simulate the actual deformation of the two-piece implant neck, which was then confirmed in a separate study [45]. Because the linear solution assumes the implant-abutment complex as a one-piece solid structure, it is likely that the increase in metal thickness will reduce deformation and the strain amplitudes. Indeed, a comparison between non-linear FE with in vitro strain-gauge analysis [104] showed high correspondence in solid structures like an implant abutment and acrylic resin. However, nonlinear FE results were higher than strain-gauge measurements under lateral loading (Figure 12 and Table 2). This outcome could be related to the fact that either the "average" of strains recorded by measuring grid of the gauges or trimming the gauge backing/bonding on convex implant body decreased it accuracy.

They also argued that the clamping force or preload of the abutment might have an influence on FE outcome, as this was not included into their FE model.

More recently, ex vivo strain-gauge measurements on human cadavers were compared with non-linear FE analysis using a partly subject-specific FE model for positioning implants and bar-retainers of each cadaver [105] (Figure 13). The results suggested statistically significant similarity in the data obtained by the two techniques. The authors argued that a strict comparative approach would probably best fit into such experimental conditions, where controlling several factors is possible. However, several factors such as variations in bone density in jawbone, angulations of implants, direction and amplitude of forces, superstructure fit, and resilience of soft tissue overlying the residual ridges might affect the quantity of strains.

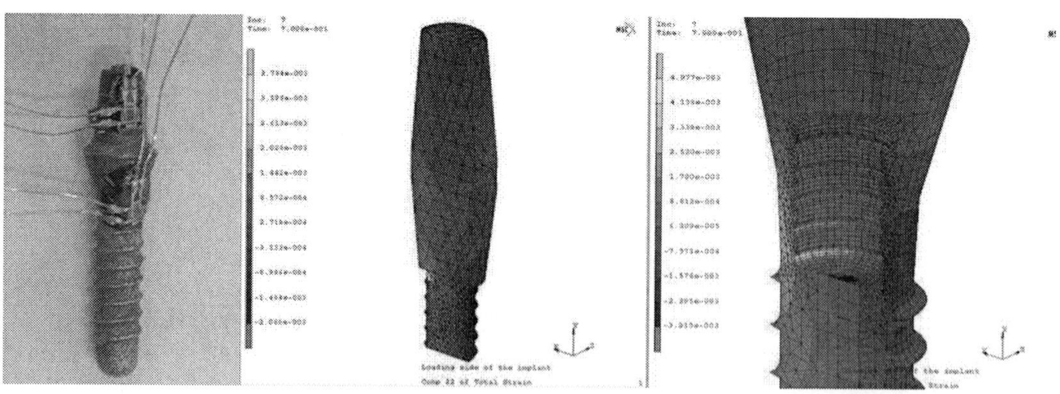

Figure 12. Miniature rosette gauges bonded on a solid abutment and the neck of a two piece implant (left), whose FE models for the abutment and the implant body appear on the right.

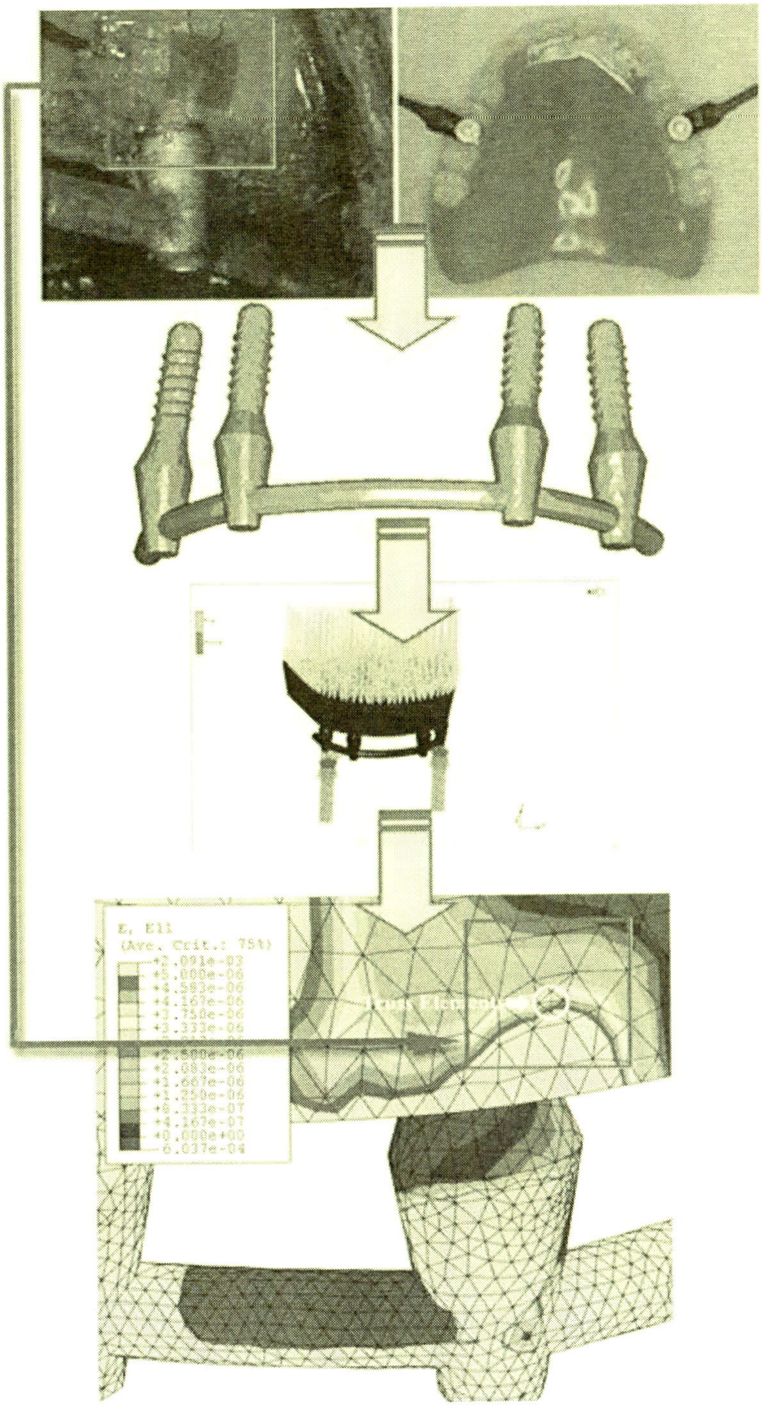

Figure 13. The ex vivo strain-gauge data on cortical bone adjacent to implants supporting bar-retained maxillary overdentures are recorded under controlled loading by means of load cells. The 3-dimensional subject-specific positions of the implants are recorded by a coordinate measuring machine and individualized CADs of arch forms are created and meshed. The subject-specific loading of the implants are simulated and strains in the area previously corresponding to the measuring grid of the gauge (red arrow) were measured at truss elements.

Table 2. Strains (X 10^{-6}) measured by in vitro strain-gauge analysis and nonlinear FE analysis on the implant, abutment and in the resin (as the bone simulant) under 75 N force application [104]

	In vitro SGA Mean strain (SD)			NL-FEA		
	Implant	Abutment	Resin	Implant	Abutment	Resin
Vertical Load	48 (6) Y -200 (21) Z	78 (24) Y -230 (45) Z	103 (52) Y -204 (86) Z	43.1 Y -200.1 Z	89.05 Y -242.5 Z	84 Y -104.3 Z
Lateral Load	545 (75) Y -1496 (123) Z	114 (27) Y -210 (42) Z	1700 (322) Y -1525 (415) Z	1048.8 Y -3395.1 Z	131.7 Y -216.1 Z	2000.6 Y -1433.7 Z

SGA: strain-gauge analysis; NL-FEA: nonlinear finite element stress analysis; Y: strain in -Y axis; Z: strain in -Z axis.

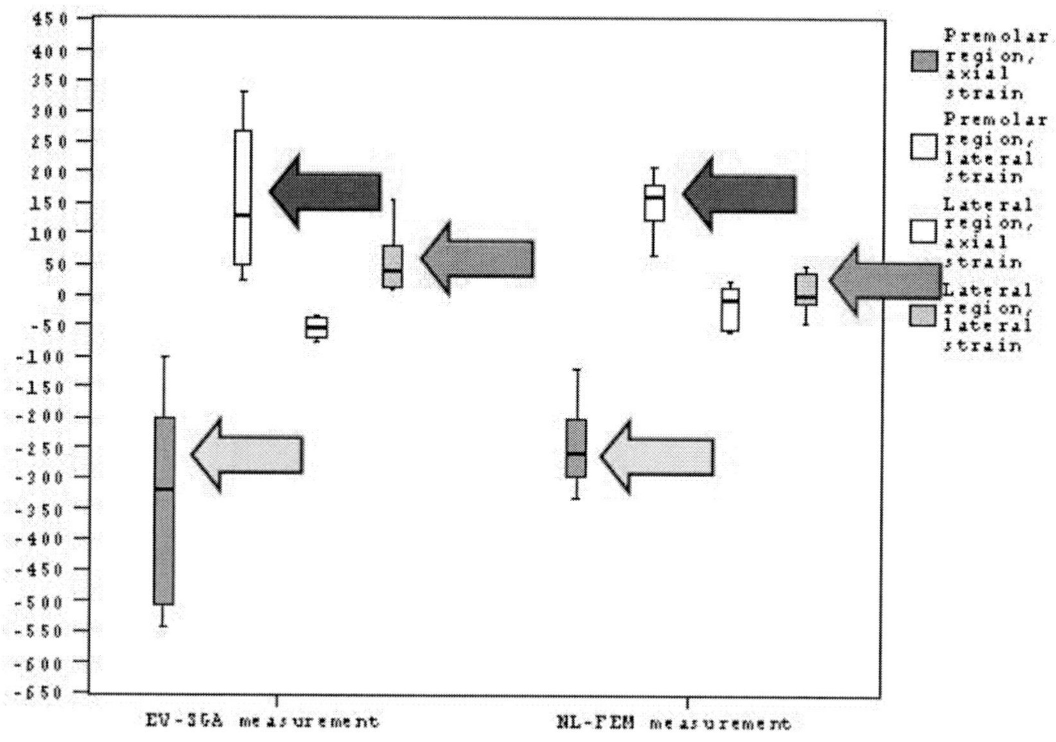

Figure 14. The reproducibility of the same data in strain-gauge measurements on bone is lower than the FE method, resulting in a range of strains that may be potentially misinterpreted when solely the means or the median is considered.

Indeed, these factors along with inter-individual differences led to differences of several hundreds of microstrains (> 900 με) in strain-gauge analysis. Therefore, it does not seem rationale to strictly compare the means of strain-gauge measurements with those of non-linear FE analysis, but one might consider comparing the "window" (data range) of strain measurements obtained by strain-gauge analysis with the means of non-linear FE analysis (Figure 14). The strain-gauge technique always suffers from random error and the results are never the same. A finite-element study based on in vitro and in vivo strain-gauges measurements on prosthesis misfit showed that the mean in vivo strain values ranged between 15-170 μm/m, which was in line with the previous findings of the authors [106,107]. In vitro strain measurements on bone around implants also show a range of strain values for a specific condition [18,20] (Figure 15, Table 3).

Table 3. Microstrains around natural teeth versus immediate oral implants supporting fixed prostheses under 100 N static load

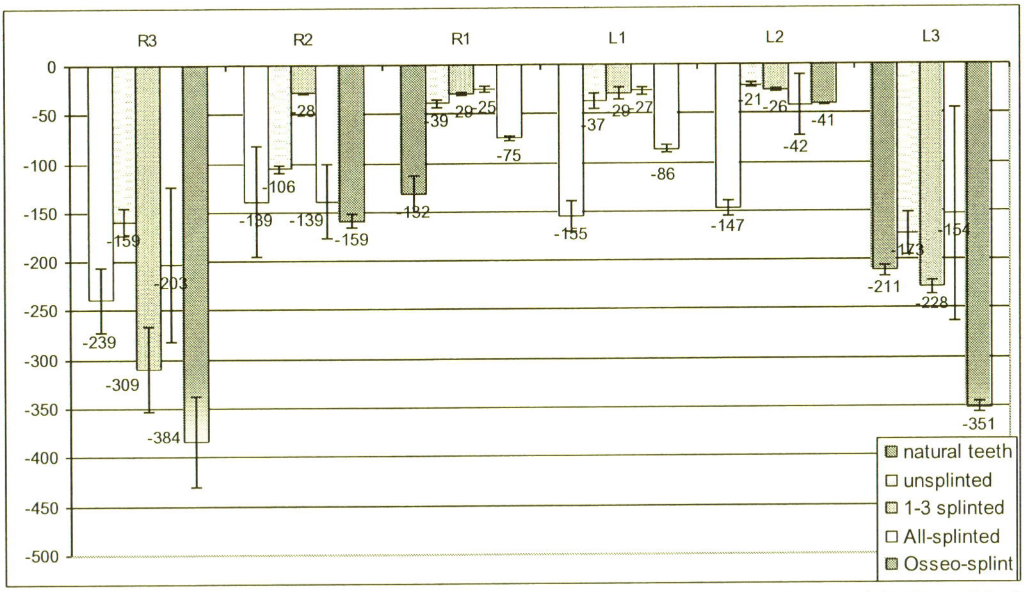

R3: right canine; R2: right lateral incisor; R1: right central incisor; L1: left central incisor; L2: left lateral incisor; L3: left canine.

Ichim and collaborates [108] explored distribution of functional strains in the mandible using numeric analysis. The FE model included the entire mandible, teeth and periodontal ligament simulation, where all contact between the components were considered as bonded. A mesh convergence test was also undertaken to confirm adequacy of mesh size. An in vitro strain-gauge analysis of the same situation was performed using linear strain-gauges for validation. The authors stated that the outcome of the two techniques was very close. In an attempt to explore root fracture during root canal therapy with FE analysis and in vitro strain-gauge analyisis, Lertchirakarn et al. [109] used a rather simplistic FE model, where the cementum layer was not modeled.

The authors claimed the FE results showed good qualitative and quantitative correspondence with strain-gauge data. Knoell [110] compared strains measured below the tooth row on the periosteal surface of cortical bone with values obtained at similar locations in the FEM. Although he stated that "reasonable agreement appeared to exist between predicted and test results", his data demonstrated the contrary when the first molar was loaded horizontally. Through regression analysis, Tajima et al. [111] found a very high correlation (r=0.94) between FE and in vitro strain-gauge analyses on surface measurement of natural teeth crowns. Likewise, Vollmer et al. [112] reported a very high correlation (r=0.992) between FE and in vitro strain-gauge analyses on surface measurements of the mandible. Alsukhun et al. [113] found 3-18% differences between FE and in vitro strain-gauge analyses on surface measurements of the mandible.

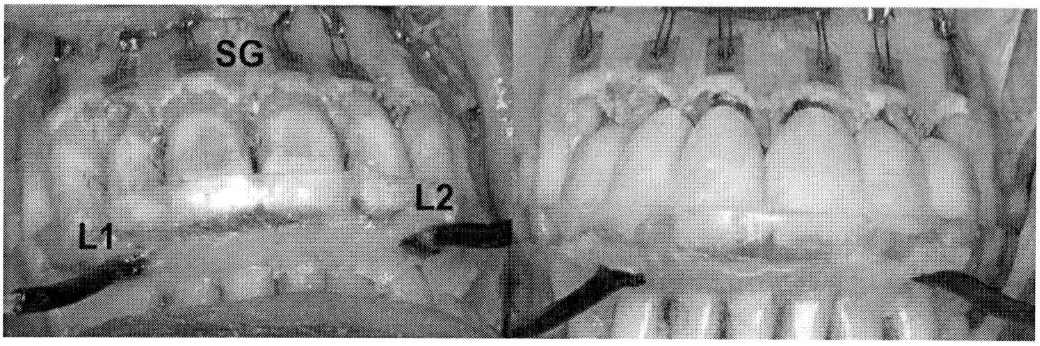

Figure 15. Linear strain-gauges (SG) bonded on the surface of cortical bone adjacent to natural teeth. Strain-measurements are undertaken under controlled-loading by means of miniature load cells (L1 and L2). The teeth are extracted, implants are placed and load-induced strains area recorded for unsplinted, splinted prosthetic versions as well as bone-implant relationships (adapted from Cehreli et al. [18]).

3.1.3. FE Analysis Versus Photoelastic Stress Analysis

In finite element part of this chapter, the impact differences in FE modeling have been already discussed. Therefore, it will not be included in this section. Comparative evaluation between FA analysis and photoelastic stress analysis in the context of implants is extremely scarse and a conclusion cannot be drawn. In search of the critical implant length and orthodontic load amplitude, Gracco et al. [114] compared the stress magnitudes around orthodontic micro implants using photoelastic stress analysis and FE analysis with contact-interaction algorithm between nodes to simulate three different stages of interface healing. They found that for moderate loads the photoelastic technique showed similar results, while the FE analysis suggested higher von Mises stresses under low loads.

3.2. Comparison of the Experimental Stress/Strain Data of Engineering Methods with in Vivo "Biological" Findings for Validation

In essence, it is not clear whether such approach really includes "validation" experiments, because it does not comprise in vivo stress/strain measurements whatsoever. This approach rather searches for a biological "*evidence*" that would correspond to the experimental stress/strain data. This is where the main problem starts, because available experimental strain data, based predominantly on resistance strain-gauge experiments, is unreliable. As stated previously, strain gradients may be measured extremely different by other techniques such as digital image correlation. Moreover, the available strain data is limited to measurements from cortical bones of a few species and predominantly long bones. It does not make sense to use strain data of the turkey ulna or draw "*scientific*" conclusions based on such data in the context of dental implant biomechanics! Today, because strain amplitudes that rule tissue differentiation and bone response to mechanical stimuli have not been measured by micro biosensors at the interface [117], this approach rather seems as a "hide and seek" desperately searching for a biological "*evidence*" that would correspond to the stress/strain data.

A good example of such approach could be found in "Numerical simulation studies in Animal experimental findings on the effect of mechanical load on peri-implant tissue differentiation and adaptation by VanDamme, Naert and Duyck" in this book. The KU Leuven Biomat Research Group has several years of experience in the field of such experiments and has reported good correspondence between biologic and numerical data. Nevertheless, one should consider that the bone chamber model does not correspond to the actual clinical situations due to several limitations. Despite allowing controlled loading experiments, more time is needed to allow bone fill into the chamber and the exerted shear forces are not directly transferred to the surrounding bone in the first weeks – as it is for implants hosted in native bone – but to the differentiating tissue. Moreover, as the bone chamber is a bone- and bone-marrow containing defect, fibrous tissue does not originate from an extrabony site like the periosteum or skin [115]. Beyond these issues, unlike the functionally adapted bone tissue, the bone tissue regenerated in the bone chamber does not compisre cortical and trabecular bone, and the site-specific mechanical properties of the "bone" in the chamber has not been thoroughly investigated. As an extension of an animal study, Jaecques et al. [41] created individualized μCT-based FE models of the guinea pig tibia with implants. They assigned different elastic modulus for different time intervals to study peri-implant stress/strain. Individualized μCT-based FE models of implants placed in animals and humans have also been constructed [42,118]. In an animal model, Duyck et al. [20] studied the biological, radiological and histomorphometric consequences of high insertion torque for placement of implants. In one minipig and two rabbit tibia, strain-gauges were bonded to measure strain amplitudes at the in vitro level. While the gauge in the minipig was relatively close to the neck of the implant, the rosette gauge on the rabbit tibia was placed between the implants at a distance that is not close to the implant neck. The authors, found that the strain levels in the minipig model was approaching to 4000-5000 με for the experimental implant, while strains less than 1000 με was recorded in the tibia. The strain data of this experiment seems rather questionable, because the closer the gauge to the neck of the implant, the higher the strains will be. In addition, the area of the measuring grid of the rosette gauge on the tibia is larger, which affects the outcome.

3.3. Theoretical Models with Tissue Differentiation or Bone Remodeling Theories

In this approach, there is not any animal or human experiment conducted in conjunction with the numeric part of the study. The literature contains an abundance of animal and clinical data that could be used for numerical coherence. In this sense, the numeric results can be compared with "available" animal and clinical data. Therefore, this approach attempts to "individualize" the experimental animal/human model of a separate study to eliminate the need for validation experiments. This author believes that this approach may have several reliable applications in the future, if highly-controlled animal/human experiments are conducted to provide in vivo biomechanical data. Let us give some examples regarding this method.

Amor and co-workers [119] implemented the mathematical model of Bailon-Plaza and van der Meulen [120], to simulate peri-implant healing around an endosseous oral implant. They attempted to correlate their time-dependent numerical findings to biological observations reported from highly-controlled animal experiments [121,122]. The peri-implant healing model developed by Berglundh et al. [121] used different species (Labrador dogs versus mice) and anatomical location (mandible versus tibia) than the fracture healing model, for which the mathematical model was originally developed [119]. The authors stated that, inter-individual and species-specific variations in some parameters might affect the numerical outcome. In addition, the amount of blood loss during haemorrhage, and the type of the healing process in peri-implant and fracture, primary healing and secondary healing respectively, are also different, and more importantly, the presence of a biomaterial, a key factor influencing the healing mechanisms, has to be considered in this approach. Akca et al. [123] used a different approach to analyze peri-implant stresses at different healing stages. They changed the mechanical properties of peri-implant elements in the numeric model to adjust for the differentiating tissue and corroborated their findings to the outcome of similar animal data. Mellal et al. [94], Eser et al. [85,86], and to some extent Chou et al.[124] created finite element models and implemented bone remodeling theories (including the lazy zone) to study time-dependent changes in stresses around load-carrying implants. To compare and corroborate their findings, a literature search was undertaken and available in vivo data was used. As a concluding remark of this chapter we may take into consideration the statements by Mellal et al.[94]:

> " ...The models applied in the present study are descriptive models. They are devoid of biologically based parameters. This issue was raised in a debate between John Currey and Rik Huiskes. While Currey argued that any model should be confirmed by biological data [125], Huiskes contended that designing a numerical model was a legitimate endeavor *per se*. Support data could (or could not) be produced at a later stage, there by validating the model *a posteriori* [126]. In this debate, Currey's standpoint essentially sets objectives for the future, while Huiskes' view appears to be more realistic. Still, models that are entirely based on biological parameters should be our prospective goal."

REFERENCES

[1] Frost, H.M.. (1983) A determinant of bone architecture. The minimum effective strain. *Clinical Orthopedics*,175,286-292.

[2] Frost, H.M. (1987) Bone "mass" and the "mechanostat" a proposal. *Anatomical Records*, 219,1-9.

[3] Frost, H.M. (1990) Skeletal structural adaptations to mechanical usage (SATMU): 1. redefining Wolff-s law: The bone modelling problem. *Anatomical Records*, 226, 403-413.

[4] Fanuscu, M.I., Iida, K., Caputo, A.A. and Nishimura, R.D. (2003) Load transfer by an implant in a sinus-grafted maxillary model. *International Journal of Oral and Maxillofacial Implants*,18,667-74.

[5] Cehreli, M., Duyck, J., De Cooman, M., Puers, R. and Naert, I. (2004) Implant design and interface force transfer: A photoelastic and strain-gauge analysis. *Clinical Oral Implants Research*,15,249-257.

[6] Wedendal, P.R. and Bjelkhagen, H.I. (1974) Dental holographic interferometry in vivo utilizing a ruby laser system. I. Introduction and development of methods for precision measurements on the functional dynamics of human teeth and prosthodontic appliances. *Acta Odontologica Scandinavia*,32,131-45.

[7] Goldstein, G.R., Wesson, A., Schweitzer, K. and Cutler, B. (1992) Flexion characteristics of four-unit fixed partial denture frameworks using holographic interferometry. *Journal of Prosthetic Dentistry*,67,609-13.

[8] Farah, J.W. and Craig, R.G. (1971) Reflection photoelastic stress analysis of a dental bridge. *Journal of Dental Research*,50,1253-9.

[9] Pezzoli, M., Rossetto, M. and Calderale, P.M. (1986) Evaluation of load transmission by distal-extension removable partial dentures by using reflection photoelasticity. *Journal of Prosthetic Dentistry*,56,329-37.

[10] Millington, N.D. and Leung, T. (1995) Inaccurate fit of implant superstructures. Part 1: Stresses generated on the superstructure relative to the size of fit discrepancy. *International Journal of Prosthodontics*,8,511-6.

[11] Fernandes, C.P., Glantz, P.O., Svensson, S.A. and Bergmark, A. (2003) Reflection photoelasticity: a new method for studies of clinical mechanics in prosthetic dentistry. *Dental Material*,19,106-17.

[12] Akca, K. and Cehreli, M.C. (2006) Biomechanical consequences of progressive marginal bone loss around oral implants: A finite element stress analysis. *Medical and Biological Engineering and Computing*,44,527-35.

[13] Thompson, D.W. In: Bonner JT, Editor. *On Growth and Form*. Cambridge: Cambridge University; 1961; 238.

[14] Bertram, J.E. and Swartz, S.M. (1991) The "Law of bone transformation": A case of crying Wolff? *Critical Reviews in Oral Biology and Medicine*,66,245-73.

[15] Lanyon, L.E. and Smith, R.N. (1970) Bone strain in the tibia during normal quadrupedal locomotion. *Acta Orthopedica Scandinavia*,41,238-248.

[16] Lanyon, L.E. (1976) The measurement of bone strain in vivo. *Acta Orthopaedica Belgica*, 42,98-108.

[17] Szivek, J.A. and Gharpuray, V.M. Strain gauge measurements from bone surfaces. In: Yuehuei, H., Draughn, R.A. *Mechanical Testing of Bone and the Bone-Implant Interface* CRC Press, 2000;20:305-318.

[18] Cehreli, M.C., Akkocaoglu, M., Comert, A., Tekdemir, I. and Akca, K. (2005) Human ex vivo bone tissue strains around natural teeth vs. immediate oral implants. *Clinical Oral Implants Research*,16,540-8.

[19] Cehreli, M.C., Comert, A., Akkocaoglu, M., Tekdemir, I. and Akca, K. (2006) Towards the limit of quantifying low-amplitude strains on bone and in coagulum around immediately loaded oral implants in extraction sockets. *Medical and Biological Engineering and Computing*,44,86-94.

[20] Duyck, J., Corpas, L., Vermeiren, S., Ogawa, T., Quirynen, M., Vandamme, K., Jacobs, R. and Naert, I. (2010) Histological, histomorphometrical, and radiological evaluation of an experimental implant design with a high insertion torque. *Clinical Oral Implants Research*,21,877-84.

[21] Lambert, K. (1971) The weight-bearing function of the fibula. *Journal of Bone and Joint Surgery*,53A,507-13.
[22] Steigman, S., Boym, K., Weinreb, M. and Michaeli Y. (1993) Dynamics of tissue changes found after mechanical load-ing of the rat incisor. II. A three dimensional longitudinal study of the histopathologic aspects. *American Journal of Orthodontics and Dentofacial Orthopedics,* 104,492-505.
[23] Wright, T.M. and Hayes, W.C. (1979) Strain gage application on compact bone. *Journal of Biomechanics*,12,471-5.
[24] Asundi, A. and Kishen, A. (2000) A strain gauge and photoelastic analysis of in vivo strain and in vitro stress distribution in human dental supporting structures. *Archives of Oral Biology*,45,543-50.
[25] Page, A.E., Allan, C., Jasty, M., Harrigan, T.P., Bragdon, C.R. and Harris, W.H.(1993) Determination of loading parameters in the canine hip in vivo. *Journal of Biomechanics*,26,571-9.
[26] Burr, D.B., Milgrom, C., Fyhrie, D., Forwood, M., Nyska, M., Finestone, A., Hoshaw, S., Saiag, E. and Simkin, A. (1996) In vivo measurement of human tibial strains during vigorous activity. *Bone*,18,405-10.
[27] Szivek, J.A., Gealer, R.G., Magee, F.P. and Emmanual, J. (1990) Preliminary development of a hydroxyapatite-backed strain gauge. *Journal of Applied Biomaterials*,1,241-8.
[28] Nicolella, D.P., Bonewald, L.F., Moravits, D.E. and Lankford, J. (2005) Measurement of microstructural strain in cortical bone. *European Journal of Morphology*,42,23-9.
[29] Nicolella, D.P., Moravits, D.E., Gale, A.M., Bonewald, L.F. and Lankford, J. (2006) Osteocyte lacunae tissue strain in cortical bone. *Journal of Biomechanics*,39,1735-43.
[30] Bay, B.K. (1995) Texture correlation: a method for the measurement of detailed strain distributions within trabecular bone. *Journal of Orthopedic Research*,13,258-267.
[31] Bay, B.K., Smith, T.S., Fyhrie, D.P. and Saad, M. (1999) Digital volume correlation: three-dimensional strain mapping using X-ray tomography. *Experimental Mechanics*,39,17–226.
[32] Brekelmans, W.A., Poort, H.W. and Slooff, T.J. (1972) A new method to analyse the mechanical behaviour of skeletal parts. *Acta Orthopedica Scandinavia*,43,301-317.
[33] Eswaran, S.K., Bayraktar, H.H., Adams, M.F., Gupta, A., Hoffmann, P.F., Lee, D.C., Papadopoulos, P. and Keaveny, T.M. (2007) The micromechanics of cortical shell removal in the human vertebral body. *Computer Methods in Applied Mechanical Engineering*,196,3025–3032.
[34] Wintera, W., Heckmann, S.M. and Weber, H.P. (2004) A time-dependent healing function for immediate loaded implants. *Journal of Biomechanics*,37,1861-1867.
[35] Chou, H.-Y., Jagodnik, J.J. and Muftu, S. (2008) Predictions of bone remodeling around dental implant systems. *Journal of Biomechanics*,41,1365-1373.
[36] Wovern, N.V. and Stolze, K. (1978) Sex and age differences in bone morphology of mandibles. *Scandinavian Journal of Dental Research*,86,478-485.
[37] Ulm, C.W., Kneissel, M., Hahn, M., Solar, P., Matejka, M. and Donath, K. (1997) Characteristics of the cancellous bone of edentulous mandibles. *Clinical Oral Implants Research*,8,125-130.
[38] Lettry, S., Seedhom, B.B., Berry, E. and Cuppone, M. (2003) Quality assessment of the cortical bone of the human mandible. *Bone*,32,35-44.

[39] Dechow, P.C., Nail, G.A., Schwartz-Dabney, C.L. and Ashman, R.B. (1993) Elastic properties of human supraorbital and mandibular bone. *American Journal of Physical Anthropology*,90,291-306.

[40] O'Mahony, A.M., Williams, J.L., Katz, J.O. and Spencer, P. (2000) Anisotropic elastic properties of cancellous bone from a human edentulous mandible. *Clinical Oral Implants Research*,11, 415–421.

[41] Jaecques, S.V., Van Oosterwyck, H., Muraru, L., Van Cleynenbreugel, T., De Smet, E., Wevers, M., Naert, I. and Vander Sloten, J. (2004) Individualised, micro CT-based finite element modelling as a tool for biomechanical analysis related to tissue engineering of bone. *Biomaterials*,25,1683-96.

[42] Limbert, G., van Lierde, C., Muraru, O.L., Walboomers, X.F., Frank, M., Hansson, S., Middleton, J. and Jaecques, S. (2010) Trabecular bone strains around a dental implant and associated micromotions--a micro-CT-based three-dimensional finite element study. *Journal of Biomechanics*,43,1251-61.

[43] Keyak, J.H., Rossi, S.A., Jones, K.A. and Skinner, H.B. (1998) Prediction of femoral fracture load using automated finite element modeling. *Journal of Biomechanics*,31.125–33.

[44] Akca, K., Cehreli, M.C. and Iplikcioglu, H. (2002) A comparison of three-dimensional finite element stress analysis with in vitro strain gauge measurements on dental implants. *International Journal of Prosthodontics*,15,115-21.

[45] Cehreli, M.C., Akca, K. and Iplikcioglu, H. (2004) Force transmission of one- and two-piece morse-taper oral implants: a nonlinear finite element analysis. *Clinical Oral Implants Research*,15,481-489.

[46] Kuhn, J.L., Goldstein, S.A., Feldkamp, L.A., Goulet, R.W. and Jesion, G. (1990) Evaluation of a microcomputed tomography system to study trabecular bone structure. *Journal of Orthopedic Research*,8,833-842.

[47] Muller, R., Van Campenhout, H. and Van Damme, B. (1998) Morphometric analysis of human bone biopsies: a quantitative structural comparison of histological sections and micro-computed tomography. *Bone*,23,59–66.

[48] Zannoni, C., Viceconti, M., Pierotti, L. and Cappello, A. (1998) Analysis of titanium induced CT artifacts in the development of biomechanical finite element models. *Medical Engineering and Physics*,20,653-9.

[49] Bernhardt, R., Scharnweber, D., Müller, B., Thurner, P., Schliephake, H., Wyss, P., Beckmann, F., Goebbels, J. and Worch, H. (2004) Comparison of microfocus- and synchrotron X-ray tomography for the analysis of osteointegration around Ti6Al4V implants. *European Cell Materials*,30,42–51.

[50] Butz, F., Ogawa, T., Chang, T.L. and Nishimura, I. (2006) Three-dimensional bone-implant integration profiling using micro-computed tomography. *International Journal of Oral and Maxillofacials Implants*,21,687–95.

[51] Jaecques, S.V.N., Muraru, L., Van Lierde, C., De Smet, E., Van Oosterywck, H., Wevers, M., Naert, I. and Vander Sloten, J. (2004) In vivo micro-CT-based FE models of guinea pigs with titanium implants: an STL-based approach. *International Congress Series*,1268,579- 583.

[52] Boyd, S.K. and Müller, R. (2006) Smooth surface meshing for automated finite element model generation from 3d image data. *Journal of Biomechanics*,39,1287-1295.

[53] Bardyn, T., Reyes, M., Larrea, X. and Büchler P. Influence of smoothing on voxel-based mesh accuracy in micro finite-element. In: Miller, K., Nielsen, P.M.F. *Computational Biomechanics for Medicine IV*. MICCAI Workshop Proceedings. 2009:78-86.

[54] Taddei, F., Martelli, S., Reggiani, B., Cristofolini, L. and Viceconti, M. (2006) Finite-element modeling of bones from CT data: sensitivity to geometry and material uncertainties. *IEEE Transactions of Biomedical Engineering*,53,2194-200.

[55] Viceconti, M., Davinelli, M., Taddei, F. and Cappello, A. (2004) Automatic generation of accurate subject-specific bone finite element models to be used in clinical studies. *Journal of Biomechanics*,37,1597-605.

[56] Keyak, J.H., Kaneko, T.S., Tehranzadeh, J. and Skinner, H.B. (2005) Predicting proximal femoral strength using structural engineering models. *Clinical Orthopedics and Related Research*,437,219-228.

[57] Keyak, J.H. (2001) Improved prediction of proximal femoral fracture load using nonlinear finite element models. *Medical Engineering and Physics*,23,165-173.

[58] Lotz, J.C., Cheal, E.J. and Hayes, W.C. (1991) Fracture prediction for the proximal femur using finite element models: Part I-Linear analysis. *Journal of Biomechanical Engineering*,113,353-360.

[59] Lotz, J.C., Cheal, E.J. and Hayes, W.C (1991) Fracture prediction for the proximal femur using finite element models: Part II-Nonlinear analysis. *Journal of Biomechanical Engineering*,113,361-365.

[60] Ota, T., Yamamoto, I., Morita, R. (1999) Fracture simulation of the femoral bone using the finite-element method: how a fracture initiates and proceeds. *Journal of Bone and Mineral Metabolism*,17,108-112.

[61] Gomez-Benito, M.J., Garcia-Aznar, J.M. and Doblare, M. (2005) Finite element prediction of proximal femoral fracture patterns under different loads. *Journal of Biomechanical Engineering*,127, 9-14.

[62] Haraldson, T., Carlsson, G.E. and Ingervall, B. (1979) Functional state, bite force and postural muscle activity in patients with osseointegrated oral implant bridges. Acta *Odontologica Scandinavia*,37,195-206.

[63] Carlsson, G.E. and Lindquist, L.W. (1994) Ten-year longitudinal study of masticatory function in edentulous patients treated with fixed complete dentures on osseointegrated implants. *International Journal of Prosthodontics*,7,448-53.

[64] Jemt, T. and Carlsson, G.E. (1986) Aspects of mastication with bridges on osseointegrated implants. *Scandinavian Journal of Dental Research*,94,66-71.

[65] Haraldson, T. and Carlsson, G.E. (1977) Bite force and oral function in patients with osseointegrated oral implants. *Scandinavian Journal of Dental Research*,85,200-8.

[66] Carr, A.B. and Laney, W.R. (1987) Maximum occlusal force levels in patients with osseointegrated oral implant prosthesis and patients with complete dentures. *International Journal of Oral and Maxillofacial Implants*,2,101-8.

[67] Mericske-Stern, R. and Zarb, G.A. (1996) In vivo measurements of some functional aspects with mandibular fixed prostheses supported by implants. *Clinical Oral Implants Research*,7,153-61.

[68] Keyak, J.H. and Falkinstein, Y. (2003) Comparison of in situ and in vitro CT scan-based finite element model predictions of proximal femoral fracture load. *Medical Engineering and Physics*,25,781-787.

[69] Weinans, H., Sumner, D.R., Igloria, R. and Natarajan, R.N. (2000) Sensitivity of periprosthetic stress-shielding to load and the bone density–modulus relationship in subject-specific finite element models. *Journal of Biomechanics*,33,809-817.

[70] Huiskes, R., Ruimerman, R., van Lenthe, G.H., Janssen, J.D. (2000) Effects of mechanical forces on maintenance and adaptation of form in trabecular bone. *Nature*,405,704-706.

[71] Prendergast, P.J., van der Meulen, M.C.H. Mechanics of bone regeneration. In *Bone Mechanics Handbook*. (Cowen, S.C., Ed.) CRC Press: Boca Raton,2001. Chapter 32, pp. 32.1-32.19.

[72] Lacroix, D. and Prendergast, P.J. (2002) A mechanoregulation model for tissue differentiation during fracture healing: analysis of gap size and loading. *Journal of Biomechanics*,35,1163-1171.

[73] Geris, L., Andreykiv, A., Van Oosterwyck, H., Sloten, J.V., van Keulen, F., Duyck, J. and Naert, I. (2004) Numerical simulation of tissue differentiation around loaded titanium implants in a bone chamber. *Journal of Biomechanics*,37,763-769.

[74] Prendergast, P.J. (2007) Computational modelling of cell and tissue mechanoresponsiveness. *Gravitational and Space Biology*,20,43-50.

[75] Eser, A., Tonuk, E., Akca, K. and Cehreli, M.C. (2010) Predicting time-dependent remodeling of bone around immediately-loaded dental implants with different designs. *Medical Engineering and Physics*,32,22-31.

[76] Eser, A., Tonuk, E., Akca, K. and Cehreli, M.C. (2009) Numeric simulation of time-dependent remodeling of bone around loaded oral implants. *International Journal of Oral and Maxillofacial Implants*,24,597-608.

[77] Hart, R.T. Bone modeling and remodeling theories and computation. In: Cowin, S.C. (Ed.), *Bone Mechanics Handbook*, second ed. CRC Press, Boca Raton, FL, 2001;pp. 31-1–31-42.

[78] Huiskes, R., Weinans, H., Grootenboer, H.J., Dalstra, M., Fudala, B. and Sloof, T.J. (1987) Adaptive bone-remodeling theory applied to prosthetic-design analysis. *Journal of Biomechanics*,20,1135-1150.

[79] Weinans, H., Huiskes, R. and Grootenboer, H.J. (1989) Convergence and uniqueness of adaptive bone remodeling. *Transactions of the 35th Annual Meeting of the Orthopaedic Research Society*, 310.

[80] Weinans, H., Huiskes, R. and Grootenboer, H.J. (1992) The behaviour of bone-remodeling simulation models. *Journal of Biomechanics*,25,1425-1441.

[81] Carter, D.R., Orr, T.E., Pyhrie, D.P. (1989) Relationships between loading history and femoral cancellous bone architecture. *Journal of Biomechanics*,22,231-244.

[82] Beaupre, G.S., Orr, T.E. and Carter, D.R. (1991) An approach for time-dependent bone modeling and remodeling-theoretical development. *Journal of Orthopedic Research*,8,551-651.

[83] Beaupre, G.S., Orr, T.E. and Carter, D.R. (1991) An approach for time-dependent bone modeling and remodeling-application: a preliminary remodeling simulation. *Journal of Orthopedic Research*,8,662-670.

[84] Mellal, A., Wiskott, H.W., Botsis, J., Scherrer, S.S. and Belser, U.C. (2004) Stimulating effect of implant loading on surrounding bone. Comparison of three numerical models and validation by in vivo data. *Clinical Oral Implants Research*,15,239-248.

[85] Carter, D.R., Fyhrie, D.P. and Whalen, R.T. (1987) Trabecular bone-density and loading history. Regulation of connective-tissue biology by mechanical energy. *Journal of Biomechanics*,20,785-94.

[86] Lin, D., Li, Q., Li, W. and Swain, M. (2009) Dental implant induced bone remodeling and associated algorithms. *Journal of Mechanical Behavior of Biomedical Materials*,2,410-32.

[87] Van Oosterwyck, H., Vander Sloten, J., Puers, R. and Naert, I. (2002) Finite element studies on the role of mechanical loading in bone response around oral implants. *Meccanica*,37, 441-451.

[88] Crupi, V., Guglielmino, E., Rosa, G.L., Sloten, J.V. and Oosterwyck, H.V. (2004) Numerical analysis of bone adaptation around an oral implant due to overload stress. *Proceedings Institution of Mechanical Engineers Part H: Journal of Engineering Medicine.*

[89] Li, I., Li, H., Shi, L., Fok, A.S.L., Ucer, C., Deulin, H., Horner, K. and Silikas, N. (2007) A mathematical model for simulating the bone remodeling process under mechanical stimulus. *Dental Materials*, 23,1073-1078.

[90] Brosh, T., Pilo, R. and Sudai, D. (1998) The influence of abutment angulation on strains and stresses along the implant/bone interface: comparison between two experimental techniques. *Journal of Prosthetics Dentistry*,79,328-34.

[91] Clelland, N.L., Gilat, A., McGlumphy, E.A. and Brantley, W.A. (1993) A photoelastic and strain gauge analysis of angled abutments for an implant system. *International Journal of Oral and Maxillofacial Implants*,8,541-8.

[92] Keyak, J.H., Fourkas, M.G., Meagher, J.M. and Skinner, H.B. (1993) Validation of an automated method of three-dimensional finite element modelling of bone. Journal of Biomedical Engineering,15,505-509.

[93] Baiamonte, T., Abbate, M.F., Pizzarello, F., Lozada, J. and James R. (1996) The experimental verification of the efficacy of finite element modelling to dental implant systems. *Journal of Oral Implantology*,22,104-1996.

[94] Iplikcioglu, H., Akca, K., Cehreli, M.C. and Sahin, S. (2003) Comparison of non-linear finite element analysis with in vitro strain-gauge measurements on a morse-taper implant. *International Journal of Oral and Maxillofacial Implants*,18,258-265.

[95] Eser, A., Akca, K., Eckert, S. and Cehreli, M.C. (2009) Non-linear finite element analysis versus *ex vivo* strain-gauge measurements on immediately-loaded implants. *International Journal of Oral and Maxillofacial Implants*,24,439-46.

[96] Heckmann, S.M., Karl, M., Wichmann, M.G., Winter, W., Graef, F. and Taylor, T.D. (2006) Loading of bone surrounding implants through three-unit fixed partial denture fixation: a finite-element analysis based on in vitro and in vivo strain measurements. *Clinical Oral Implants Research*,17,345-350.

[97] Heckmann, S.M., Karl, M., Wichmann, M.G., Winter, W., Graef, F. and Taylor, T.D. (2004) Cement fixation and screw retention: parameters of passive fit – an in vitro study of three-unit implant-supported fixed partial dentures. *Clinical Oral Implants Research*,15,466-475.

[98] Ichim, I., Kieser, J.A. and Swain, M.V. (2007) Functional significance of strain distribution in the human mandible under masticatory load: numerical predictions. *Archives of Oral Biology*,52,465-73.

[99] Lertchirakarn, V., Palamara, J.E. and Messer, H.H. (2003) Finite element analysis and strain-gauge studies of vertical root fracture. *Journal of Endodontics*,29,529-34.

[100] 110. Knoell, A.C. (1977) A mathematical model of an in-vitro human mandible. *Journal of Biomechanics*,10,159-166.

[101] Tajima, K., Chen, K.K., Takahashi, N., Noda, N., Nagamatsu, Y., Kakigawa, H. (2009) Three-dimensional finite element modeling from CT images of tooth and its validation. *Dental Materials Journal*,28,219-26.

[102] Vollmer, D., Meyer, U., Joos, U., Vègh, A. and Piffko, J. Experimental and finite element study of a human mandible. *Journal of Craniomaxillofacial Surgery*,28,91-6.

[103] Al-Sukhun, J., Kelleway, J. and Helenius, M. (2006) Development of a three-dimensional finite element model of a human mandible containing endosseous dental implants. I. Mathematical validation and experimental verification. *Journal of Biomedical Materials Research A*,80,:234-46.

[104] Gracco, A., Cirignaco, A., Cozzani, M., Boccaccio, A., Pappalettere, C., Vitale, G. (2009) Numerical/experimental analysis of the stress field around miniscrews for orthodontic anchorage. *European Journal of Orthodontics*,31,12-20.

[105] Vandamme, K., Naert, I., Geris, L., Vander Sloten, J., Puers, R. and Duyck, J. (2007) Histodynamics of bone tissue formation around immediately loaded cylindrical implants in the rabbit. *Clinical Oral Implants Research*,18, 471-480.

[106] Qin, Y.X., McLeod, K.J., Guilak, F., Chiang, F.P. and Rubin, C.T. (1996) Correlation of bony ingrowth to the distribution of stress and strain parameters surrounding a porous-coated implant. *Journal of Orthopedic Research*,14,862-70.

[107] Cehreli, M., Sahin, S. and Akça, K. (2004) Role of mechanical environment and implant design on bone tissue differentiation: current knowledge and future contexts. *Journal of Dentistry*,32,123-32.

[108] Matsunaga, S., Shirakura, Y., Ohashi, T., Nakahara, K., Tamatsu, Y., Takano, N. and Ide, Y. (2010) Biomechanical role of peri-implant cancellous bone architecture. *International Journal of Prosthodontics*,23,333-8.

[109] Amor, N. , Geris, L. , Vander Sloten, J. and Van Oosterwyck, H.(2009) Modelling the early phases of bone regeneration around an endosseous oral implant. *Computer Methods in Biomechanics and Biomedical Engineering*,12,459 -468.

[110] Bailon Plaza, A., van der Meulen, M.C.H. (2001) A mathematical framework to study the effects of growth factor influences on fracture healing. *Journal of Theoretical Biology*,212,191-209.

[111] Berglundh, T., Abrahamsson, I., Lang, N. and Lindhe, J. (2003) De novo alveolar bone formation adjacent to endosseous implants: a model in the dog. *Clinical Oral Implants Research*,14,251–262.

[112] Abrahamsson, I., Berglundh, T., Linder, E., Lang, N.P., Lindhe, J. (2004) Early bone formation adjacent to rough and turned endosseous implant surfaces. An experimental study in the dog. *Clinical Oral Implants Research*,15,381-92.

[113] Akça, .K, Eser, A. and Canay, S. (2010) Numerical simulation of the effect of time-to-loading on peri-implant bone. *Medical Engineering and Physics*,32,7-13.

[114] Chou H.-Y., Jagodnik J.J. and Muftu, S. (2008) Predictions of bone remodeling around dental implant systems. *Journal of Biomechanics*, 41, 1365-1373.

[115] Currey, J.D. (1995) The validation of algorithms used to explain adaptive remodeling in bone. In: Odgaard, A. and Weinans, H., eds. *Bone structure and remodeling*, Singapore: World Scientific.9-13.

[116] Huiskes, R. (1995) The law of adaptive bone remodeling: a case for crying Newton? In: Odgaard, A. and Weinans, H., eds. *Bone structure and remodeling*, Singapore: World Scientific.15-24.

In: Biomechanics of Dental Implants
Editor: Murat Çehreli
ISBN 978-1-62100-780-7

Chapter 11

Treatment Planning for Implant-Supported Fixed and Removable Prostheses

Regina Mericske-Stern

1. Introduction

The placement of implants is a reconstructive, preprosthetic surgical intervention, and therefore different from most goals in oral surgery as there are: extraction of decayed teeth, treating an infection, removing a pathological process from soft or hard tissues. Thus, implants are part of the final prosthetic-reconstructive treatment, which encompasses a functional, esthetic and social rehabilitation. The patient's functional status, oral conditions, interjaw relation and the correct assessment of individual needs determine the goal of implant-prosthetic treatment. Dentists tend to base the selection of the prosthetic design on the number of implants that can be placed meaning that such implant dentistry is prevalently bone driven. Other criteria have to be considered: esthetic appearance, facial morphology, lost hard and soft tissues, biomechanics such as stability of the prostheses, complications and adjustments required. A variety of prosthetic solutions are available to restore the completely edentulous jaw. More recently specific computer assisted methods have been developed such as computer guided planning and surgery, CAD CAM fabrication of prostheses and processing of new materials like high strength ceramics. This evolution of technologies will translate into a better predictability of treatment outcomes. It enables processing of biologically well accepted materials and simultaneously will enhance more uniform quality and passive fit of the prosthesis.

2. Development of Treatment Concepts for the Edentulous Jaw

Implants have changed prosthodontics more than any other innovation in dentistry Replacement of teeth with removable prosthesis in the edentulous jaw is considered to be a" restitutio ad similem", while Implants may provide a feeling of "restitutio ad integrum".

Implant prosthodontics means restoring function, esthetics, and providing technology; biology and technology are combined. Branemark and co-workers seminal work had one primary goal: to restore the edentulous jaw by means of fixed prostheses supported by "titanium roots". This should approach what is meant by "restitutio ad integrum". This treatment goal could be achieved more often in the mandible than in the maxilla for several reasons. Within this treatment concept the overdenture prosthesis was not considered a viable solution, and was regarded as inferior to fixed prostheses. The systematic and scientific approach of developing the concept of osseointegration and implant dentistry is documented in the first comprehensive overview in 1985 which represented the Gold standard at that time for this field of dentistry. The book chapters, particularly describing the treatment of the edentulous maxilla, reveal an insufficient understanding of planning of proper implant placement related to the prostheses. Likewise, it is obvious that well developed laboratory technologies were not available at this time and that there was a lack of sophisticated planning tools. Further, it must be considered that adjunctive techniques to optimize implant position and improve the implant bed (e.g. local bone regeneration, local grafts, sinus floor elevation and complete grafts) were not yet known or practiced and have yet not been considered. The technical aspects of retention (screw retention) and fabrication of prostheses in the early period of implant dentistry posed many problems. Passive fit was difficult to achieve, particularly with large frameworks of precious or non-precious alloys. The selection of secondary components (abutments, analogs, screws, impression copings and healing caps) was limited and was subjected to frequent complications. Later implant companies introduced new implant designs and changed the connection between implant and superstructure, offering a variety of components. This facilitated fabrication of prostheses and but still did not always fulfill the demands of prosthetic designs. It also led to an abundance of components and was confusing. This has significantly changed in the past years. Gradually, due to the frequent complexity of providing fixed prostheses, implant overdentures were introduced. First clinical reports and case series in the late eighties presented successful results with a reduced number of implants– mostly 2 implants – for mandibular overdentures. Early studies reported a high failure rate (biological failures) for maxillary overdentures, which is explained by the specific indication given for this treatment modality; i.e. compromised situations with osteoporotic bone and advanced jaw atrophy. This allowed for placement of only a limited number of small sized implants. Thus, overdenture treatment resulted in a selection of risk patients who often had experienced failures with fixed prostheses and therefore received overdentures supported by the remaining implants. Increased success was achieved with careful patient selection and technical protocols of overdenture fabrication in various studies. Today virtual implant planning and additional surgical techniques allow for optimized prosthetically driven implant placement and implant rehabilitation. Over time the clinical experience in practicing implant dentistry increased, particularly when practicing prosthodontics and resulted in an improved understanding of prosthetically-driven implant planning and placement.

A more recent breakthrough in implant-prosthetic technology has been the computer assisted analysis, virtual planning and CAD CAM technology. Biological research laid the scientific basis for the concept of osseointegration in the late 1960s and 1970s. The establishment of prosthetic concepts and technological development followed in the 1980s and 1990s. Since then a rapid and broad evolution in implant dentistry has occurred with an exponential increase in publications (Figure 1).

Only a few studies have evaluated the benefits of implant treatment on the basis of controlled randomized clinical trials. Prosthetic methods in general and related to implants are not evidence based. They rely on clinical experiences, patients' demands, technical considerations and reports on maintenance. Biomechanical aspects of implant prostheses to be considered for the edentulous jaw are:

- fixed or removable
- number / distribution of implants and retention / fixation of prostheses
- passive fit, cementation and screw retention
- choice of material, design of prostheses
- occlusal scheme

While some of these topics can be disputed on the basis of relevant literature, for others we miss evidence. However, from clinical experience, well-designed clinical concepts have evolved and the benefit of the patients concerned appears to be high and obvious.

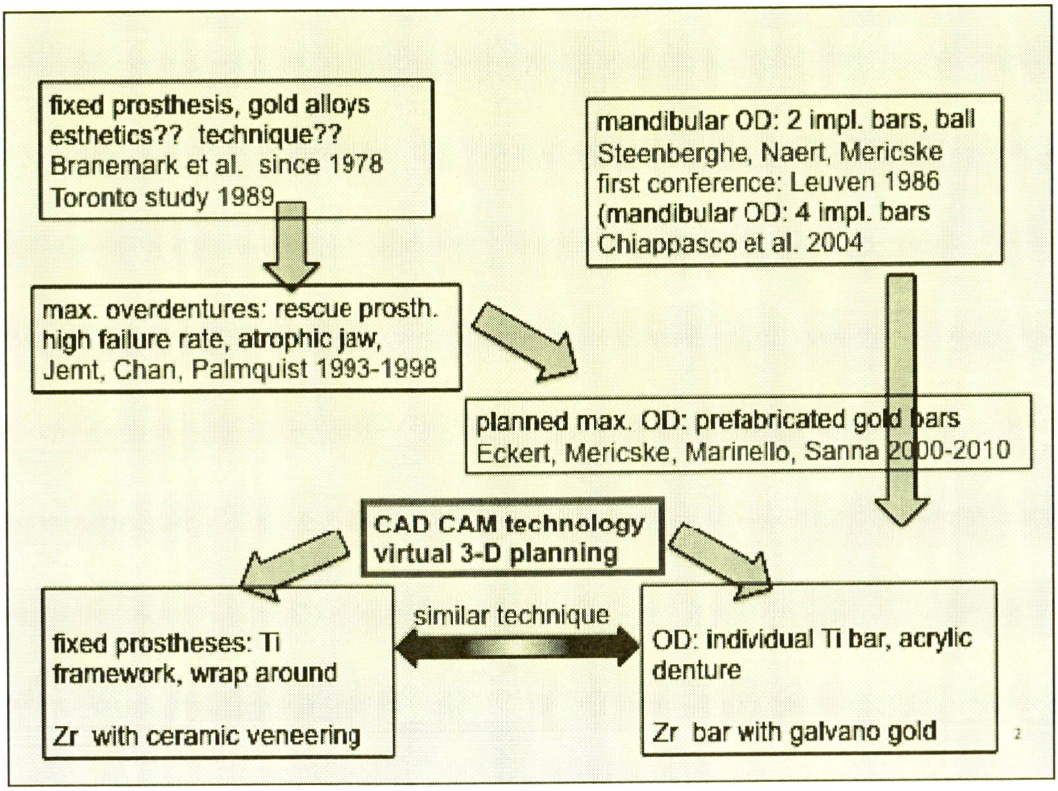

Figure 1. Overview on treatment concepts for the edentulous jaw from 1980 to 2010. Modern CAD CAM technology individualized and simplified the fabrication of frameworks and bars for the edentulous jaw. The technology is comparable for both types of prosthesis, fixed and overdentures.

3. Aspects of Treatment Planning

Goals of implant rehabilitation are: restoring function comprising mastication and phonation, restoring esthetic appearance including aspects of facial morphology, providing the best emergence profile with respect to the planned prosthetic design and to enable good oral hygiene (Figure 2).

Careful, accurate prosthetically driven implant planning today must be the first step to achieve these goals. A frequently disputed question is, which prosthetic design and technology should be selected; and whether fixed or removable prostheses should be recommended for patients with edentulous jaws. In general, the following criteria will determine the treatment planning of the edentulous jaw:

- The prosthetic design will depend on the distribution of the implants over the arch, their location and their number and vice versa
- The natural dentition or type of prosthesis in the opposing jaw will influence design considerations
- The intermaxillary relationship has to be considered
- The occlusal scheme is influenced by all these factors
- Esthetic and morphological aspects have to be involved

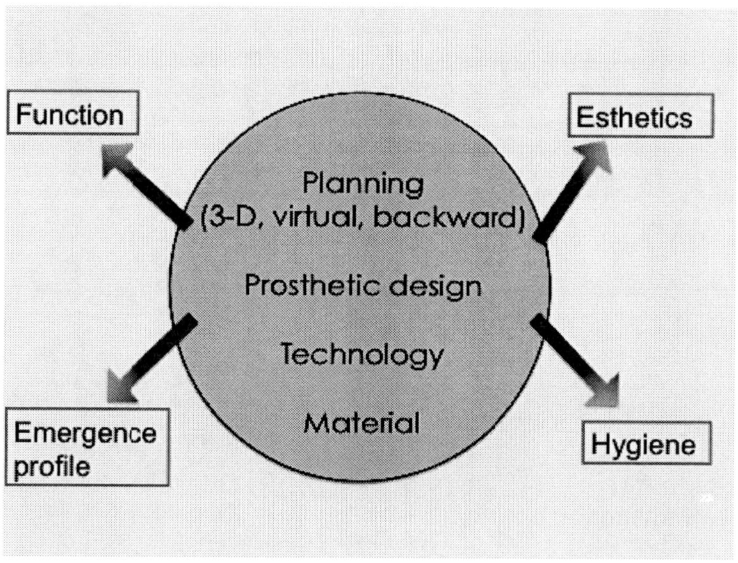

Figure 2. components of treatment planning and outcomes of implant prosthetic rehabilitation.

Prosthodontic treatment of the edentulous patient or jaw with implants is demanding, because it usually implies a complete rehabilitation of oral structures and facial morphology. A preliminary overview on treatment outcomes in the edentulous jaw obtained from the recent literature is given in Table 1.

Treatment planning is the first and most important step in clinical implantology with respect to surgical, prosthodontic and laboratory procedures. There is a broad range of planning requirements meaning that the planning tools and documents may differ significantly, according to minimum or maximum solutions. While restoring the mandible often offers a broad range of options - fixed prostheses with different design, removable prostheses with different attachment and retention systems - the maxilla is much more demanding, requires more analysis and planning steps and may be limited in the options. One single implant may stabilize a mandibular overdenture, while up to 10 implants are suggested for fixation of prosthesis in the edentulous maxilla. This often comprises elective surgical procedures like SFE and local GBR to allow for the placement of multiple implants.

A clinical extra- and intraoral examination of the patient and casts are always required, which provides the clinician a first perception of the space and interjaw relation for implant-prosthetic rehabilitation. Well-fitting, well-designed existing dentures can directly be used for planning and presurgical analysis. Or, a new setup has to be made if extended implant overdentures and/or fixed bridgework are planned. The orientation index obtained from the tooth setup is utilized to analyze dimensions of tissue volume and to compare the prospective implant and tooth axis (Figure 3).

Table 1. Parameters and outcomes of implant prostheses in the edentulous jaw

Parameter	Outcome
Survival of implants	
longevity, edent. jaw	mand > max implants
	Fixed p. > OD
immediate implants	interforaminal safe (good evidence)
Physiological impact	
mastication, nutrition	mand OD > CD
biting forces	mand OD > CD
bone remodeling	not clear
phonation	edentulous maxilla OD > fixed
Psychological impact	
satisfaction, well being, QoL	mand OD > CD
	long bar, single attachments: similar
	max. OD almost equal to fixed prosthesis
handling, hygiene	individual preferences
	OD > fixed in elderly patients
esthetics	maxilla: OD provides better lip support
Cost	
direct	implants > natural teeth
	OD > CD
	OD < fixed prosthesis
indirect (service, maintenance)	fixed < OD, not clear
	ball anchors > bars> rigid bars
investment of time	OD < fixed, not clear

> better or more frequent; < less frequent; OD: overdenture, CD: Complete denture.

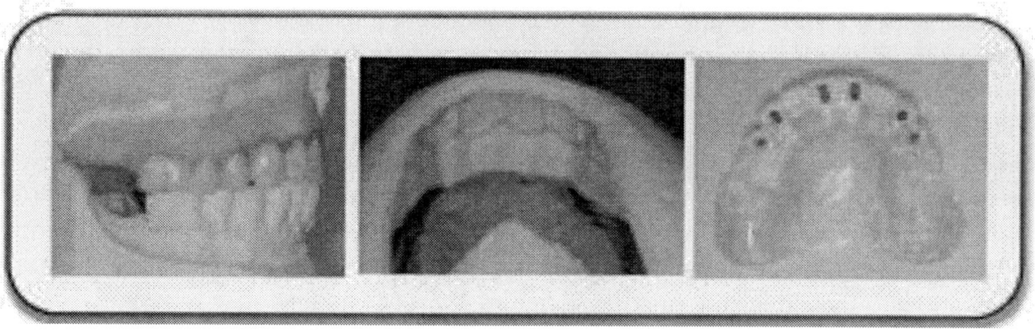

Figure 3. Analysis of dimensions. Tooth setup: space and relation between cast (jaw) and tooth setup (prospective prosthesis) by means of analog method is analyzed. An orientation index of the tooth position is fabricated from silicon material. The tooth setup can be converted into an acrylic splint for analog or digital radiographic analysis.

The tooth setup is further used to fabricate templates with markers for radiographs, to fabricate surgical guides, and to fabricate temporary dentures if necessary. For a long time period panoramic radiographs and 2-D analysis of CT scan were used together with radiographic templates and markers to determine the prospective implant position, altogether leading to good treatment outcomes. Modern computer software today has a great impact on planning in implant dentistry. Along with computer assisted 3-D analysis a backward planning was introduced with high predictability of the envisaged therapeutic goal. This comprises optimized implant axes, choice of retention and fixation, of design and material for the prostheses (Table 2).

Table 2. Treatment planning: tools and documents

Documents / Parameters	Overdenture		Bridgework	
	Mandible	*Maxilla*	*Mandible*	*Maxilla*
Minimum, simple conditions				
Intraoral diagnosis	++	++	++	++
Casts alone	++	+	-	-
Casts with setup	---	++	++	+++
OPT with markers	+	++	++	++
Complex situations				
Intraoral diagnosis	++	++	++	++
Mounted casts with set up	++	++	+++	+++
Virtual implant planning	+	++	++	+++
Surgical guide	+	++	+++	+++

– not necessary; + useful; +++ recommended, strongly recommended or necessary.

According to the complexity of the case, the parameters and planning tools are selected.

Planning of the edentulous maxilla with fixed prosthesis requires the most in-depth analysis. Today, a computer assisted 3-D backward planning should be considered whenever possible (Figure 4).

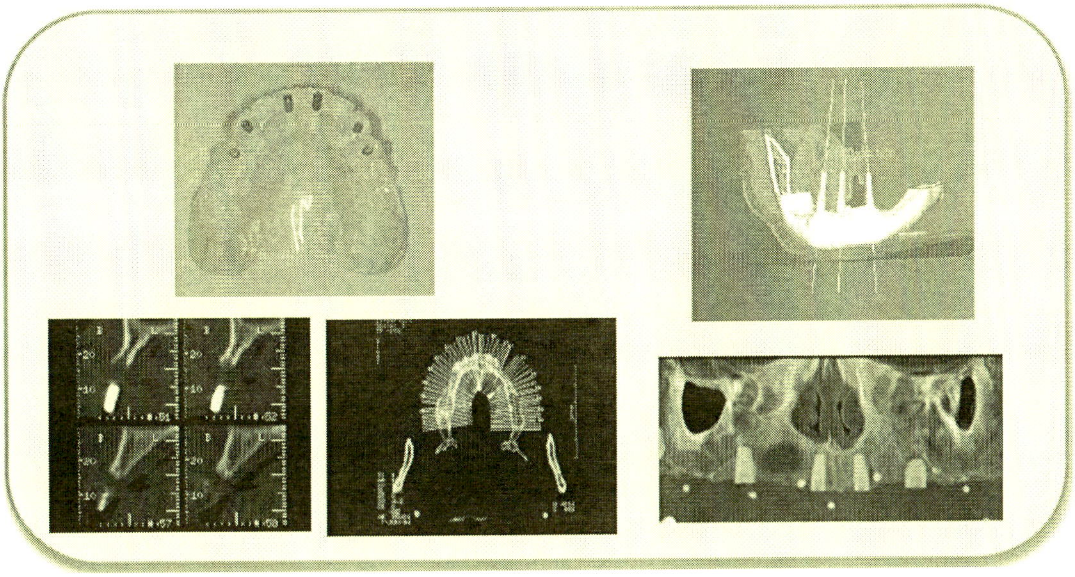

Figure 4. conventional planning and 3-D planning software. Radiographic analysis left top: Acryl splint of complete denture with measuring pins used as landmaks for analog radiographic analysis, bottom: CT-can slices and overview right: computer assisted planning with implant placement and 3-D or 2D visualization of prospective implant position.

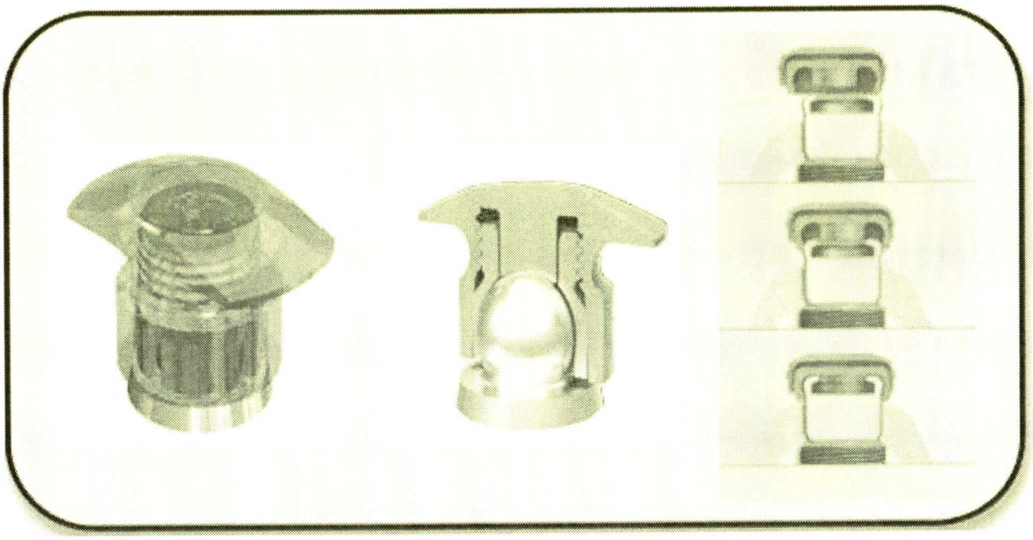

Figure 5. Ball anchors and locators for overdenture retention. The most frequently used single attachments. Left: Dalla Bona plus. The inner part of the female retainer can be activated or exchanged chair-side if the retentive force is lost. Right: The Locator. This is an inverse retentive system. Plastic inserts are engaging and non-engaging for dysparallel implant axes. Various colors means indicate retentive forces.

4. Overdentures

4.1. Indications for Overdentures

The overdenture adopted its design from the complete denture. This is particularly true for the mandible, whle in the maxilla a horseshoe design with no or minimal coverage of the palatal mucosa is aimed. Single retention devices like ball anchors (Dalla Bona type) and locators are popular (Figure 5). Handling is simple and their use does not require complex technology for overdenture fabrication.

For bar fabrication the implant systems offer gold and titanium copings to whichprefabricated gold bars can be soldered and laser welded. Round clip bars, the egg shaped or U-shaped Dolder bar are most frequently used. Various female parts either simple clips or long sleeves are available. The inner walls can be activated or the long sleeves contain plastic inserts of different retentive force, indicated by different colors. They can easily be exchanged. The bars are fixed with screws on titanium abutments or, depending on the implant system, directly at the level of the implant shoulder. Prefabricated Dolder bars and their female parts are also available in titanium material for soldering and laser welding. Recently a new stress-free (SF) round clip bar made from titanium was released to the market. Its advantage is that it can be adjusted directly in the mouth of the patients without the need of laboratory completion. No heat processing, no solder joints are required and stress is avoided. This SF-I bar is suggested for use when doing immediate loading. Furthermore, it should reduce costs (Figure 6).

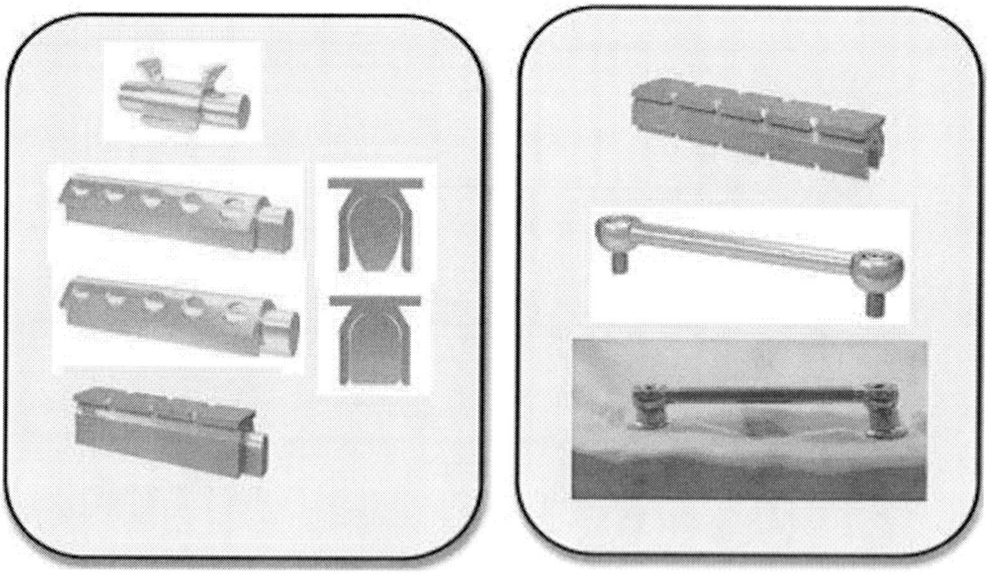

Figure 6. Various types of prefabricated bars for overdenture retention. Various types of prefabricated bars. Left: round clip bar, Dolder bars with egg- or U-shaped profile and corresponding female parts. The Dolder bars are available in Gold- or Titanium alloys for soldering / welding. A female retainer with plastic insert is also available (bottom), which can easily, be exchanged chairside, if retentive force is lost. Right: SF-I bar, the stress free bar, that can be directly mounted in the mouth of the patients and is also intended for immediate loading.

With soldered and laser-welded bars over time the solder joints may exhibit microfractures and corrosion products. The CAD CAM fabrication of milled titanium bars reduces or eliminates such technical problems and should become the standard of bar fabrication. CAD CAM planning favors a highly individual design (Figures 7 and 8).

4.2. Mandibular Overdentures

The mandibular overdenture has become a preferred treatment modality, particularly for elderly and maladaptive patients who exhibit problems with wearing mandibular complete dentures. It is considered to be a geriatric concept. The patints are mostly edentulous in the maxilla as well and do not ask for fixed prostheses. They had ben edentulous for a long time or lost or will loose the last remaining residual dentition in the mandible. A large body of literature on mandibular overdentures is available. Preferred topics are the number of implants and the retention mechanism with regard to anchorage of the overdenture and technical complications. Further treatment outcomes are chewing function and satisfaction as compared to complete dentures.

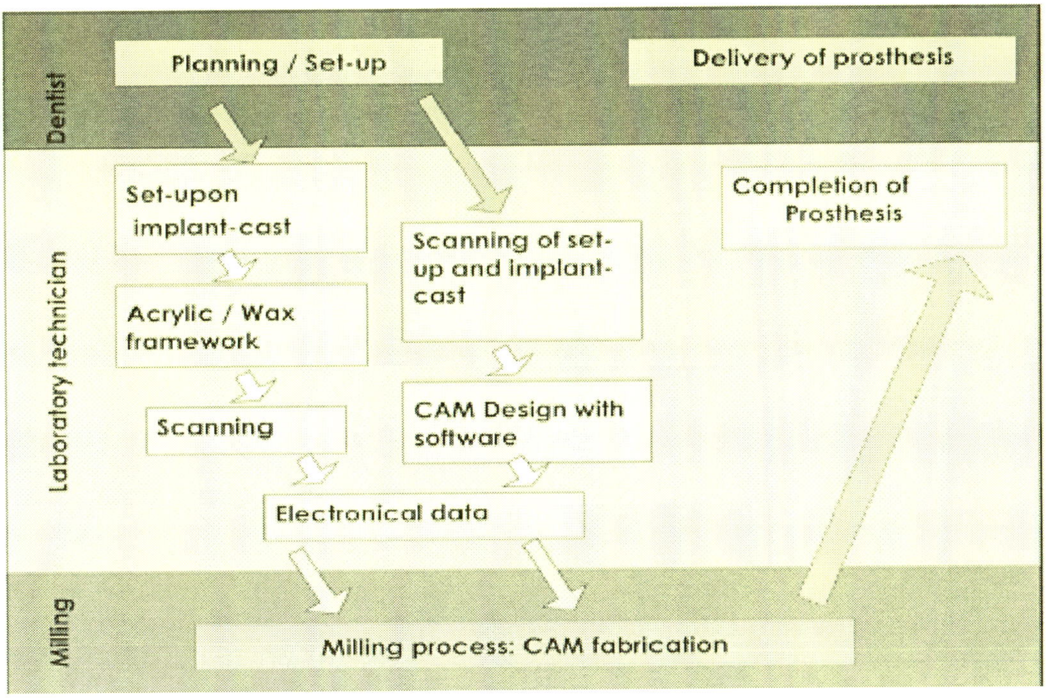

Figure 7. CAD CAM fabrication process. Flow chart of the CAD CAM fabrication process. Either the bar / framework is prepared in wax / resin pattern according to the setup, then scanned and the virtual design finalized by the software. Or the setup and the master-cast are scanned and the entire framework is designed virtually. On the basis of the elctronic data the milling process is perfomed and the prosthesis can be completed.

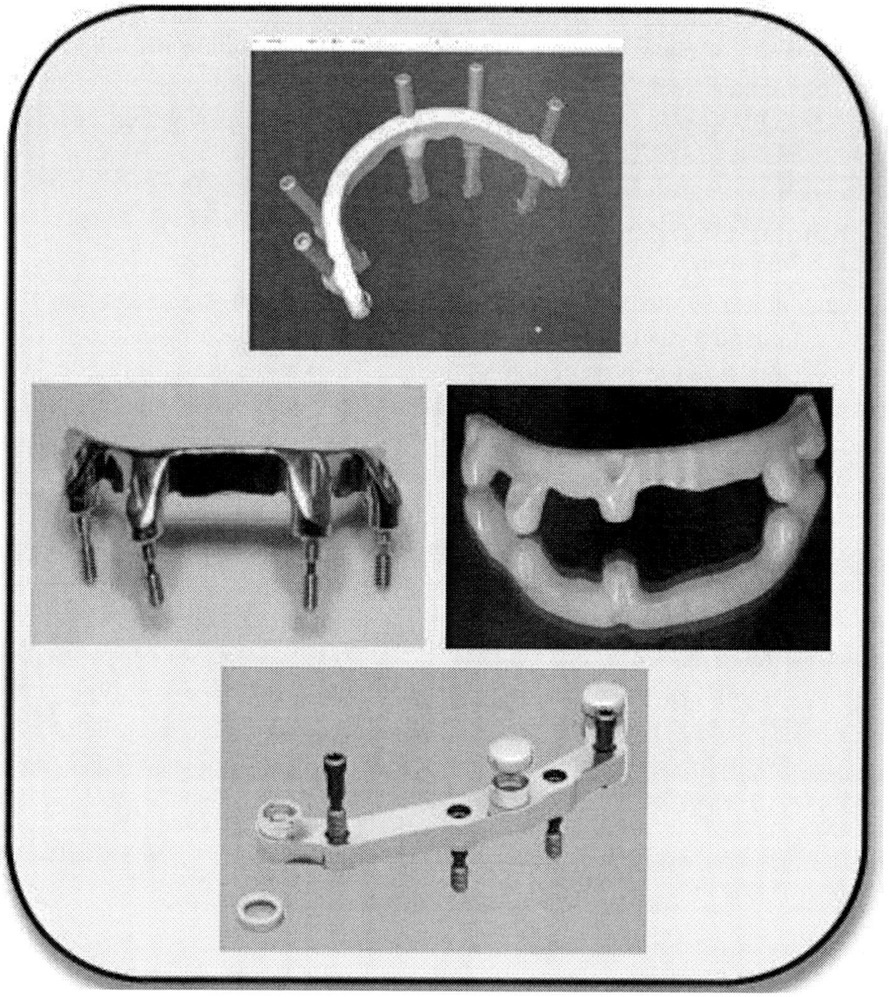

Figure 8. CAD CAM fabricated bars. Top: virtual design, left: titanium bar, right: Zirconia bar, bottom: Titanium bar with inbuilt Locator attachments.

4.3. Maxillary Overdentures

While most patients asking for mandibular overdentures are completely edentulous in both jaws, the maxillary implants are indicated for patients who have natural teeth in the opposing mandible or fixed or removable prostheses supported by implants and teeth. Thus, this is a younger segment of elderly patients. Some of them have no experience with wearing a removable prostheses and first of all ask for fixed prostheses. Studies published in the 90-ties exhibited a surprisingly high failure rate for maxillary overdentures, i.e., over 20%. This failure rate was significantly higher in comparison with fixed prostheses or mandibular implants. A critical analysis of the treatment outcomes revealed that the indication for overdentures was often given in an emergency situation, meaning that overdentures were a substitute for failing fixed prostheses. They became a rescue were prescribed if adequate placement of implants to support fixed prostheses was not possible. Otherwise, in properly planned overdenture an increased survival was found. The marginal bone surrounding the implants was maintained at the same level as with fixed prostheses, also in ridges with advanced atrophy.

Overdentures have the advantage that they compensate for deficits of the jawbone. Thus, adverse morphological effects can be more easily eliminated with the utilization of overdentures instead of fixed prosthese and esthetic problems are more easily solved. Maxillary overdentures gradually became more popular but he number of publication on this topic is still low as compared to the mandible.

4.4. Number / Distribution of İmplants

4.4.1. Mandible

Four interforaminal implants with a connecting bar for denture support were often and still today are suggested to obtain cross-arch stabilization and to better maintain osseointegration. So far, no study has confirmed the biological and biomechanical superiority of 4 implants as compared to 2 implants for mandibular overdenture fixation and one study found no advantages with 4 implants. Two mandibular implants were proposed in the late eighties and 10 years later the majority of studies on mandibular overdentures were based on 2 and occasionally 3 interforminal implants. It was even proposed as the standard of care for the edentulous mandible although this statement was disputed controversially. In fact, the number and distribution of implants cannot be discussed without taking into account the anchorage device and retention system (Figure 9).

4.4.2. Maxilla

4.4.2.1. Anterior Concept

In many situations patients have lost posterior teeth long ago and exhibit better bone quality and quantity in the anterior jaw segment (between canines or first premolars). Four to 5 implants can be accommodated for a bar support and more invasive surgical procedures like bone augmentation and sinus floor elevation can be avoided. This results in an anterior cross arch splint.

4.4.2.2. Posterior Concept

Less frequent is the situation that patients have advanced anterior atrophy and may profit from posterior implants, if sufficient bone is available. If atrophy of the maxilla does not allow for implant placement either in the anterior or posterior segment, a relative augmentation via sinus floor elevation is a treatment option. Such posterior implants are connected on each jaw side separately by a bar segment or in selected cases, if technology enables it, an anterior segment is also added. This results in a full cross arch splint (Figure 10).

The better the implants are distributed over the entire arch in symmetrical position the more the overdenture is purely implant supported and nearly resembles a fix prosthesis from a biomechanical point of view (Figure11).

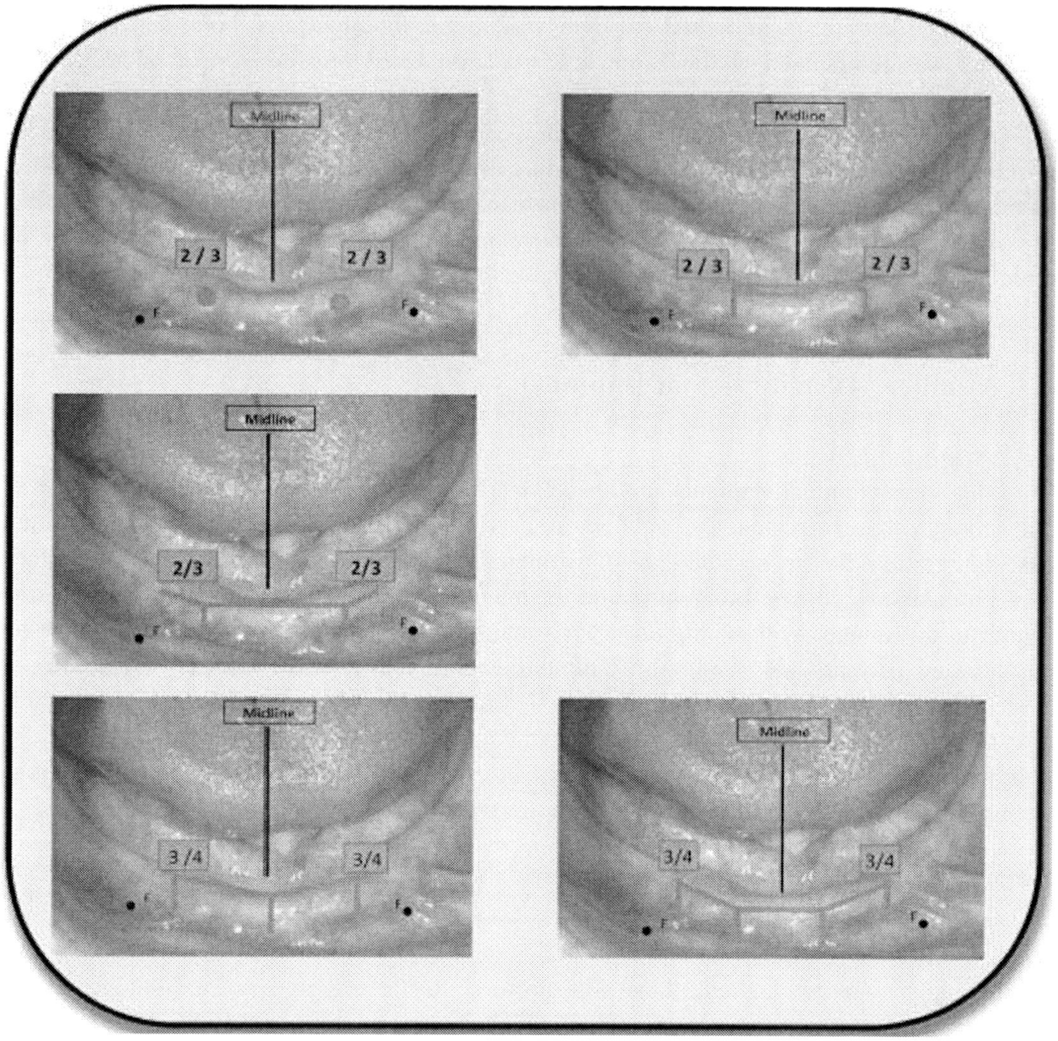

Figure 9. Various interforaminal implant locations and anchorage devices for overdentures. Top: two interforaminal implants with ball anchors or a simple bar. Middle: two implants with a rigid bar and short distal extensions Bottom: 3 to 4 implants with a large anterior cirvatur. Blls and bars can be mounted Altogether, the indication depends on the jaw morphology, location of the foramen and shape of the anterior curvature. Single anchors are not advised with advanced atrophy.

4.4.2.3. Comparison with the Mandible

The situation with maxillary overdentures is different from the mandible. For maxillary overdentures the support by only 2 implants is not suggested as a final treatment concept, but may be a solution in rare, special cases. A full palatal coverage of the prosthesis is mandatory. A connecting bar is not feasible between 2 implants since it would produce the effect of a hinging axis and interfere with the anterior space. Sufficient prosthesis stability is not provided. Thus, a minimum of 4 implants is usually recommended. This enables a bar design that follows properly the anterior curvature of the jaw. One study showed that two telescopes supporting the overdenture exhibit a high risk of failure and telescopes should not be recommended unless multiple implants, i.e. preferably 6 implants are well distributed in the anterior and posterior jaw segments for denture support. Single implants might be preferred since techniqual requirements are simple and less expensive. Due to the specific anatomical features of the maxilla a divergence in the implant axis is observed. This hinders the vertical insertion path for the prosthesis if connected to multiple single ball anchors or locators, although providers ensure proper fit with up to 40° of divergence. A well-positioned bar on four to five implant, which follows the anterior curvature or individually cast telescopic abutments will overcome non-parallel axes and may solve the problem. The telescopes twith a removable bridgework has appropriate implant support and is an equivalent to a fixed prosthesis. Today a 3-D computer assisted analysis and virtual planning is strongly recommended and should become the state of the art. Prosthetically driven implant planning becomes fully predictable and digitally produced surgical splints ensure parallel implant placement. This results in an optimum bar design and high quality of the overdenture.

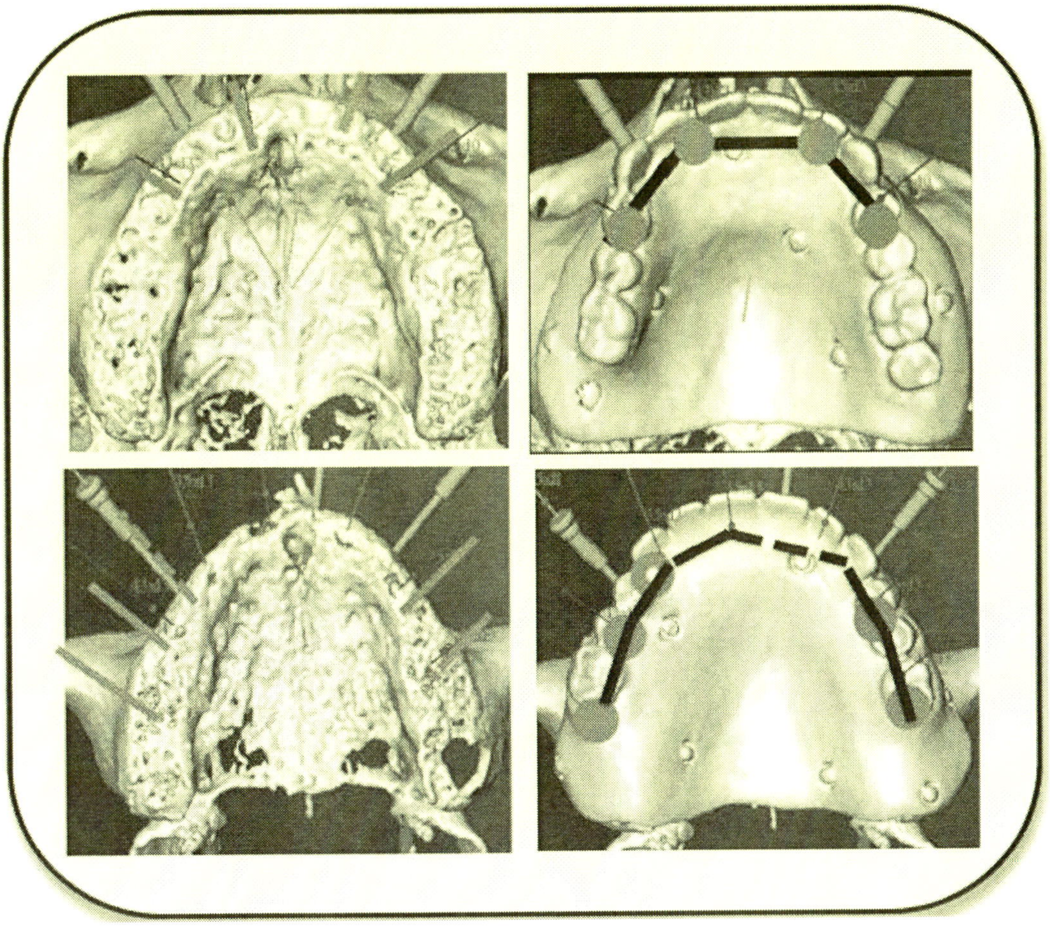

Figure 10. Anterior and posterior concept. Virtual Planning: Top: Planning of a bar overdenture on 4 implants with the anterior concept. Bottom: Planning of a bar overdenture with posterior concept. According to the individual situation in the anterior arch bar segment are as short cantilevers or the bar is completed in one segment.

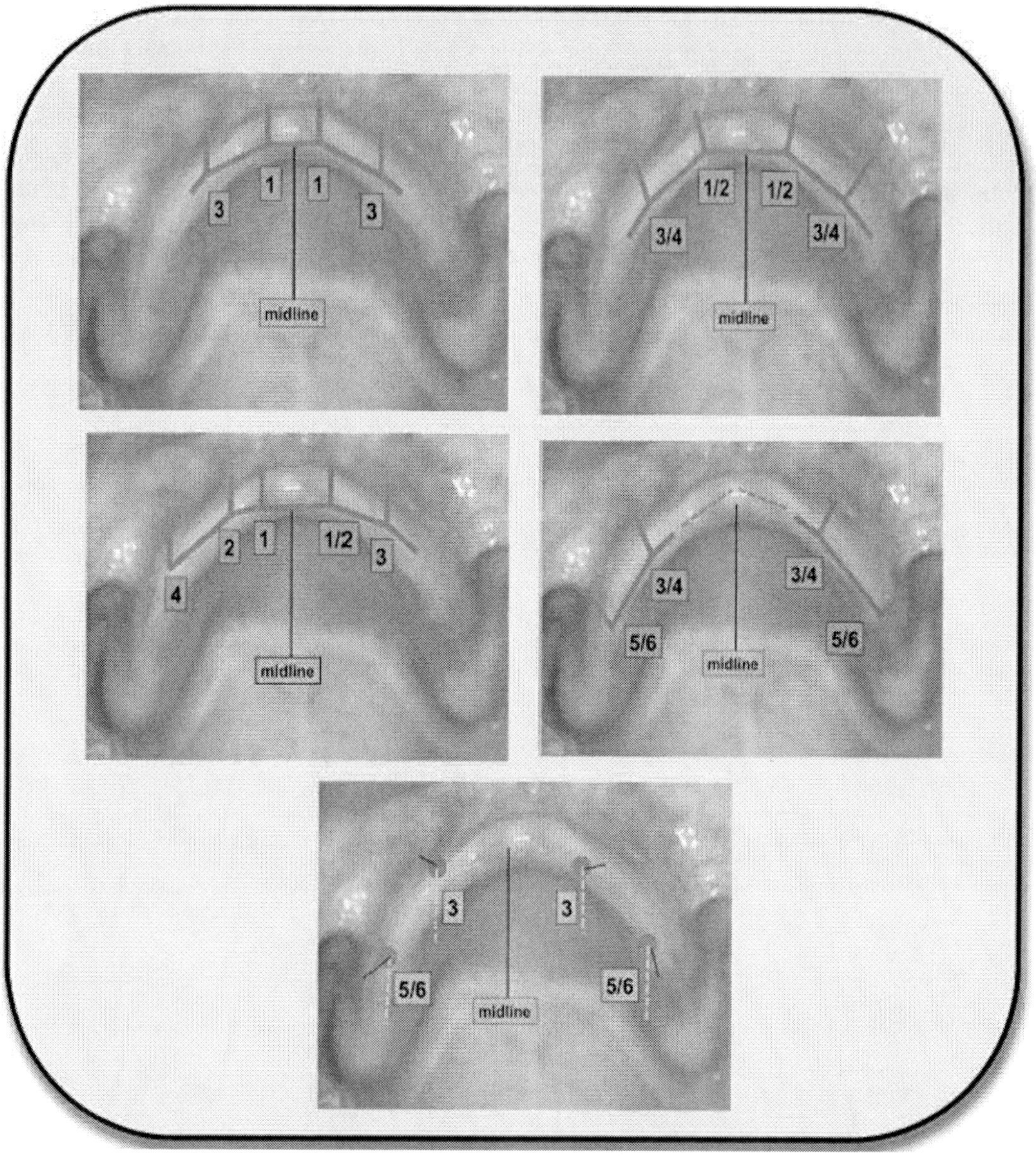

Figure 11. Maxillary implants for overdenture support. Anterior / posterior concept and single anchors Top: With a regular distribution of four implants rigid stable bars provide for good stability. Ball anchors are not recommended, particularly in the situation left insufficient stability of the overdenture is produced. All materials can be used: soldered gold bars, CAD CAM fabricated Titanium / Zirconia bars. Middle left: this anterior concept with an irregular distribution occurs with less favorable jaw conditions. It is important that the bar has to follow closely the jaw contour and the tooth set up. All materials can be used: soldered gold bars, CAD CAM frabricated Titanium / Zirconia bars. Middle right: With theposterior concept the implants are located in the area between canines and first molars. Two bar segments with a rigid design are used. With CAD CAM fabrication, a connecting anterior segment may be considered, depending on the size of the anterior curvature. Bottom: Single retention devices can be recommended: 1. If the implants are well distributed over the entire arch, 2. if parallel, perpendicular implant axes are obtained. Divergent axes are a frequent problem and single retention devices will not sufficiently compensate for.

4.5. Retention of Prostheses

4.5.1. Mandible

Equally, a major topic in overdenture rehabilitation, disputed with controversy, is the use of bars (= splinting the implants) or single standing anchors (without a splinting effect). Furthermore, the retention mechanism, i.e. resilient and stress free vs a rigid anchorage mechanism is discussed on the basis of clinical aspects and laboratory testing. With two interforaminal implants the overdenture has mostly a combined tissue and implant support. It is believed that two single ball anchors or rotational bars connecting 2 implants provide the best resilient anchorage. This should avoid overload of the implants and avoid a negative impact on the crestal bone. However, 3-D in vivo force measurements with different anchorage systems and retention mechanism clearly demonstrated that loading forces on the implants were highly similar in force magnitudes and force directions. All measurements were performed with opposingccomplete denture in the edentulous maxilla. Likewise, these forces are significantly lower as compared to single tooth replacement and short FPP in dentate patients. Furthermore the vertical / axial force component was dominating, but always accompanied by horizontal force components. The more implants are placed the more the overdenture is purely implant supported regardless of whether a resilient or rigid anchorage system is used. Tangential axes compete with the transversal axis and the rotational effect gets lost with multipl implants.

In case of advanced atrophy a resilient retention mechanism will transfer functional loads onto a small crest of the remaining jaw bone. This should be avoided. Rigid bars, comprising short cantilevers relieve the tissues, provide stability and avoid pressure of the denture base onto the foramen mentale. There is a lack of studies to prove a negative effect of a rigid bar, if connecting only 2 implants. A long-term study observed no differences in crestal bone level changes with resilient bars or single ball anchors. Two long-term studies with an observation time up to 24 years exhibit a high survival rate (> 80%) of the overdentures and implants (> 90% after 20 years). Around 70 % of the anchorage devices were rigid bars connecting two implants. The crestal bone level remained stable with little resorption. Another study observed more technical problems if the overdenture was retained by a round clip bar, as compared to single ball anchors . This study compared two resilient retention systems. In fact, the small clips are prone to mechanical complications. Although patients are highly satisfied, frequent maintenance with ball anchors has been reported. Another long-term study with up to 15 year data reported that both resilient retention (clip bar, single ball anchors) required more maintenance and were more often changed than rigid bars. Apart from rigid U-shaped bars, occasionally telescopic attachments were also used for overdenture support by two implants without negative affect on bone or prosthesis stability. Such telescopes are individually cast and need parallel orientation. Gold-alloys, titanium and / or more recently milled zirconia are the materials of choice. The seating of the secondary crown directly in the mouth of the patients will enhance stress free fit.

It is important to summarize that under real clinical conditions the shape, size, curvature and degree of atrophy play an important role in stability and maintenance of implants and prostheses. These factors must be included in the decision making. The fugures show various biomechanical aspects of retention devices in the edentulous mandible and clinical considerations (Figure 12).

4.5.2. Maxilla

The discussion about the retention mechanism i.e. stress breaking and resilient versus rigid, is not relvant and nearly absent in the literature of the maxilla. With multiple implants, the maxillary overdenture becomes mostly implant-supported regardless of the type of retention device and bar design.

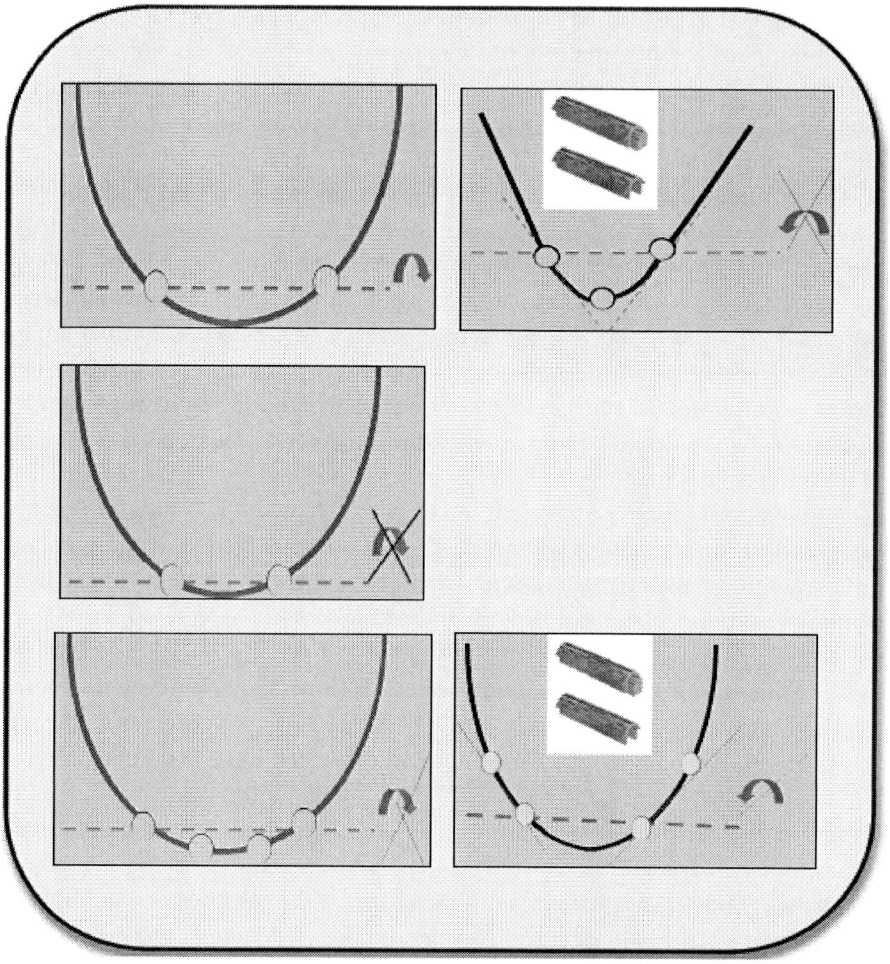

Figure 12. Biomechanical aspects of overdentures support and retention with interforaminal implants. Biomechanical consideration of retention and support: Top left: Normal situation with two mandibular implants in anterior position. Resilient types of bars or ball anchors can be used. With round clip bar, egg-shaped bar or ball anchors a slight rotational movement is produced. Top Right: Mandible with a narrow anterior curvature: 2 implants in the location of lateral incisors or canines produce a strong hinging axis. The implants cannot be connected by means of a bar an additional anterior implant is necessary. With tangential axes, the hinging movement disappears. Bars with a rigid design or ball single attachment can be mounted. Middle: Two implants in anterior position. The hinging movement is limited or absent, a rigid bar can be placed. Bottom left: With single ball anchors the hinging movement is limited, a vertical translation is possible. Four implants in close anterior location must be avoided with bars. The single segments are too short to provide good stability. The female retainers are prone to complications. Bottom right: Mandible with large anterior curvature: 2 implants in the location of lateral incisors or canines produce a light hinging axis. If four implants are placed, the hinging movement disappears and a rigid bar design or four single attachments can be mounted.

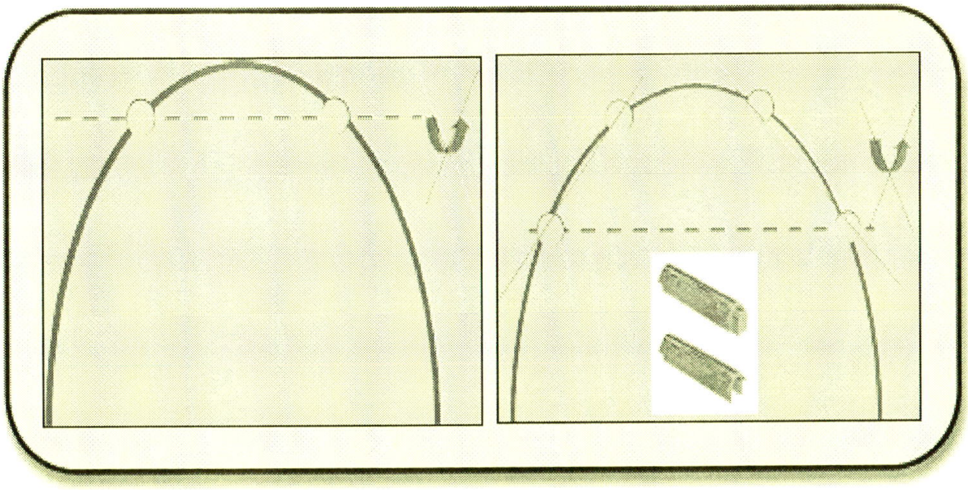

Figure 13. Biomechanical aspects of overdentures with interforaminal implants. Biomechanical consideration of retention and support: Left: Two implants in the canine area or in slightly anterior location produce the effect of a hinging axis. This is not comfortable for the patient. The implants cannot be splinted by a bar, it would interfere with the space of the tongue. Right: Four implants with tangential axes hinder a hinging movement and resilient anchorage. A rigid bar rigid provides good stability.

For the patients' comfort the rotational effect of a transversal axis should be avoided and is limited with multiple implants. Therefore, rigid bars can be recommended as standard procedure and a cast metal framework will reinforce the denture base to ensure stability and stiffness. Four to six implants connected by a rigid bar provide high stability and a maximum of support. Female bar retainers are not soldered to the cast metal framework, but rather are mounted in the acrylic denture base. This facilitates prosthetic services like tightening or renewal of retainers. In vivo force measurements with maxillary implants connected by a rigid bar or supporting full-arch bridgework have been shown to result in mostly identical force patterns for both types of prosthesis and the force magnitudes with overdenturs were insignificantly lower. If the bar was not splinted in the front area (posterior concept), this had only minimal, insignificant impact on the force pattern. The overdenture itself acts as a secondary splint. From this study the authors concluded that bar overdentures supported by multiple implants resemble fixed prostheses with regard to the force pattern. Similar judgment was made by the patients, if they assessed in a cross over study design a long bar with fixed prosthesis. In comparison with mandibular overdentures the force pattern of the maxilla was different, exhibiting higher forces in all dimensions. Particularly high forces were measured in horizontal direction at the anterior implant, with an anterior directed force vector. But still, the vertical axial force component showed the highest force magnitude (Figure 13).

One question remains. Why can stability and functional comfort of maxillary overdentures be different, although the same concepts of treatment are applied. The following figures illustrate tha the extension of the bar but als the sagittal – and of course not only the sagittla – jaw relations have an impact on the area of occlusal support. The following figures is an abstraction aimed to explain differences in support and stability (Figure 14).

4.5.3. Summary

4.5.3.1. Guidelines for Mandibular Overdenture Fabrication

2 intraforaminal implants, preferably of standards size (diameter approx. 4 mm and minimum length of 8mm) are recommended. The distance between the implants should not be less than 15-25mm, depending on size and curvature of the anterior jaw. Splinting with a clip-bar (resilient retention) or u-shaped bar (rigid retention) is possible as well as the use of single anchors, like ball attachments and locators. Some implant manufacturers will not provide single anchorage devices for use with small diameter implants. Three or 4 implants are recommended in case of reduced implant length or diameter. However, the distance between the implants should be at least 12 to 15 mm to allow for a reasonable length of the bar segments. If 4 or more implants are placed, fixed cantilever prosthesis can be considered.

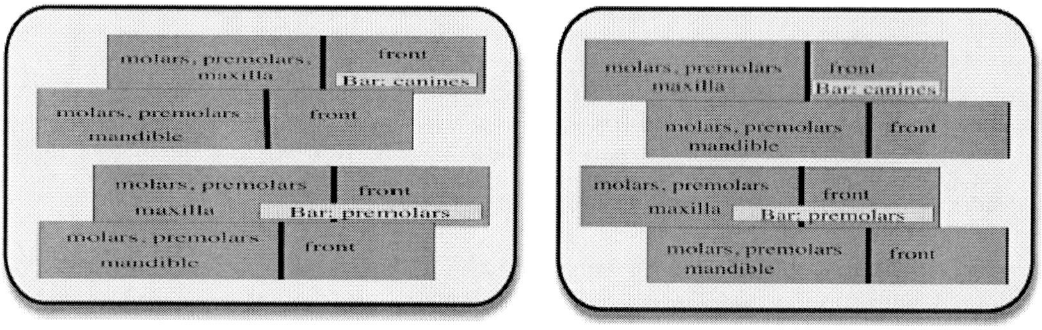

Figure 14. Functional aspects of anterior bars supporting an overdenture. Left: Functional aspects of anterior bars in sagittal class ll configuration. Top: front bar extended between canine teeth. The bar does not support the posterior units of occluding teeth are not supported. Bottom: anterior with extension between premolars. The bar slightly supports the most anterior unit of occluding teeth Right: Functional aspects of anterior bars in sagittal class lll configuration Top: front bar extended between canine teeth. The bar does support the posterior units of occluding premolars. Bottom: anterior with extension between premolars. The bar supports the posterior units of occluding teeth (premolars and also molars) Higher stability and comfort is provided than in class ll. Front teeh contacts are stable.

Traditionally, prefabricated gold bars soldered to gold copings and fitting on titanium abutments are in use. CAM fabricated milled titanium or zirconia bars with direct connection on the implant level are advantageous and eventually might become the standard.

4.5.3.2. Guidelines for Maxillary Overdenture Fabrication

The placement of only two implants in the location of the canines is not considered to be a standard procedure, exhibits an increased risk of failure and may not satisfy the patients demand of stability. A full complete denture design is mandatory. A minimum of 4 to implants evenly distributed over the anterior arch is a frequent solution. In specific individual situations the posterior concept is applied. Standard size and reduced diameter implants can be mixed. A splinting bar is recommended and the overdenture is designed as a horseshoe. Telescopic retention with 6 welldistributed implants is an optional solution but leads to higher technical investment. Utilization of 4 ball anchors or locators is technically less demanding, but proper engagement of the female parts is best ensured only with parallel implant axes. The use of standard size diameter implants is strongly recommended and some implant providers will not allow to combine single anchorage devices with small diameter implants. It is suggested that the denture base be reinforced with a metal cast framework. This increases initial cost but may reduced cost in the maintenance phase since fractures and other complications can be limited.

5. Fixed Prostheses

5.1. Overdenture or Fixed Prosthesis

One advantage of overdentures is that their utilization may be more consistent, with optimum placement of the implants with regard to the remaining bone structures. Congruence of the prosthetic tooth position and implant location is not necessary for overdentures. The rehabilitation with fixed prosthesis was a priority goal in implant prosthodontics and the first long-term results were presented in the Toronto study. In these early publications restorations for the edentulous mandible predominated as compared to limited reports on the maxilla. The prostheses were designed around a metal-framework from precious-alloy with acrylic veneering. The so-called "wrap-around" technique with acrylic denture teeth and denture base material to compensate for lost hard and soft tissues was also applied. These prostheses were supported by a minimum of 4 mostly 5 to 7 implants, and were screw retained, with a hybrid design (where the prosthesis material was not in contact with the alveolar mucosa). The implants were placed in the interforaminal respectively anterior region of the jaw, thus avoiding the mental nerve or sinus, but cantilevers and shortened dental arches became necessary. Loading patterns of these types of prostheses were investigated and demonstrated maximum loading forces on the distal implants adjacent to the cantilevers. The reason for the limited number of maxillary cases was due to the advanced atrophy and anatomical conditions that often hindered the placement of posterior implants. The prostheses at that time exhibited long cantilevers with the risk of and cantilever fracture. After the early experiences with fixed prostheses, efforts were made to fabricate porcelain fused to metal fixed prostheses with a crown design with the aim to improve esthetics and prosthesis quality. This included posterior implant support to avoid long cantilevers. The frameworks became large, were of heavy weight and misfit was frequent. Thus a segmentation was preferred. A recent study again reported on the clustering effect with multiple implant failures in the edentulous maxilla. In studies about implant survival or success the design of the prosthesis is often not clear and the authors focus more on bone level changes than on prosthetic aspects. Long-term results and comparative studies with different designs of fixed prostheses in the edentulous jaw are not available. But, it appears that bone quality and patient related factors have the greatest impact on bone remodeling capacity around maxillary implants. RFA measurements on maxillary and mandibular implants, identified thoroughly lower ISQ values in the maxilla at different time points, immediately after surgery or some years after loading. Posterior implants and elderly postmenopausal women exhibited the lowest values in the maxilla. This would justify to use cross arch fixation. Today with modern CAD CAM technologies, the hybrid fixed prostheses is taken up again with a titanium or zirconia framework and optimized design. Likewise, PFM frameworks can be replaced by CAD CAM milled zirconia or titanium. This design allows for a limited facial support if necessary and large one-piece frameworks can be processed. The frameworks are of high precision and light weight. In general, more implants are required for support of fixed prostheses than for overdenture, Reduced bone volume limits the indication for fixed prostheses unless additional surgical procedures are applied. This is particularly true for the maxilla, and implies a more specific patient selection and more individual planning than it is necessary for the mandible. Here in many situations both options – fixed and removable- are a viable solution with interforaminal implants. A standard surgical and prosthetic protocol can often be utilized.

Studies have reported that the patients' choice is rather dependent on individual preference, e.g. ease of oral hygiene procedures. Speech analysis in crossover testing with maxillary overdenture or a fixed prosthesis observed better phonation for some sounds with overdenturs. The design of the fixed prostheses however is not clear. Requirements of facial support can be fulfilled and problems with phonetics and oral hygiene are often better resolved with overdentures. In fact, hygiene procedures are mostly facilitated with removable prostheses; however, under maxillary overdentures soft tissue hyperplasia may develop. Table 3 below shows roughly some clinical parameters for decision making, particurlaly related to the edentulous maxilla.

Table 3. some parameters for decision making: fixed or removable prosthesis

Extraoral	Fixed	Removable	Intraoral	Fixed	Removable
Lipline	low	high	Ridge (shape)	vertical	buccal inclination
Tooth display	little	distinct		convex	buccal concavity
Facial support	no need	necessary	Intermax.	dist. $\leq$ 10 mm	> 15 mm
lip support					
			Intermax. relation	neutral	skelettal III
				deep	crossbite
				overbite	
			Mucosa	thick	thin

The fabrication by CAD CAM technologies for fixed prostheses follows the same principles as for bars. The CAD CAM method exhibits much more similarities in the procedures for overdenture and fixed prosthesis than with standard fabrication (Figure 15).

5.2. Mandible

For the mandible, guidelines and concepts are similar as for the maxilla, but aspects of esthetics and morphology have a lower, secondary impact on the design of the prosthesis. The implant axes have a more vertical direction and facilitate the fabrication of the prosthesis, which respects requirements of tooth position, alignment and occlusal aspects in relation to the maxilla. There is some concern that mandibular overdentures may encourage posterior jaw resorption. In contrast, vertical bone apposition in the posterior jaw underneath cantilevers was observed with fixed cantilever prostheses. As a result, it was recommended to provide younger edentulous patients with fixed mandibular prostheses. With a minimum of 4 to 5 implants a fixed cantilever prosthesis can be fabricated. With exclusively interforaminal implants a pronounced anterior curvature of the jaw is required (Figure 16).

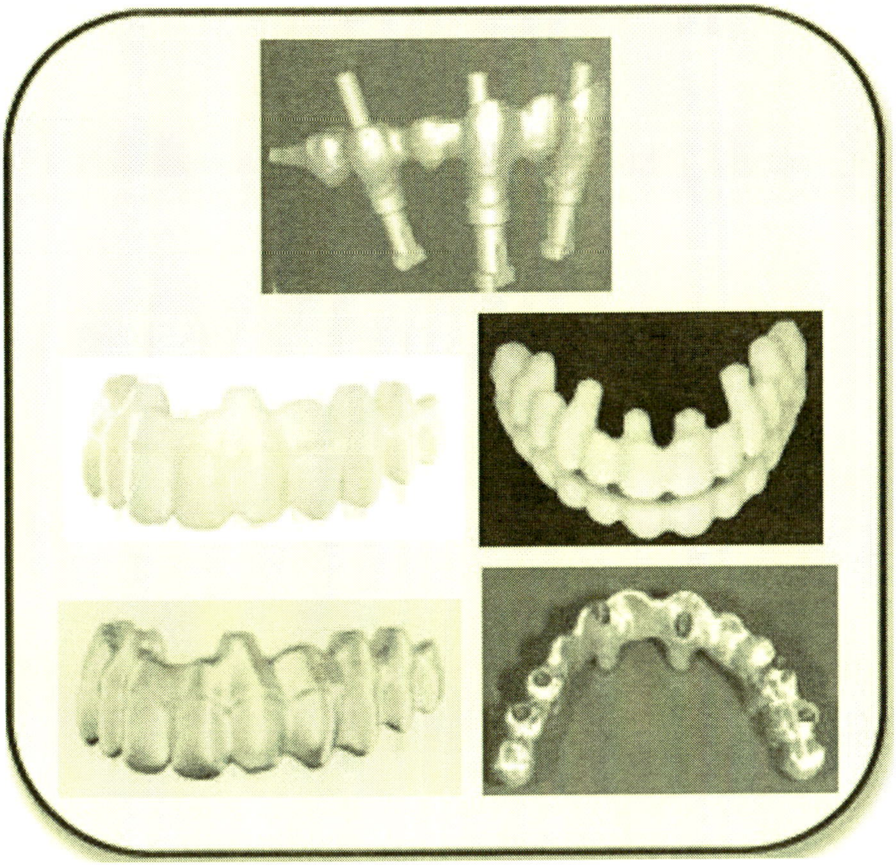

Figure 15. CAD CAM fabricated frameworks for fixed prosthesis. Top: virtual design, Middle: Zirconia framework with or without buccal flange material, Bottom: Titanium framework with or without buccal flange material

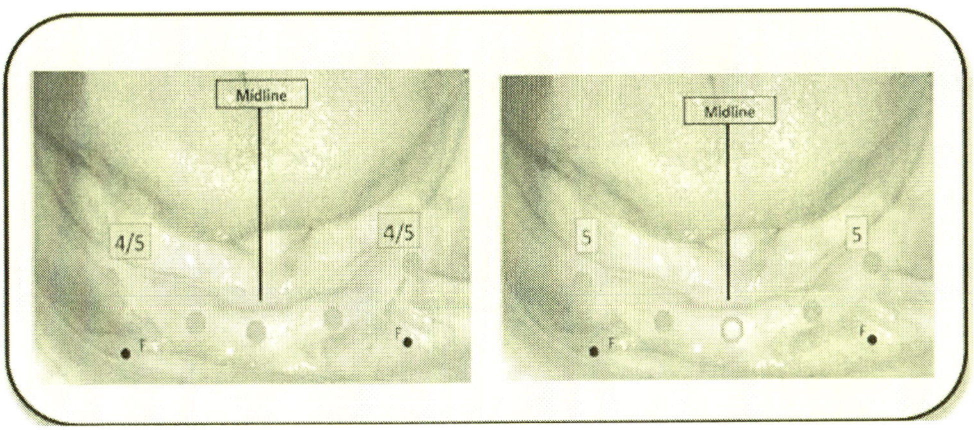

Figure 16. Mandibular interforaminal im plants for overdenture support. Distribution of 4 to 5 implants for fixed cantilever prosthesis with a one-piece framework. This type of prosthesis is only indicated with a good anterior curvature. Left: either 4 to 5 implants are well distributed in the interforaminal jaw segment or the distal implants are tilted to obtain a larger support. Right: r two implants can placed in posterior position of the foramen mentale , two tothree implants in the anterior position.

5.3. Maxilla

Special attention has to be paid if implant-prostheses are planned for the edentulous maxilla. A younger segment among patients with edentulous maxilla will ask for fixed bridgework. The treatment of the edentulous maxilla will require more elective procedures than are necessary for the mandible, particularly with respect to the following criteria:

- degree of atrophy of the residual jaw
- tissue volume dimensions
- facial morphology and esthetics
- function and phonetics

The dental status in the opposing mandible influences biomechanical considerations regarding loading and occlusion. The evidence from the literature on prosthetic aspects is minimal, mostly survival rates of implants are reported or maintenance was compared between fixed and removable prostheses.

5.3.1. Esthetic Appearance and Facial Support

Both are a topic of major concern when all teeth are lost in the maxilla. Clinical experience today shows that the soft tissue may be managed successfully in case of one single-tooth replacement and resembles a fully natural gingival architecture. Recreating a well-contoured soft tissue border around implants over an entire dental arch has not yet been documented to be practicable. Dimension of tissue atrophy and its implication for prospective tooth position and implant axis determine the choice of prosthesis design. Compensation for lost hard and soft tissue becomes difficult and is a problem that requires special attention in the planning phase. The intermaxillary distance between the incisal edge of the lower teeth and the maxillary jaw should not exceed 10 to 15 mm, otherwise the teeth will become too long, facial support is not sufficient and an overdenture would be a better indication. A low lip-line i.e. no gummy smile is advantageous. Complex skeletal, alveolar and occlusal conditions may determine the choice of a removable prosthesis. Today, analysis on the basis of computer assisted backward planning can help to predict the best solution with regard to the tooth position in relation to the edentulous jaw and loss of tissue. Measurements on 50 patients revealed that up to 60 % of all patients would benefit from facial support by a denture flange and for many their planning data revealed that fixed prosthesis of a crown design was not feasible (Figure 17).

For an optimum and esthetic crown design of the fixed prosthesis only minimum atrophy, a regular distribution of the implants and healthy thick soft tissue are the best prerequisites (Figure 18).

5.4. Number and Distribution of İmplants

The number of implants to be placed depends on the type of prosthesis and the choice of prosthetic design. Conversely, the number of implants that can be placed with respect to anatomic-morphologic conditions will determine to a certain degree the type and design of prosthesis. Case series and single case reports demonstrate that the number of implants to support the prostheses varies widely from 6 to 10 or even 12 in the edentulous jaw. Moreover, depending on individual oral conditions, shortened dental arches or short cantilevers may be considered, if a fixed prostheses is placed and implants in posterior position will be. Anecdotal reports recommend replacement of each single tooth root by one implant. A symmetrical distribution of the implants over the arch may favor the distribution of functional forces and facilitate the prosthesis design. One-piece frameworks oppose segmentation of frameworks. For segmentation, more implants are required.

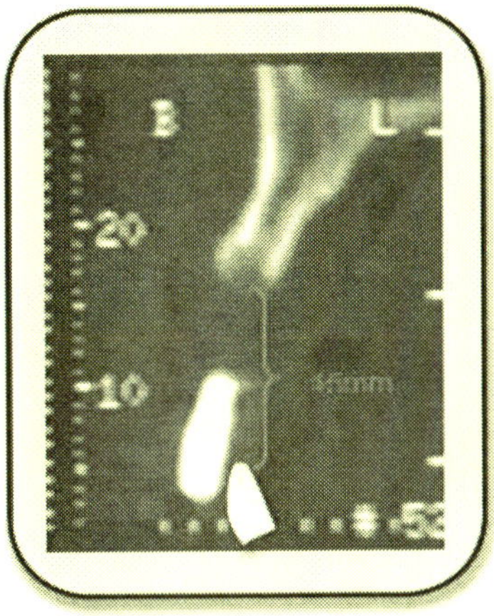

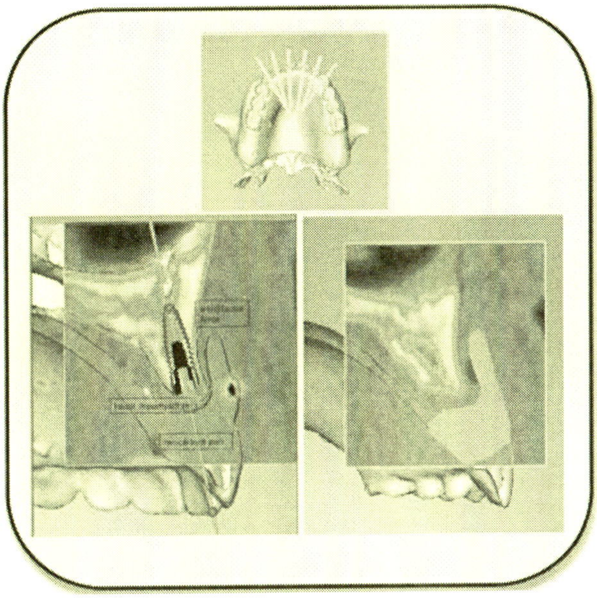

Figure 17. Decision making: fixed or removable prostheses. Treatment planning for the edentulous maxilla: Left: slice of CT-scan. It shows the jawbone and the position of the front teeth. A rough measure says, that an OD should be preferred, if the vertical distance between the incisal edge of the mandibular front teeth and the jawbone is around 15mm ore more. Right: Virtual implant planning with specific software. Different measuring parameters are used for decision making whether fixed or OD is indicated. The right picture shows the amount of tissue that is currently replaced by the denture base material. It can be visualized in all 3 dimensions and in relation to the tooth position.

5.4.1. Mandible

A screw-retained fixed cantilever prosthesis is a favorable alternative to overdentures in the mandible, especially for a younger segment of the older population. It is retained be 4 to 5 interformainal implants. The selection of this prosthetic design is based on the morphologic condition of the mandible. Prerequisite is that the implants can be distributed over a anterior curvature and are aligned in an anterior line. A tilted position of the two implants was also proposed, the so called all on 4 method, which increases the supporting area. The design of the cantilever prosthesis is consistent with perio-prosthetic requirements and the esthetic appearance is not impaired. However, greater manual skills are necessary for daily hygiene procedures than with removable overdentures. Good prosthesis support can be achieved if it is possible to locate 4 implants in the area of the canines and second premolars, adding one posterior cantilever. An exclusive interforaminal or mixed anterior / posterior location of 4 implants does not allow for a segmentation of the framework. Today one-piece CAD CAM titanium frameworks with acrylic veneering and wrap around technique provide a prosthesis of high precision, stability and lower costs as compared to the PFM technique. Instead of CAD CAM fabricated Titanium, Zirconia frameworks with porcelain veneering can also be produced (Figure 19).

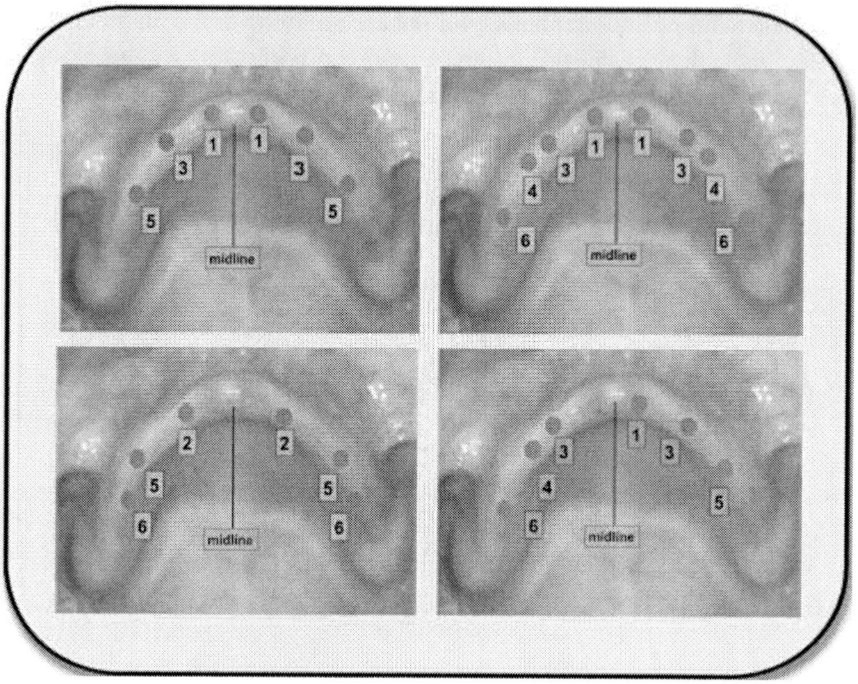

Figure 18. Maxillary implant for support of fixed prostheses. Distribution of implants for fixed prostheses; Top: With a minimum of 6 or 8 implants in symmetrical distribution an optimum framework is obtained, With 6 implants short cantilevers can be added or a SDA is an option. All materials can be used: PFM, CAD CAM Titanium or Zirconia framework. Bottom left: Regular distribution without support of implants in the location of the central incisors. A fixed prosthesis with CAD CAM frameworks in on piece is recommended. Telescopes are an option. Such a telesope prosthesis is removable but has a prevalent crown design. Bottom right: With an irregular distribution a CAD CAM titanium framework in one piece with the wrap around technique appears to be the best solution.

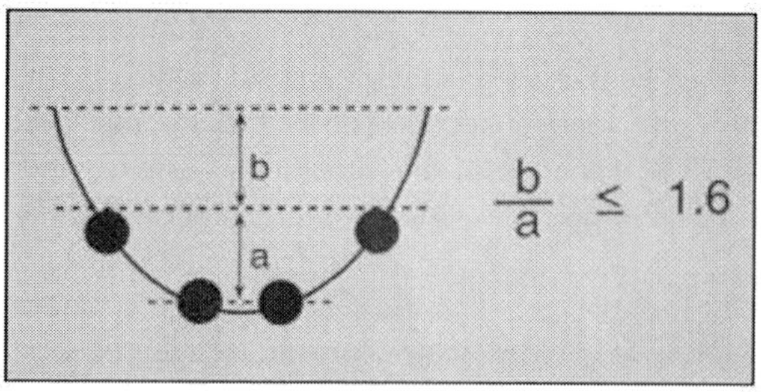

Figure 19. Design of cantilever fixed prosthesis on interforaminal implants. Geometry of cantilever prostheses: This mathematical formula was proposed to calculate the length of cantilevers. This was proposed for frameworks of cast metalalloys. The length depend on the anterior vurvature of the jaw. Data for milled framworks are not available.

If implants are planned for prosthesis support without posterior cantilevers, >4 implants are required. A full cross-arch connection from molar to molar position however should be avoided to prevent stress and strain from distortion of the mandible during opening and closing movements. A segmentation of the framework is indicated. Standard procedures with PFM are possible for all indications as well, but usually with this technique a segmentation is preferred to obtain higher precision of fit.

5.4.2. Maxilla

Based on early experiences with screw retained cantilever prostheses supported by implants in mostly anterior position were not much suggested due to esthetical and morphological implications. In fact, some speech problems were observed and the metal structures of the denture or the implant shoulder could not always be hidden. With increasing experience in implant dentistry, better surgical techniques to compensate insufficient bone structures and increasing demand regarding esthetics the fabrication of fixed prostheses with a crown design became the treatment goal. However, full-arch fixed prosthesis of a crown design from PFM are technically complex, precise fit is not and the achieved and creating natural soft tissue contour with papillae is difficult. For placement of this type of full-arch fixed prosthesis the most important prerequisite is the congruence of implant location and tooth position. If proper screw access from the palatal side is not obtained angled and individually cast abutments may overcome the implant alignment problems and the prosthesis is cemented. The minimum number of implants is 6 but clinicians mostly recommend to 10 implants. The implants are, whenever possible, symmetrically distributed over the entire arch, allowing for pontics. Studies have shown, that view many adjacent implants ma not be favorable particularly in the front segment.

If implants with "non-standard" size are selected this should be compensated for by including implants of standard lengths and diameters. Due to the maxillary resorption pattern, which often is irregular, a symmetrical distribution of the implants is hindered. Likewise long teeth and insufficient lip support may result. Therefore fixed prosthesis with the wrap-around technique more easily overcomes anatomical and aesthetic problems.

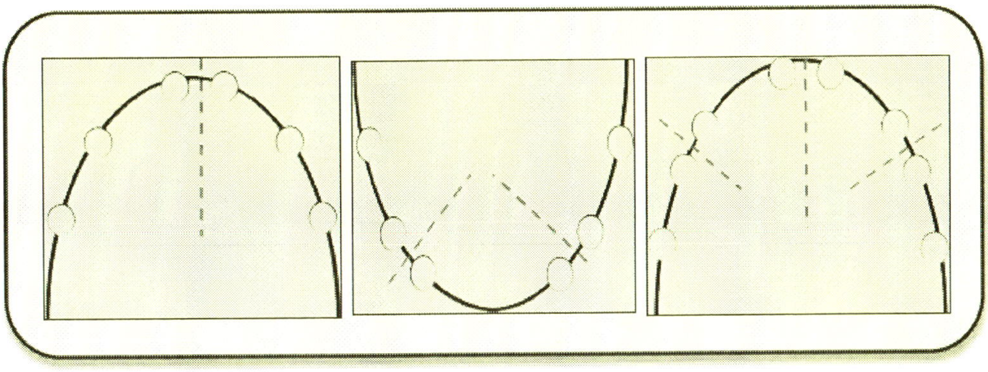

Figure 20. symmetrical distribution of implants over the entire arch. Segmentation of frameworks; Favorable, symmetrical distribution of implants which allows for segmentation of the frameworks. All types of frameworks can be fabricated (PFM, CAD CAM Titanium or Zirconia). In the maxilla, a onepiece CAD CAM framework is also possible, segmentation is not necessary.

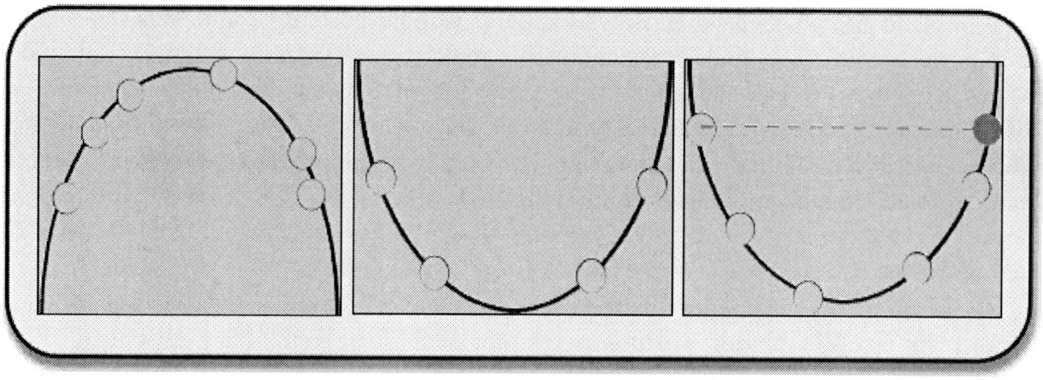

Figure 21. asymmetrical distribution and / or reduced number of implants. Segmentation of frameworks; Irregular distribution or reduced number of implants: a one-pieceCAD CAM milled framework from Titanium is a favorable solution. Cantilevers to replace the first molars can be added. Mandibular posterior implants: a full splinting framework connecting the molars should be avoided.

Today with the use of virtual planning and CAD CAM technology, a better design is achieved that partly compensates for loss of tissue. 5 to 7 implants distributed over the arch are required without the need of full symmetry. Short distal cantilevers (one molar) can be added. This prosthesis has a one-piece framework with precise, stress-free fit. Segmentation is not necessary and cross-arch stabilization is maintained. This type of prosthesis is slightly more consistent with optimum implant placement in relation to the remaining bone in comparison with a crown design. A full congruence of prosthetic tooth position and implant is not necessary but preferred and screw access from the palatinal side must always be provided. While hyperplasia—particularly in the maxilla—of the soft tissues is more often observed with an overdenture, the change to a fixed bridgework will lead to shrinkage of the soft tissue. The figures 20 and 21 show the implant distribution in the maxilla and mandible with and without segmentation.

5.5. Fixation of Fixed Prosthesis

It is obvious that the retention and anchorage devices - bars and single anchorage -for overdentures are screw retained, retrievable by the dentist. It is suggested, that extended cross arch fixed prostheses are retrievable as well. This enables monitoring of the single implants with the prosthesis removed, repair and changes of the prosthesis in the laboratory. While screw retention is more likely to be feasible in the mandible, with perpendicular implant axes and good access to the screw holes, this may be sometimes difficult in the maxilla. If divergent implant axes are observed abutments are indicated for cementation. With regard to the occlusal surface of the prosthetic teeth, cementation is technically more simple since no access holes for the occlusal screw has to be built in. A problem that occurred at various instances is, that the abutment becomes loose under the cemented prosthesis and cannot be tightened anymore. Provisional cements are suggested with the option that the prosthesis remains retrievable. One disadvantage is that in some cases it may not be possible to remove the prosthesis even though provisional cement has been used.

5.6. Summary

Jaw atrophy and interjaw dimension strongly determine the selection of the type of prosthesis, particularly in the edentulous maxilla. A minimum of four implants for the mandible and 6 implants for the maxilla are required. If implants with "non-standard" size are i.e. reduced diameter are selected this should be compensated for by including implants of standard lengths and diameters.

Traditionally, PFM reconstructions are used with screw retention or cementation on abutments. Today modern CAD CAM technology brought new options with millded titanium and zirconia. Cross arch stabilization with one-piece framework and direct screw access at the implant level or seating on abutmnts becomes possible with clinically well-acceptable precision. A summary of prosthetic indications, number and size of implants of implants and design considerations is given in Table 4.

Table 4. Implant prostheses in the edentulous jaw

Indication	Outcomes, recommendations, problems
Edentulous maxilla	
2 implants	OD: not standard, full palatal coverage
4-5 implants	OD: palatal free, bar preferred to single anchors, metalframework, facial support provided Single anchors: parallel axes
5-7 implants	Fixed cantilever with wrap around, CAD CAM framework Titanium, one piece screw retained, facial support provided
6 – 10 implants and good bone	FPP ceramometal: segmented, screw retained or cementation on abutments, symmetrical distribution suggested CAD CAM Titanium, Zirconia, one piece, screw retained
space available:	≥15mm between maxillary crest and mandibular teeth: OD suggested or wrap around
sagittal class lll / need for lip support	– OD or wrap around prosthesis suggested
Edentulous mandible	
Interforaminal position	
2 -3 implants	OD: bar or single anchors (curvature!) possible
4 implants	OD: rigid bar preferred to single anchors, length of bar segments!!
4-5 implants	Fixed cantilever prosthesis with wrap around technique CAD CAM framework Titanium screw retained
Anterior / posterior	
≥6 implants, good bone	FPP segmented (screw retained), symmetrical distribution suggested
	CAD CAM Titanium,/ Zirconia, no splinting from molars to molars!!
4 implants in position of teeth 35,33,43,45	SDA or short cantilevers for first molars, one piece framework, PFM, CAD CAM TitaniumZirconia

6. PASSIVE FIT

6.1. Cementation or Screw Retention

One topic of studies was the type of fixation of fixed prostheses, namely cementation on abutments or screw retention, mostly via abutments. The measurements were conducted with heat processed cast metal alloys supported be only few implants. The conclusion was that passive fit is better obtained by cementation but the gap between abutment shoulder and superstructure is larger, which may evoke an inflammatory reaction of the tissues. Veneering of the framework additionally affects its fit. The tightening sequence with screw retention otherwise affects stress development and misfit. Due to heat processing of cast metal-alloys precise fit decreases with larger frameworks and stress is induced, if the superstructure is fixed to the implants. Subsequently technicians suggest that frameworks are preferably segmented. This otherwise hinders cross arch stabilization which was an early rule for implants in the edentulous jaw and advised be researchers. Most implant manufacturers offer a broad choice of prefabricated and custom made abutments, which enable cementation. Cementation appears to be user-friendly and is popular. Moreover, no access holes in the occlusal surfaces are present, which could hinder optimum design of the occlusal contacts. Thus, technicians and dentists master the cementation technique more easily. The use of angled abutment for axes correction clinically disguises non-axial functional loads. The visual perception of the connection between implant and superstructures implies that of an axial load transfer. It is so far not known, whether such non axial connection and load transfer has a negative impact on the bone or over time, may lead to fatigue of the materials. However, as mentioned previously, the problem of axis correction becomes much better under control with computer assisted 3-D analysis and virtual implant planning. Therefore, today strong prosthetically driven implant planning favors screw retention for large frameworks.

6.2. Passive Fit, Stress, Misfit

Passive fit is a frequently disputed aspect of implant-prostheses and regarded from different perspectives. It is the aim of the dentists, the technicians and manufacturers of implants and superstructure components to achieve passive fit. The goal must be to mount superstructures and frameworks to implants, if not fully stress-free, though with as little stress as possible. Passive fit means to be a protection of the bone by avoiding stress concentration particularly in the cervical portion of the jawbone, but it also means a protection of the superstructure and a prevention of technical complications. While evidence with regard to crestal bone behavior around the implant under stress is less clear, stress and strain comprising undue loading conditions, favor technical complications and failure of the prostheses. However, clinically passive fit is not really measurable and cannot be reached. Various studies analyzed biomechanical aspects, stress and strain development and fit or distortion of prostheses by means of experimental, laboratory methods such as finite element analysis or stress gauge measurements and complex 3-D measuring eqquipments. Behavior of superstructures with regard to design, material, fabrication process, number of implants and fixation mechanism with or without abutments on the implants was investigated. All these parameters play a role in the development of misfit. Technologies like spark erosion or Cresco Ti and laser welding were developed and suggested for optimized fit of the superstructures. Clinically misfit is explored by a probe, by thin disclosing matarial a d a feeling of rocketing or strain when positioning the superstructure or tightening the screws Cross arch screw tightening may reduce misfit and is suggested (Figure 19).

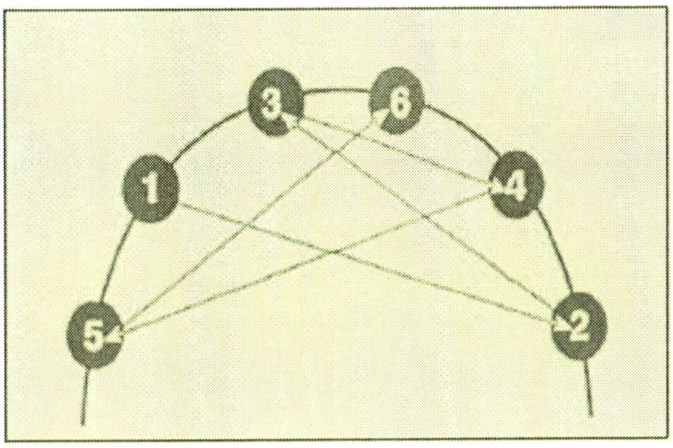

Figure 19. Screw tightening of large frameworks. Cross arch tightening reduces misfit.

Radiographically a gap may be visible. If such signs of misfit are clinically detectable, imprecision of the frameworks has reached a degree that is clinically not acceptable. Particularly ith extended, large cross-arch frameworks from gold alloys misfit may become visible and clinically detectable. Such frameworks are of significant weight as compared to modern milled materials. Thus, segmentation in smaller pieces is preferred, which requires up to 10 implants. Obviously passive fit will not be achieved by our traditional techniques and therefore the primary goal in clinics is to minimize misfit. A typical sign of stress and misfit is loosening of abutments and occlusal screws. One important step forward in the direction to reduce misfit is the CAD CAM fabrication of bars and frameworks, mostly Titanium or Zirconia. This modern technology of designing and milling processes enhances quality and precision of fit. Just recently, the CAD CAM fabrication of large cross-arch frameworks fitting on multiple implants became possible. Another step forward in this direction is that these frameworks – and also bars - are sitting directly on the implant-shoulder without interposition of an abutment. This configuration of framework connection may have biomechanical and technical advantages with only one interface and only one screw for fixation. Likewise, this connection between implant and superstructure appears to promote higher stability with a significant reduction of screw loosening. As of today this method of framework fabrication is not possible with all implant types and CAD CAM systems. Implant manufacturers offer implant systems with an internal or external often hexagonal connection or flat to flat contact surface. There is not sufficient evidence to show which connection is advantageous. Under load it appears that a minimal opening and development of a gap between abutment and superstructure have to be expected. One study showed high precision of one piece CAD CAM frameworks with minimal misfit in all dimensions. The axial misfit appears to be more important then lateral misfit. The axial distortion (gap between implant shoulder and framework) appears to introduce the highest misfit and tightening of the prosthetic screws increases stress. With a direct connection at the implantlevel the tightening torque is higher than at the abutment level. High torque increses the prelaod at screw joint and increases the stress level. Otherwise, the loosening of abutments is a sign of misfit at the interface abutment / superstructure. So far it is not possible to suggest the best type of connection i.e. design of the implant shoulder, with or without abutment. However it appears that the more components are introduced in a system and the more complexe these are (angled, small screws, geometry) the more sensitive becomes the laboratory processes, which in turn may increase misfit and imprecision. It could not be demonstrated in the animal model that static load of implants has negatively influenced the osseointegration and bone remodeling proccess

7. Occlusion

The occlusal contacts during mastication are of limited time duration while nonchewing function with tooth to tooth contact is often observed during prolonged time periods, particularly in light and heavy bruxists. Therefore occlusal contact schemes are of importance, not exclusively with regard to chewing function. Principles of therapeutic occlusion apply to two basic conditions:

1. Occlusal scheme for complete dentures: This often comprises bilateral guidance and lingualized occlusal contacts. An anterior free-way space of 1 to 2 mm is observed to hinder tilting of the upper denture from strong front teeth contacts. Characteristics are: Cusp-to-fossa contacts in centric occlusion and simultaneous guidance on working and non-working sides.
2. Occlusal scheme for fixed prostheses with natural teeth: This often comprises freedom in centric, with a canine-protected lateral guidance on the working side and no balancing contacts. Front teeth contacts and anterior guidance are provided. The anterior path from a most retrusive contact position to maximum intercuspidation (1 to 2 mm) is without interferences.

These empirical occlusal schemes were eventually adopted with minor modifications for implant prostheses. In implant prosthodontics a specific evidence-based occlusal philosophy has not yet been developed. However, there are a few specific guidelines, which may favor optimum load distribution onto the implants and ensure stability of the dentures. The greater the number of implants placed and the greater the rigidity of the prosthetic connection are, the more the occlusal scheme may resemble freedom in centric. From a biomechanical point of view, however, a balanced occlusal guidance as utilized with complete dentures might favor equilibration of occlusal loads onto the implants due to simultaneous contacts on the working and non-working sides.

1. With all types of implant- reconstructions, occlusal contacts and guidance of buccal cusps on premolars and molars should be avoided.
2. The occlusal scheme has to favor the jaw with the weaker, less stable type of reconstruction.

3. Vertical implant axes in relation to the occlusal plane, symmetrical distribution of the implants over the arch and (primary) splinting may favor appropriate load transmission and reduce bending moments.

These guidelines are of importance when teeth are in contact otherwise than when chewing. It has a protective function particularly with regard to parafunctional habits like grinding. The following figures represent possible combination of implant retained and supported prostheses in the edentulous jaws.

7.1. Occlusal Concept of Complete Dentures

Indications for a bilaterally balanced occlusion are a combination of mandibular overdentures supported by a few implants occluding with a complete denture in the opposing jaw. The same scheme can be recommended for mandibular overdentures occluding with maxillary overdentures. It hinders minor loosening and tilting of the overdentures. With anterior bars in both jaws however front teeth contacts and biting on front teeth is possible without destabilization of the overdenture. If a bar supprted overdenture opposes natural dentition or fixed implant prostheses, balancing contacts as built up with complete dentures, may also contribute to the denture stability. This is particularly true for the anterior overdenture concept in the maxilla. Contacts of the non-working side on posterior madibular cantilevers must strongly be avoided.

7.2. Freedom in Centric

Indications are multiple implants supporting bridgework and occluding with fixed prostheses or natural teeth. Clinicians often debate the lateral guidance of the working side and if an implant should be placed in the location of the canine. While a canine-protected lateral guidance is easy for the technician to build up, a group function could protect the implant in the position of the maxillary canine. Full lateral guidance on a single implant should be avoided, e.g. single implant crown on maxillary canine. However, in the completely edentulous jaw primary or secondary splinting is provided and can be considered a protection.

For an even distribution of the implants over the maxillary arch and particularly if segmentation of the frameworks is desired a canine implant is mandatory and alateral guidance on the implant cannot be avoided.

7.3. Overdentures and Complete Dentures (Figure 20)

Occlusion: for all combinations balanced, lingualized occlusion and bilateral guidance.
Material: prevalent custom made acrylic teeth

7.4. Overdentures in Both Jaws, or Mandibular Cantilever Prosthesis on İnterforaminal İmplants (Figure 21)

1. maxilla, anterior overdenture concept, mandibular overdenture / or cantilever prosthesis:
    - Occlusion: bilateral guidance, canine and front teeth guidance is possible, Cave: no lateral guidance on distal cantilevers , centric contacts on cantilevers releaved
    - Material: prevalent custom made acrylic teeth
2. maxilla, posterior overdenture concept, mandibular overdenture / or cantilever prosthesis:
    - Occlusion: bilateral guidance not necessary, canine and front teeth guidance is possible, Cave: no lateral guidance on distal cantilevers , centric contacts on cantilevers releaved
    - Material: prevalent custom made acrylic teeth

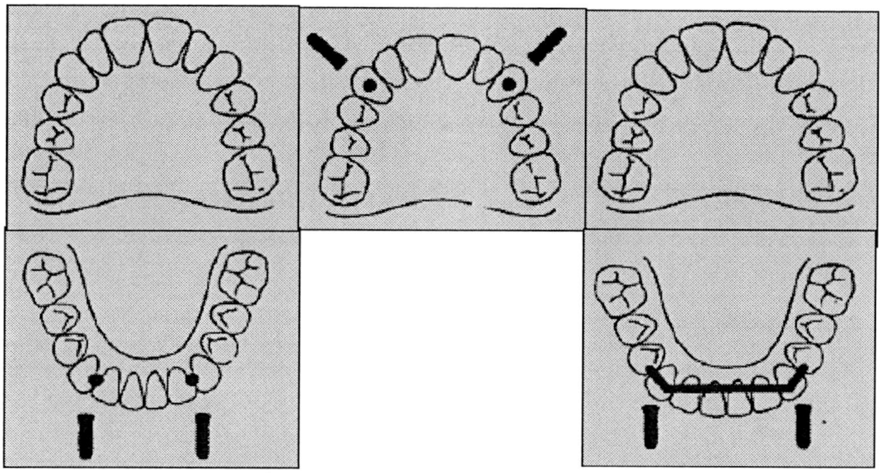

Figure 20. Overdentures and complete dentures.

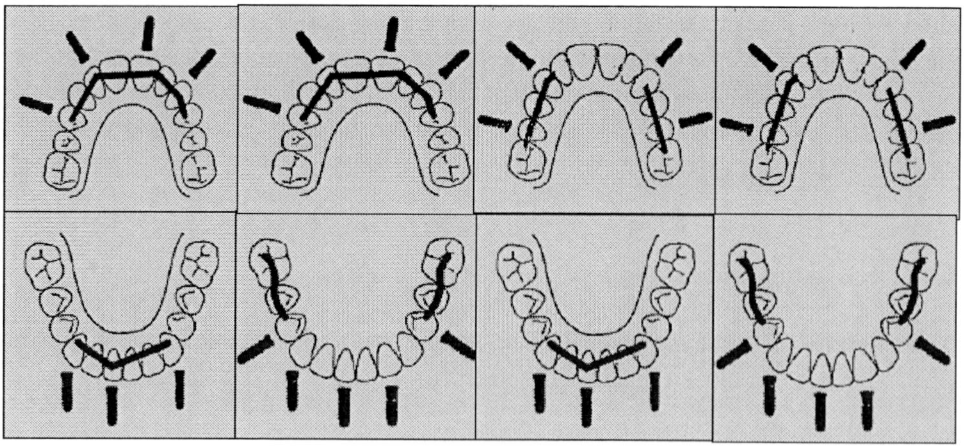

Figure 21. Overdentures in both jaws, or mandibular cantilever prosthesis on interforaminal implants.

7.5. Overdentures in Maxilla, Mandibular Fixed Prosthesis on İnterforaminal and Posterior İmplants (Figure 22)

1. maxilla, with extended bar, mandible with natural teeth (or implants with fixed prosthesse):
   - Occlusion: bilateral guidance not necessary, canine and front teeth guidance is possible, avoid contacts on buccal cusps of maxilla
   - Material: maxilla: custom made acrylic teeth, mandible: enamel, acrylic material, ceramics
2. maxilla, with anterior overdenture concept, mandible with implants and fixed prosthesis:
   - Occlusion: bilateral guidance recommended, light canine and front teeth guidance is possible, avoid contacts on buccal cusps of maxilla
   - Material: maxilla: custom made acrylic teeth, mandible: enamel, acrylic material, ceramics
3. maxilla, with posterior overdenture concept, mandible with implants and fixed prosthesis:
   - Occlusion: no bilateral guidance required, canine and front teeth guidance is possible, Cave: front teeth of maxilla are not supported by implants, no strong contacts exclusively on front teeth, avoid contacts on buccal cusps of maxilla
   - Material: maxilla: custom made acrylic teeth, mandible: enamel, acrylic material, ceramics

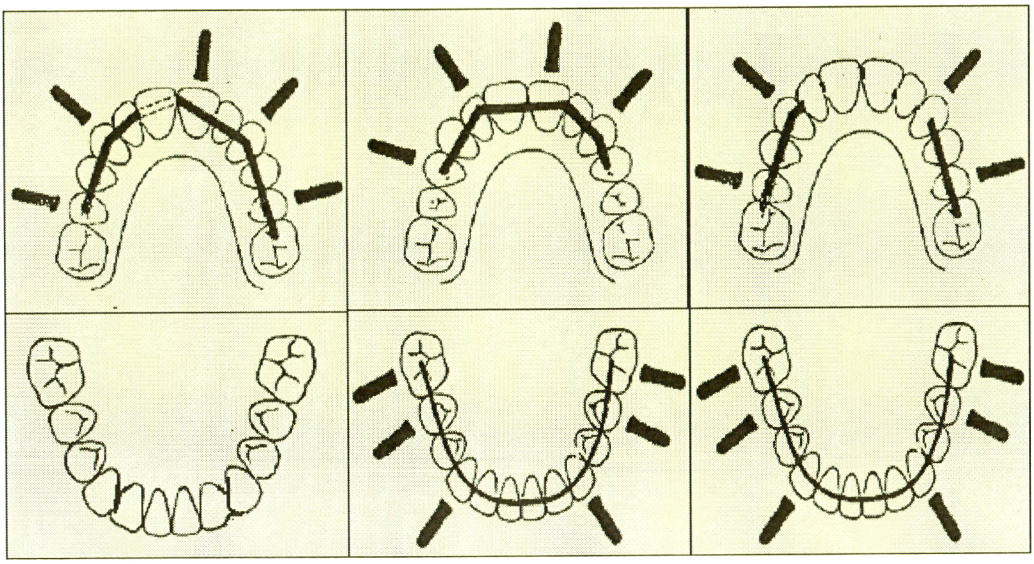

Figure 22. Overdentures in maxilla, mandibular fixed prosthesis on interforaminal and posterior implants.

7.6. Fixed Prosthesis in Maxilla with Posterior Support, Mandibular Fixed Prosthesis or Natural Teeth (Figure 23)

1. maxilla: fixed prosthesis with posterior support, mandible with natural teeth
   - Occlusion: canine and front teeth guidance, avoid contacts on buccal cusps of maxilla
   - Material: maxilla: acrylic tooth material, or PFM or Zirconia with ceramic veneering, mandible: enamel, composite, ceramics
2. maxilla: fixed prosthesis with posterior support, mandible with fixed prosthesis and posterior support
   - Occlusion: bilateral guidance recommended, canine and front teeth guidance, avoid contacts on buccal cusps of maxilla
   - Material: maxilla: acrylic tooth material, or PFM or Zirconia with ceramic veneering, mandible: maxilla:
     - acrylic tooth material, or PFM or Zirconia with ceramic veneering, avoid Zirconia reconstructions in both jaw. With Zirconia in maxilla, acrylic material in mandible recommended. Cave: PFM heavy wight of frameworks!!
3. maxilla: fixed prosthesis with posterior support, mandible with fixed cantileever prosthesis

   - Occlusion: no bilateral guidance, canine and front teeth guidance, no strong contacts on cantilecer. Centric stops releaved by thickness of shimstock foil, avoid contacts on buccal cusps of maxilla
   - Material: maxilla: acrylic tooth material, or PFM or Zirconia with ceramic veneering, mandible: maxilla: acrylic tooth material
   - Best recommendation: in both jaws titanium framework, wrap around technique, acrylic tooth material

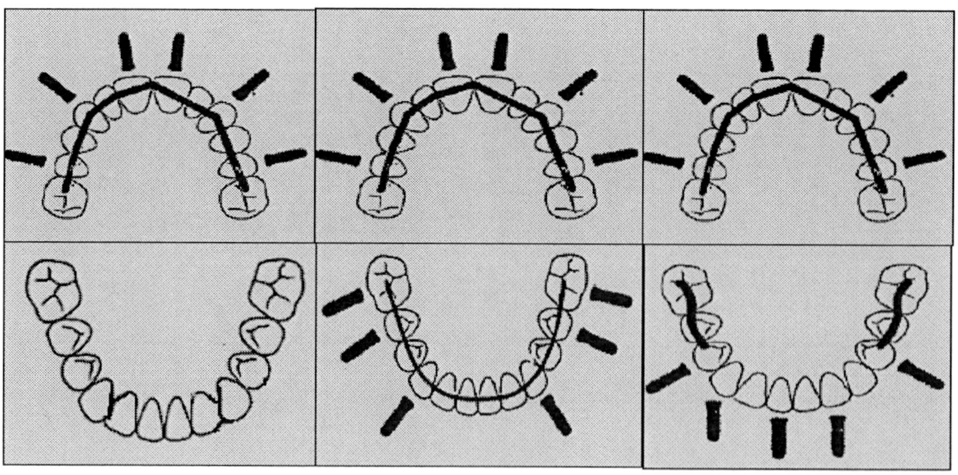

Figure 23. Fixed prosthesis in maxilla with posterior support, mandibular fixed prosthesis or natural teeth.

7.7. Occluding Material

Basically occluding surfaces of the same material in contact appears favorable. Resin tooth material will exhibit more wear over long time, but nowadays the material quality has significantly improced. Ceramic veneering may exhibit chipping and fractures under undue loading. A gold alloy surface is a favorite to occlude with all type of oposing material, but today not frequently used. The occluding material of overdentures and fixed CAD CAM titanium prosthesis with the wrap around technique consists of resin teeth.

Various configurations are observed:

- Resin / Resin
- PFM PFM
- Resin / PFM
- Resin / Zirconia
- Resin / Natural teeth
- PFM / Natural teeth
- Zirconia / Natural teeth

Not recommended: Zirconia / Ziorconia

Conclusion

Prosthetic principles have not changed basically, over many years. Scientific and clinical evidence from clinial practice brought new directions in implant dentistry. The future of implant dentistry is strongly influenced by new technologies Prosthetically driven implant dentistry is gradually being accomplished and has more and more translated into clinical concepts. Backward planning is the trend and is enabled by modern software. In this respect the complete denture is the basis of indepth knowledge and remains a highly important planning instrument, even though clinically the treatment with full dentures will eventually disappear. CAD CAM technologies will determine the development of material processing, material properties, the quality of laboratory fabrication and design of superstructures by adequate software. More safety, higher precision and predictability of outcome is achieved. However, this does not mean that prosthodontics are or become again technically dominated. High standards of planning, of quality of material and techniques comprise biological principles in implant dentistry.

In: Biomechanics of Dental Implants
Editor: Murat Çehreli

ISBN 978-1-62100-780-7

Chapter 12

IMPLANT SUPPORTED FIXED PARTIAL PROSTHESES: CURRENT CONCEPTS AND FUTURE DIRECTIONS

Kıvanç Akça

1. GENERAL VIEW: PROSTHODONTICS AND DENTAL IMPLANTS

Dental implants have been considered, researched and developed specifically for patients with edentulous mandible whom generally experience functional insufficiencies and psychological problems with conventional tissue supported complete dentures. Use of dental implants in clinical practice literally followed definition of biological interaction between implant and bone, namely called as osseointegration. Long-term successful results for both fixed and removable prostheses supported by implants placed between the mental foramens established the scientific basis for fundamental and permanent changes in the principles of prosthetic dentistry. In the following years, the same principles were applied successfully to unilateral distal edentulous patients, bilateral distal edentulous patients, patients with extended edentulous areas, and single tooth loss. Consequently implants have become the most attractive field in dentistry in terms of scientific research and clinical practice. In conclusion, when evaluated scientifically, implants are found to hold an indispensible place among the clinical applications of modern dentistry (Figure 1). This advancement, which took roughly 30 years to solidify, has also been presented in a variety of consensus reports, workshops and guidelines. All the developments in this filed to date can generally be classified under two major categories: concepts and applications. Today's clinical application principles of dental implantology have also emerged as a result of scientific evaluation of concepts and applications. Although dental implantology has indeed made great strides in such a short time to occupy a place in routine clinical proactie, the current technology is not likely to be the endpoint of this field due to the nature of science. As a result, a third developmental classification can potentially be added for potential future developments.

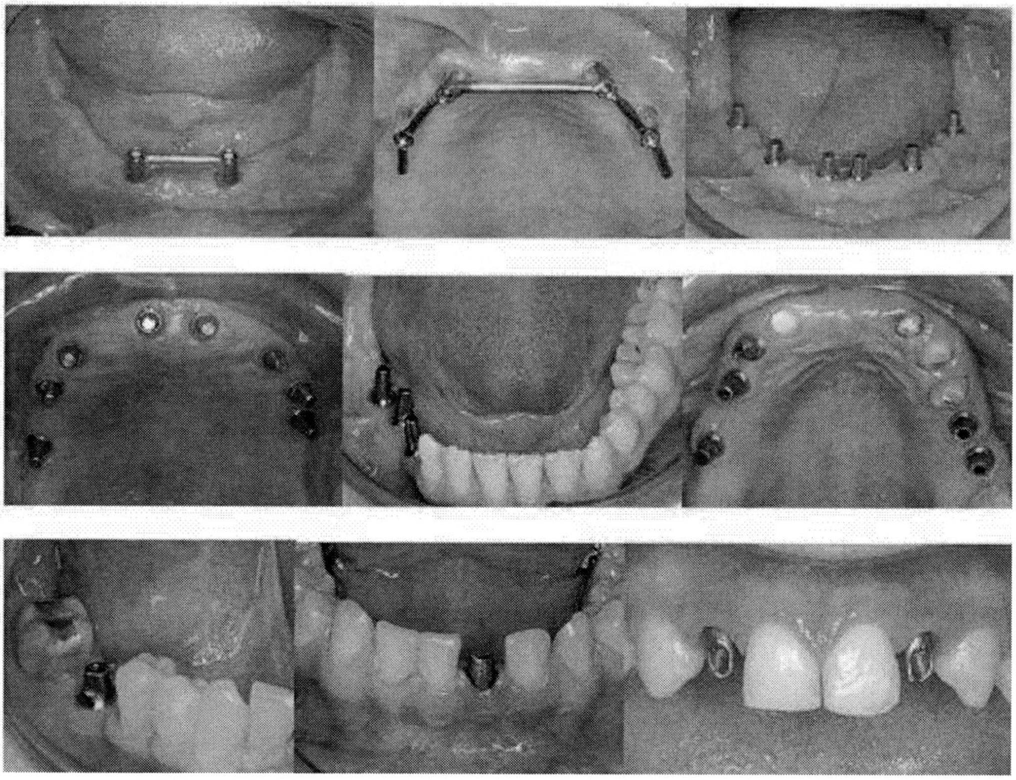

Figure 1. Dental implants successfully used to support mandibular and maxillary overdenture, (a and b), mandibular and maxillary fixed prostheses (c and d), fixed partial prosthesis for unilateral distal edentulism (e), fixed partial prostheses for long span edentulims (f), single tooth missing tooth in functional area and mandibular anterior region (g and h) and in esthetic zone(i).

2. Development of Concepts

In the period following the identification of osseointegration, conceptual developments have generally focused on the implant-bone interface as well as extending the use of implants in the treatment of all missing tooth/teeth cases. In fact, it was the requirements of prosthodontic treatment that guided these two main research avenues. Even though prosthodontic needs provide the foundation for dental implantology, achieving the biological and mechanical retention of the implants in the bone into which they are surgically placed is one of the most important factors in serving these prosthodontic needs. Because of this relationship, the implant-bone interface is one of the most highly studied topics in the conceptual field. Studies in this context have aimed at decreasing the time required or otherwise intentional stimulation for osseointegration. The common aim in both cases is to complete the treatment with implant supported prostheses in the shortest time by minimizing the edentulous period. The conceptual categories, which emerged from surgical procedures and have gained acceptance in today's clinical application, include immediate, early and late placement of implants following extraction sockets. Similarly, these categories also involve prosthodontic loading time of implants as immediate, early and late following implant placement. As such, conceptual processes that minimize even eliminate edentulous period have become the subject of scientific research and have shaped the principles of today's clinical applications (Figure 2).

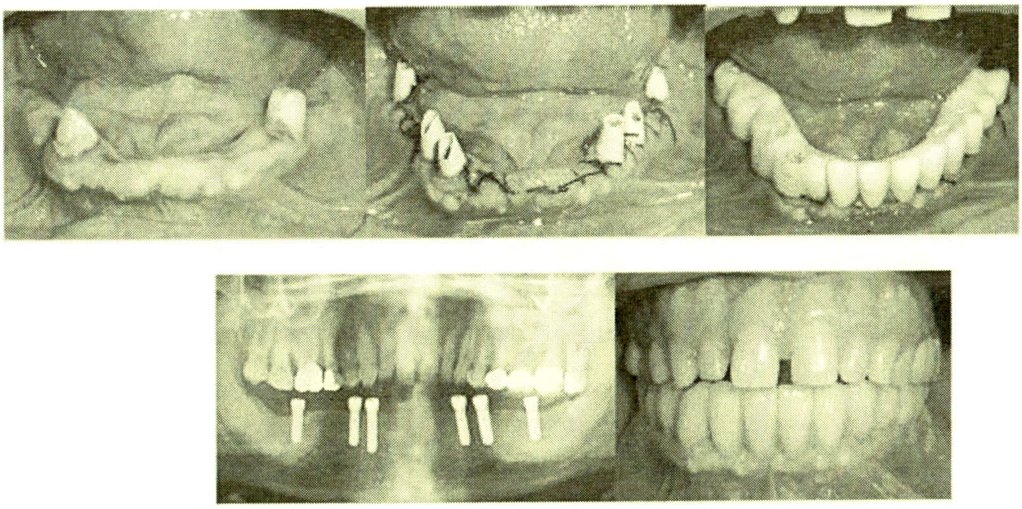

Figure 2. Elimination of edentulism with immediate placement and immediate loading of dental implants placed in mandible to support fixed prosthesis. Existing mandibular arch with #34 and #43 teeth (a), temporary abutments to support immediately loaded fixed prostheses (b and c), radiographic follow-up of two months (d) to start permanent restorations (e).

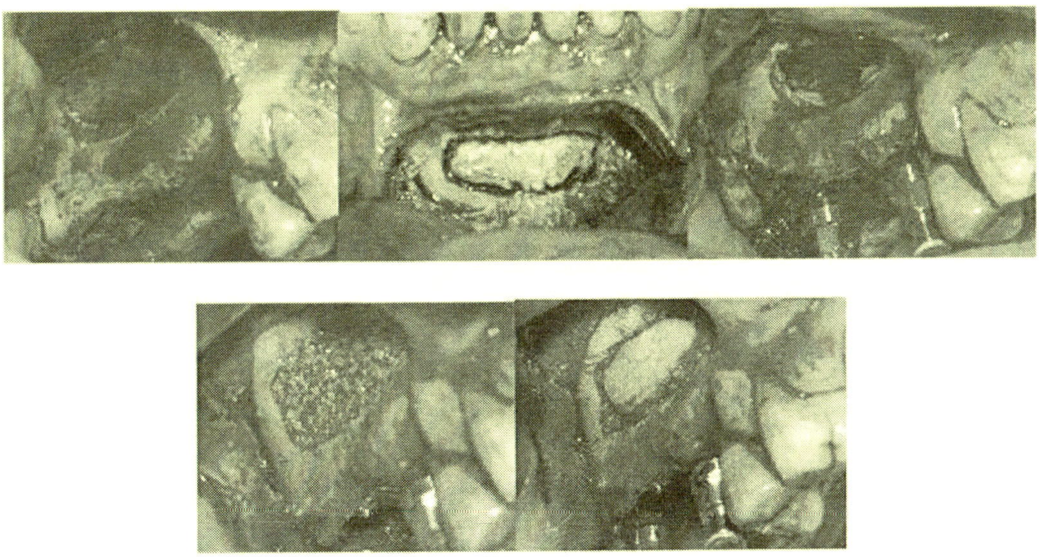

Figure 3. Advanced surgical approach to restore unilateral maxillary distal edentulism with implant supported fixed partial prostheses (a). Intraoral autogenous bone grafting (b) to support simultaneous implant placement with sinus elevation procedure (b to e).

The second main topic in the development of concepts is improvement of existing bone tissue when they may cause problems in the surgical placement of implants due to insufficiencies. This field, again, aims at satisfying prosthodontic needs. The concepts have emerged from the development of advanced surgical techniques and related biological materials, and, in general, the resulting methods can be classified as guided tissue regeneration methods, bone grafting methods and sinus elevation methods (Figure 3).

2.1. Development of Applications

Clinical applications that developed following, but more often simultaneously with, the aforementioned concepts have mostly evolved with the goal of meeting the daily-life expectations of patients. In this sense, treatment of tooth losses and associated aesthetic requirements with implants has come distinctly to the forefront. Accordingly clinicians were placed under pressure to fulfill this clinical expectations and therefore clinical practice by trial and error actually eclipsed scientifically supported applications. Therefore, both biological and mechanical failures were often observed during this period (Figure 4). Conversely, today, dental implantology has completed its developmental steps for clinical practice, and common application principles have largely taken shape. In reaching this point, the correlation of preclinical studies with evidence-based clinical application principles has played a crucial role in understanding the biomechanical characteristics of implants and bone integrity. In short, through evaluation of applications that follow conceptual developments from the standpoint of preclinical and clinical studies, the guidelines for implant supported prostheses in clinical practice of contemporary dentistry have become quite clear.

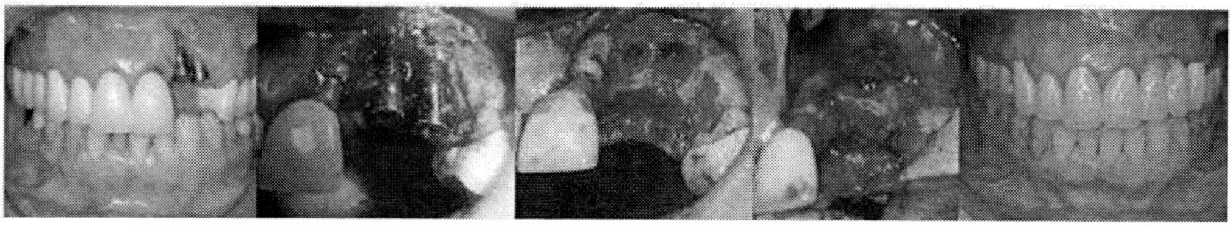

Figure 4. Implant failure with regards to periimplant-soft (a) and hard (b) tissues due to clinical application with lack of scientific support. Treatment of the case with teeth supported fixed partial prostheses following removal and advanced surgical hard/soft tissue grafting procedures (c to e).

2.2. Future Developments

The fact that successful treatment of tooth loss with implants, as emphasized briefly above, has perhaps become the most important and established field in dentistry does not indicate that a developmental plateau has been reached. The technology today will face a scientific body with different views, and changes will ensue together with future developments. However, they will be mostly limited to the applied implant and associated materials along with their application technologies. In this context, especially dental ceramics and developments related to CAD/CAM technology will lead research and development (Figure 5).

In addition to these, it is also highly likely that developments will take place to allow implant supported prosthesis design with lesser application of advanced surgical techniques. As a result, correlation of *in vitro* experimental studies that feed preclinical animal studies with evidence-based clinical studies will be important for the development of clinical applications in implant dentistry.

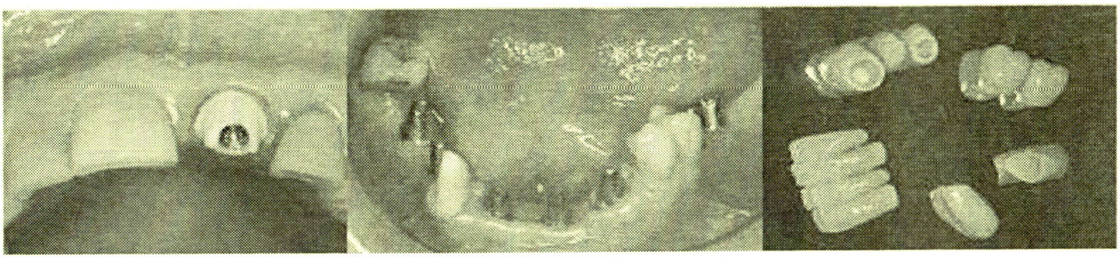

Figure 5. CAD/CAM technology used in implant supported prostheses for customized abutment production (a) and the restorations themselves (b to c).

There is no universal classification for implant supported prostheses in dentistry. However, classification based on missing teeth may be meaningful when arranging implants in order of usage if implants are considered as providing support and retention to the prostheses used in the treatment of teeth loss. A classification of dental implant supported prostheses according to teeth loss status is given below:

- Complete Edentulism Cases
  - Implant supported removable prostheses
  - Implant supported fixed prostheses
- Partial Edentulism and Single Tooth Missing Cases
  - Implant supported fixed partial prostheses
  - Implant supported single tooth restorations

In this section, clinical application principles that are accepted as in the foundation for scientific support in the treatment of partial edentulism will be discussed in detail in relation to scientific studies and new directions. In this context, it should be kept in mind that scientific studies that form the basis for clinical application are referred to as preclinical and clinical scientific studies can be classified under the biomechanical heading.

3. Factors Contributing to the Success of Implant Supported Fixed Partial Prostheses

The two main factors needed to achieve ultimate long-term success with implant supported fixed partial prostheses are an implant-based prosthesis system and adequate hard/soft tissues to house the implants.

Scientific studies related to implant-abutment-prosthesis complex have focused on implant design, implant geometry and the implant-abutment connection. The most explanatory description in implant design comes from the relation of the implant to prosthesis transition to the bone tissue in vertical direction. Implants that meet with a prosthesis placed vertically above the bone level are referred to as soft tissue implants, while those that interface with a prosthesis at the level of the bone are referred to as bone level implants. Thus, commercially available implants for clinical applications by dentists are referred to either as bone level or soft tissue level implants (Figure 6).

For several years, both implant designs have been researched for their prosthodontic advantages and disadvantages in scientific preclinical and clinical studies. In these studies, the most widely studied topics are load transfer to the peri-implant tissues by the prostheses and the effects of those forces, especially on the bone in the cervical region (Figure 7 and 8).

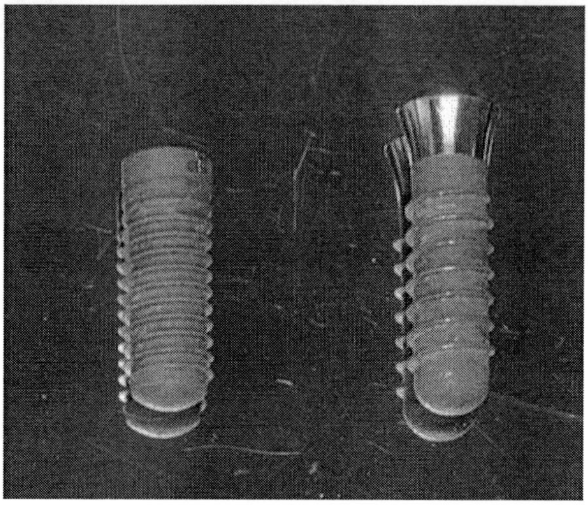

Figure 6. Bone- (left) and soft tissue (right) -level implant designs.

Although there are some studies that show both implant designs to have clinically similar long-term bone stability, loss of approximately 1.5 mm of cervical bone, specifically within the first year following placement of prosthesis, is deemed acceptable for bone level implants in success criteria. In contrast, bone loss within the first year is not included in the soft tissue level implant success criteria. The reasons for this difference have been widely investigated scientifically. Such studies have focused on the effects of mechanical forces concentrated in the peri-implant region and the biological effect of the microgap at the implant-prosthesis interface. The physiological effects of the forces on the bone tissue in the peri-implant region have been studied only with *in vitro* experimental models, algorithmic simulations, animal studies and *ex vivo* experimental conditions. Because conducting similar studies in clinics is impossible for ethical reasons, the *in vivo* effects of forces in the peri-implant region can only be understood by evaluating results from clinical studies conducted within the scope of evidence-based applications in dentistry. Moreover, the biological effects of the microgap at the implant-prosthesis interface have been scientifically researched only in animal studies and to some extent in clinical studies. The observed difference in cervical bone level between bone and soft tissue level implants within one year following the placement of prosthesis can be explained mostly biologically after evaluating all the available studies. In other words, it is concluded that the association of the implant-prosthesis microgap with the bone level in vertical direction is what determines the early stage bone physiology. Because of this, soft tissue level implants are thought to be more suitable for early stage stabilization of bone in the cervical region. However, for aesthetic reasons that are particularly important in the maxillary anterior region, clinicians feel uncomfortable about the transmucosal part that remains above the bone tissue and thus may become visible in the soft tissue implants (Figure 9).

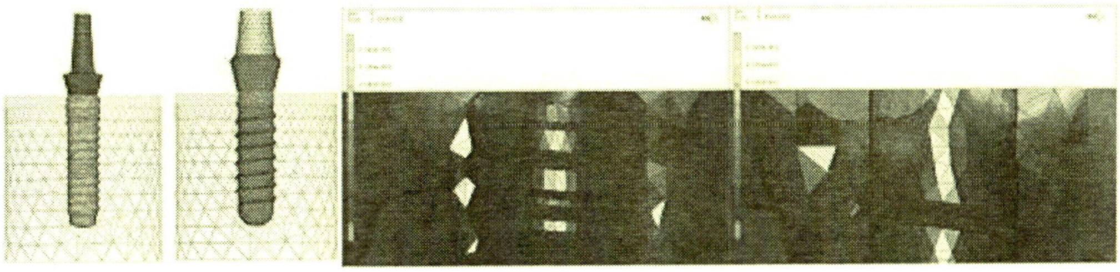

Figure 7. Quantitative and qualitative comparison of load transfer of bone- and soft tissue-level implants in an mathematical simulation (a to c).

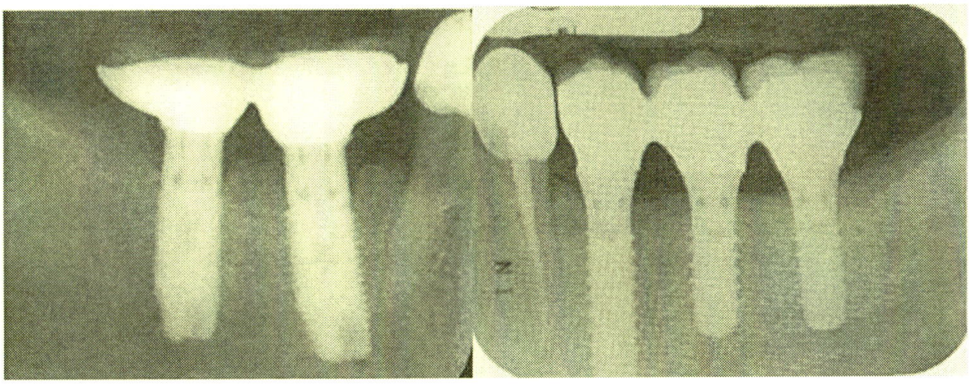

Figure 8. Radiographic comparison of bone- and soft tissue-level implants in clinical practice (a and b).

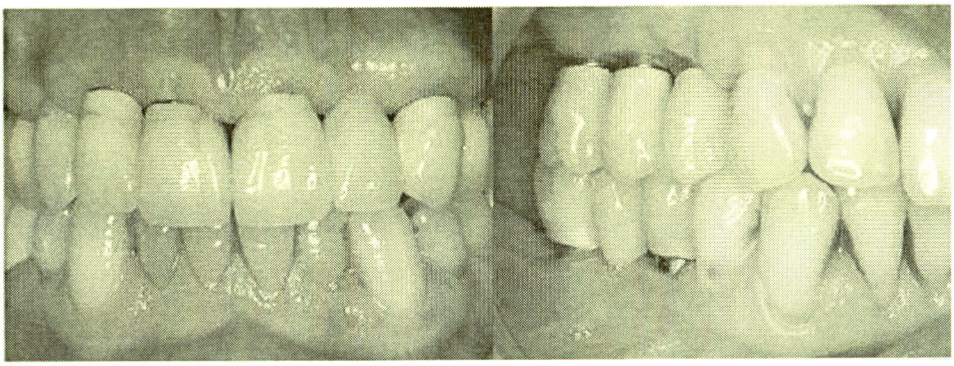

Figure 9. Exposure of metallic transmucosal part of tissue-level implants to support fixed prostheses in esthetic zone (a) and functional area (b).

To assess this inadequacy of soft tissue level implant design, the implant-abutment-prosthesis interface has been scientifically reevaluated in the last decade. Studies have concentrated on the horizontal position of the implant-abutment interface in accordance to implant long axis at bone level. In studies where this relationship shifted towards the implant center, results were so positively encouraging that scientific studies speedily changed their focus to move in this direction. Limited but meaningful scientific data with this redefinition of the implant-abutment connection for bone level implants demonstrated that they have similar early-term peri-implant bone stabilization to soft tissue level implants (Figure 10). Similar biomechanical studies have been performed on the peri-implant region to reevaluate relevant results from this new concept, and rather than mechanical factors related to force distribution, a mucosal barrier was biologically shown to form in this region as a result of relocation of the microgap in the bone level implants towards to the implant center (Figure 11). As a result, going forward, both implant designs can be used with equal success rates on implant supported fixed partial prostheses. Preference regarding which implant design will be used is influenced by the expertise of the clinician. However, it is recommended that decision be made according to the edentulism, the surgical protocol and the prosthesis design. For example, while bone level implants are preferred for fixed prostheses in the maxillary anterior region, soft tissue level implants should be preferred for removable prostheses in the treatment of edentulous patients (Figure 12).

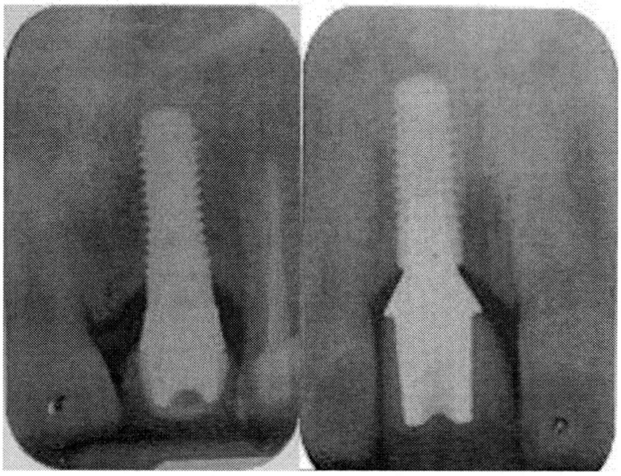

Figure 10. Similar periimplant bone stability with tissue level implants (a) and bone-level implants with laterally placed abutment connection (b).

Implant geometry has been studied primarily under two main topics: macro and micro. As characteristics of macrogeometry, implant shape, and design and the number of the threads on its surface are evaluated. Parallel-sided or tapered cylindrical implants were scientifically demonstrated by several studies to be the only reasonable choice in the early years of implantology. Nevertheless, the physical structure and number of threads still remain highly studied subjects in the macro level definition of the implant surface. The primary factor in these studies is the idea that threads are effective for mechanical and biological stabilization of the bone tissue surrounding the implant following its placement. On the other hand, it has been demonstrated mostly in preclinical experimental studies that different designs and numbers of threads have similar outcomes in the intraosseous stabilization. In similar fashion, it has been shown that implants with different thread geometry also have similar long-term clinical results. Microgeometric characteristics, on the other hand, are related to the implant surface properties. The matter of highest importance related to microgeometric specifications is the time required to achieve osseointegration. Without question, scientific studies have demonstrated that rough surfaces formed particularly by subtractive procedures from the implant surface positively influence osseointegration time. Therefore, in the treatment principles for implant supported fixed prostheses, implant geometry is more important in saving time biologically than it is for prosthodontic effect.

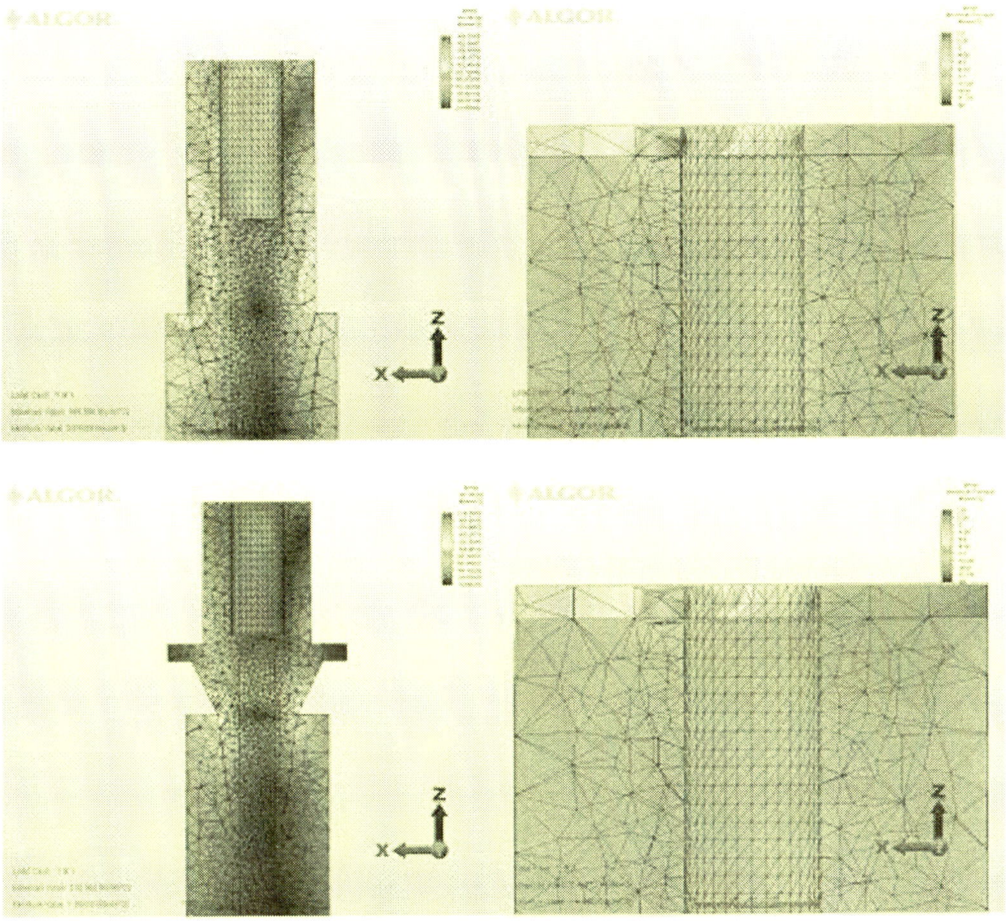

Figure 11. Comparative force transmission to periimplant bone tissue with regular butt-joint (a and b) and lateralized-joint (c and d) abutment connection in bone-level implant design.

The combined effects of implant macro- and micro-geometry at the bone level in peri-implant region have been subject of several scientific studies. In this vein, preclinical studies have mostly evaluated forces transmitted by the implant body to the bone surrounding the designed cervical region. For this purpose, strain-gauge, photo-elastic and mathematical models have been used in researches (Figure 13). However, there are obvious deficiencies in these experimental studies in terms of the accurate integration of biological factors into the models. Because the evaluation of the effect of biological factors in the living tissues is restricted for ethical reasons, the validity of results from experimental studies should again be evaluated according to the principles of evidence-based dentistry in clinical applications. When these kinds of such studies are examined carefully, noticeable differences are observed in some areas between preclinical studies and clinical results. Based on clinical findings, it can be concluded that stabilization of bone in the cervical region related to macro level differences in cervical region geometry does not correlate well with the type of implant.

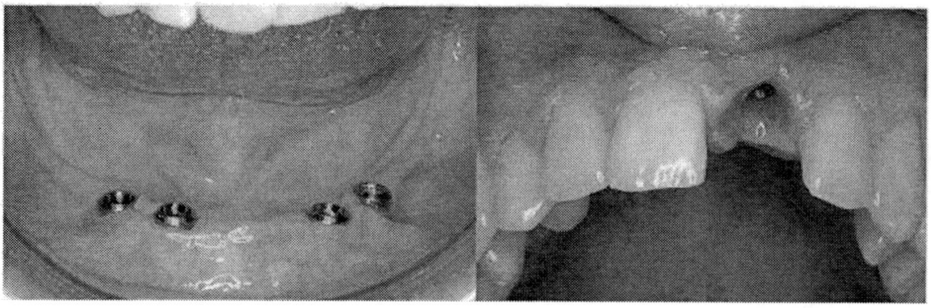

Figure 12. Soft-tissue level implants to support removable prostheses (a) and bone level implant to support single-tooth restoration (b).

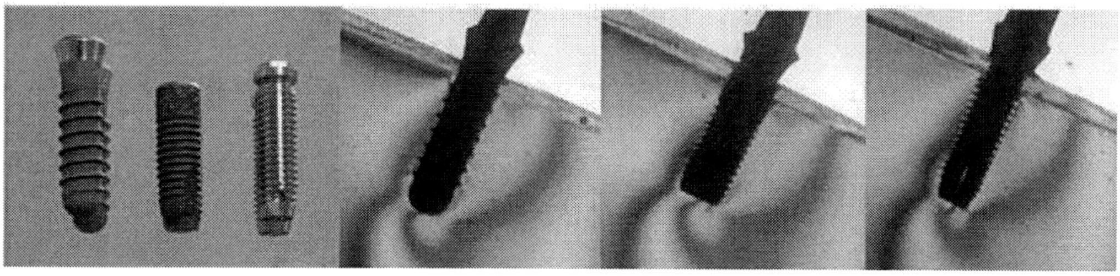

Figure 13. Soft-tissue and bone level implants with different neck designs (a) used to evaluate load transfer in a photoelastic stress analysis (b to d).

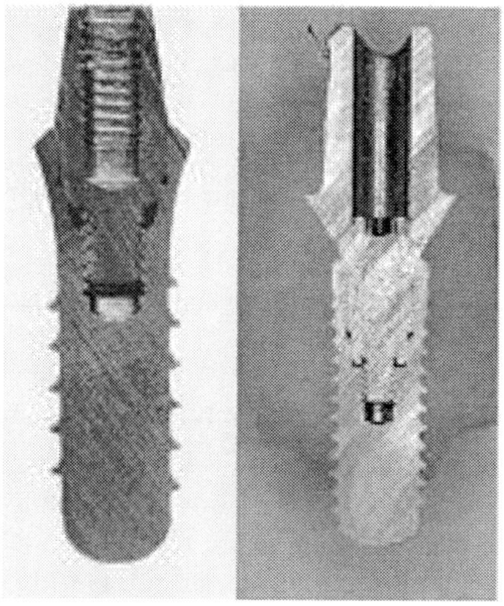

Figure 14. Internal conical abutment connections within a soft tissue (a) and bone (b) level implant.

Deterioration of implant-abutment stability subsequent to oral functions initially results in mechanical complications such as breakage of the screw and/or the abutment. Thus, implant-abutment interface has been, in particular, mechanically evaluated especially in *in vitro* experimental studies. The most important subject of evaluation in these studies has been the behavior of this interface under applied mechanical forces. This is probably the most important topic where biological factors do not play a primary role, and there is thus much less doubt on the validity of *in vitro* results. The validity of preclinical studies has been confirmed by high congruence with the clinical results regarding the implant-abutment interface stabilization. In this manner, the mechanical effectiveness of abutments that can be referred to as internal conical connection has been confirmed with both preclinical studies and clinical results (Figure 14).

Apart from the implant-prosthesis relationship, another equally important factor in the success of implant supported fixed prostheses is the integrity of the hard and soft tissues that will house the whole system. Again, we can also evaluate all the components here both biologically and mechanically. In clinical applications, mechanical factors are closely related to functional requirements, while biological factors are more closely related to aesthetic requirements.

Functional requirements correlate most with overcoming the reduced chewing ability due to teeth loss in partially edentulous patients. In other words, functional requirements relate to restoration of the lost maxillomandibular occlusal relationships. This, in turn, involves posterior mandibular and maxillary teeth. Although the hard tissue that will house the implants in both regions has similar anatomical limitations, there are distinct differences in the respective surgical solutions. While the maxillary sinus forms an anatomic boundary in the upper jaw, the lower jaw contains the mandibular canal. Both of those anatomical formations influence the implant length that will be placed into the bone in the respective region. In cases of extreme limitations, conditions can be improved with advanced surgical techniques to facilitate the use of an optimal implant lenght. It is difficult to plan *in vitro* studies and determine their clinical validity regarding the success of the implants used for bone conditions that are thus enhanced using advanced surgical techniques (Figure 15). This is because we do not yet fully understand the biomechanical properties of the regenerated bone. Thus, evaluations aiming toward clinical applications can be deduced partially based upon results from preclinical animal studies. However, the validity of these conclusions will be in question because of species-specific tissue variations. Under these conditions, clinical applications have been shaped by the results from clinical studies within evidence-based dentistry application principles. The unknown information in these situations is the kind of reaction that the regenerated bone tissue will have when faced with normal functional forces in the long term. In addition, the inability to perform clinical studies on this topic compounds the situation. In situations like these, in general, changes are made to the number and diameter of the implants that will be used as supports for fixed partial prostheses prepared for clinical applications (Figure 16).

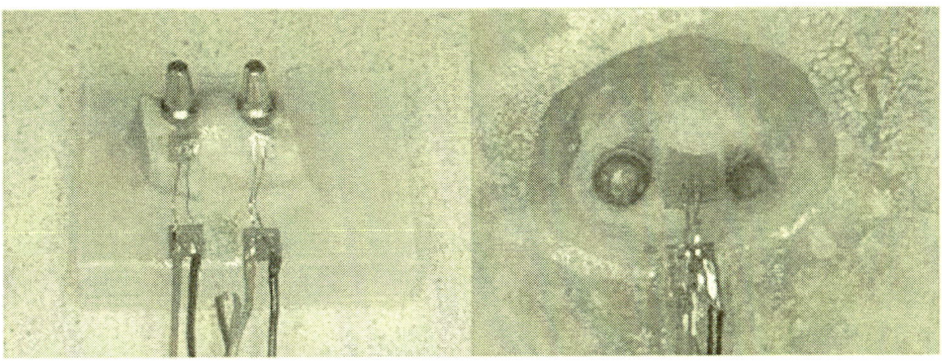

Figure 15. Strain-gauge analyses in a simulation of advanced surgical approach applied in maxillary posterior region (a and b).

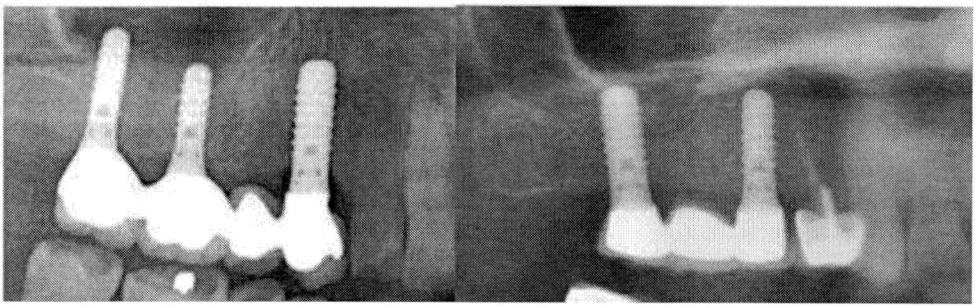

Figure 16. Lack sufficient knowledge about the biomechanical bone behavior of regenerated bone, number of placed implants placed with such advanced surgical procedures are increased (a) in comparison to cases with implants placed in native bone (b) in posterior region.

However, aesthetic requirements are related to elimination of the changes to the lower facial region as a result of tooth/teeth loss. Losses in the maxillary anterior region mostly negatively affect aesthetics in partial edentulous cases. Although this region traditionally contains bilateral canine teeth and incisors, premolar teeth are also included as they become visible when smiling. Satisfaction of aesthetic requirements is harder here compared to fulfilling functional requirements due to subjective factors. In the definition of aesthetics for implant supported fixed partial prostheses, the seemingly simplest, yet hardest to achieve, criterion is imitation of the natural teeth. Because of the anatomic structure of the related region, positioning of the hard tissues following tooth/teeth loss that will house the implant towards to palate is the first physiological process. Additionally, the presence of periodontal disease or regional trauma further complicates this situation. This is because loss of the existing bone occurs not only in the horizontal direction but in the vertical direction as well. As the natural position of the lost teeth and the surrounding soft tissues are used as criterion for aesthetic requirements, advanced surgical procedures become inevitable for successful outcomes. In addition to the morphologic structure of the region involved rehabilitation of the missing hard and soft tissues with implant supported prostheses requires a high level of knowledge and expertise. Possible complications that occur in these particular cases, in turn, bring about detrimental consequences. Research on this aesthetic region is currently, in part, taking place in preclinical studies that are only limited to animal models. Treatment principles, however, are generally established in accordance with results from clinical studies (Figure 17).

4. Planning of Implant Supported Fixed Partial Prostheses

Among the factors that affect the implant treatment plan in today's clinical applications, the implant number and dimensions have been the subjects of several scientific studies. There are plenty number of *in vitro* experimental biomechanical studies related to this topic. Even though these biomechanical studies have guided clinical applications, there exist clinical studies that are included in the evidence-based dentistry application principles that have provided more acceptable results. This is due to lack in clinical support for the results of *in vitro* experimental studies. For example, in *in vitro* experimental studies where basic engineering knowledge is put forth, it has been emphasized that using an implant for each lost tooth to be replaced should reduce the forces applied to the peri-implant region, thereby increasing the success rate. In contrast, results from clinical studies suggest that two implants placed into native bone are adequate for the treatment of three missing teeth (Figure 18).

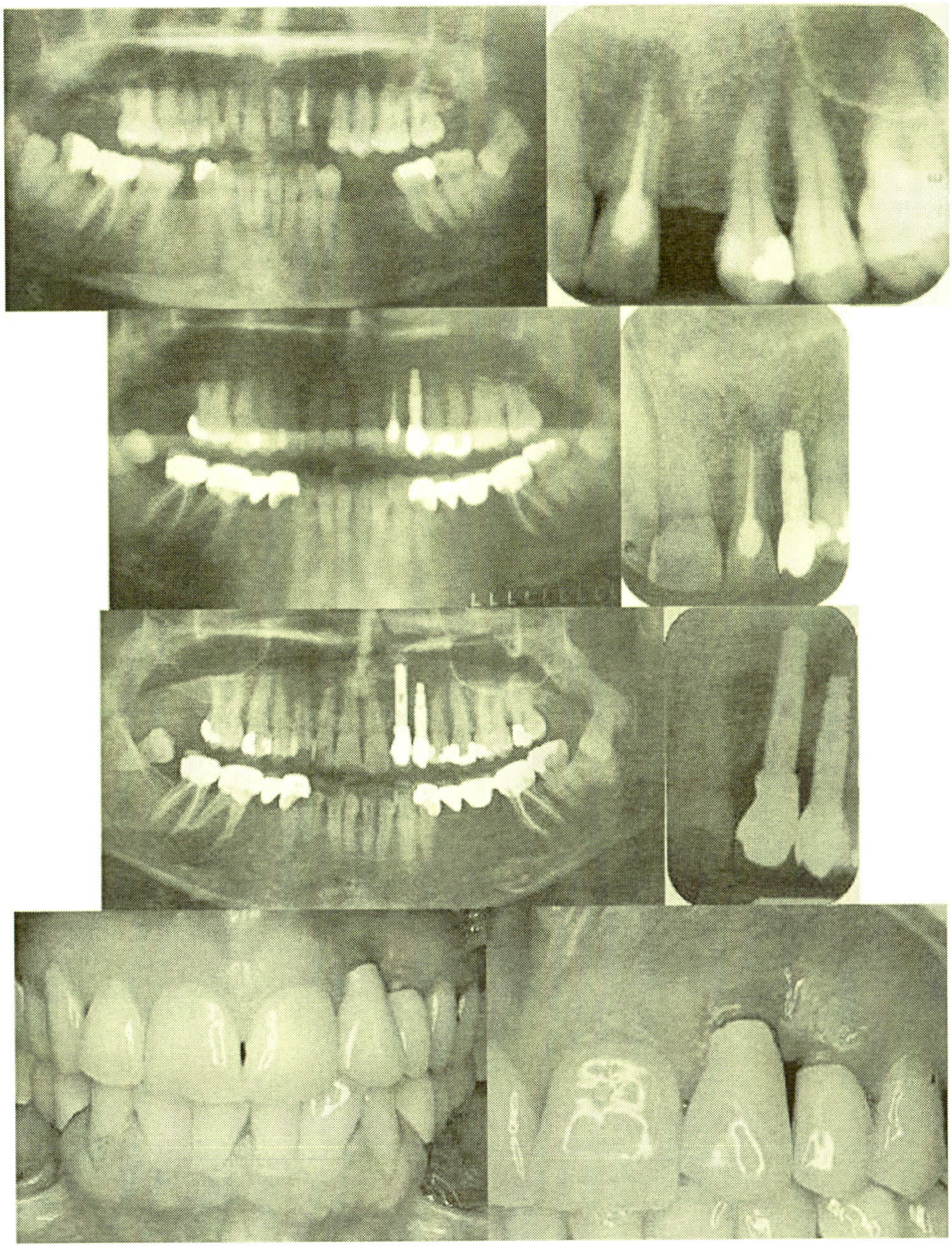

Figure 17. Specifically failures related with high expectations of patients came one step forward in constitution of treatment principles implant treatment in esthetic zone when compared to related scientific support. Trial and error based 12-year follow learning of a case in anterior region (a to h). Conventional surgical placement of #23 was followed by immediate placement of #22 in 8-years of period. Resulting with two adjacent implants at unilateral position with reduced esthetic outcomes.

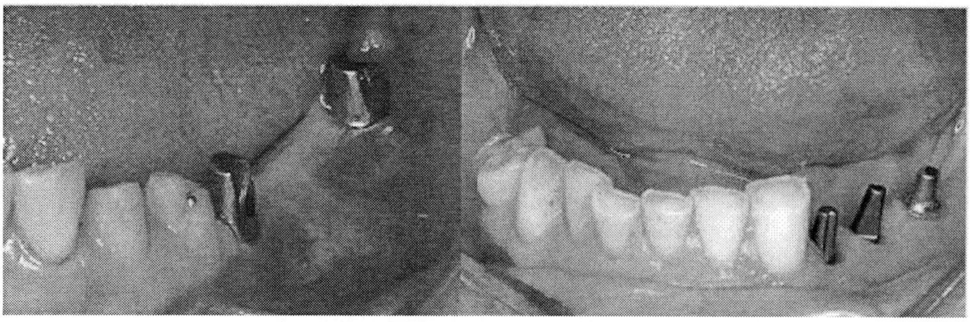

Figure 18. Two regular diameter rough surfaced implants with at least 10mm length are biomechanically sufficient to support three missing occlusal units (a), whereas three implants may be referred in existence of parafunctional activities (b).

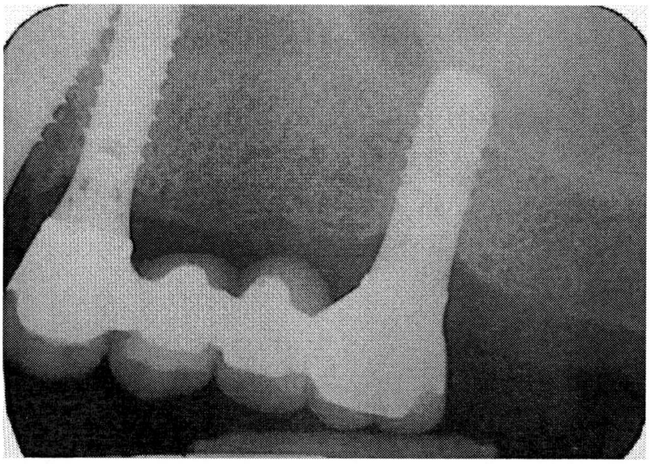

Figure 19. Use of wide diameter implants may reduce the requirement of additional implant placement for replacement of missing four occlusal units in cases with no other modifying factors.

In similar *in vitro* studies, application of widest and longest possible implant has been reported to increase success by decreasing the forces exerted at the peri-implant region. Unfortunately, again these findings were not supported by clinical studies. For determination the implant height, biological factors have been found to become more relevant relative to the implant surface, and rough surfaces with a 10-13 mm implant height have been shown to provide optimal conditions. Rather than following the clinical study requirements and using the widest possible implant in each case that bone will permit, planning of the implant diameter has been found, from clinical studies, to be an important factor. For example, while a fixed partial prosthesis can be supported by two wide implants placed at either end, in the case of four missing posterior teeth, this number decreases to three missing posterior teeth with standard diameter implants in routine clinical applications (Figure 19). Because aesthetic requirements are very important in the maxillary anterior region, differences were found compared to the purely functional region in the planning of implant supported fixed partial prostheses.

These differences originate from results from clinical rather than preclinical studies. In the functional region, if standard diameter cannot be employed for loss of two adjacent teeth because of existing bone insufficiencies, two narrow-diameter implants with two splinted crowns can provide sufficient support. Similarly, juxtaposition of three narrow-diameter implants along with three splinted crowns can complete the treatment for loss of three teeth in the posterior region. Results from preclinical studies revealed that narrow-diameter implants did not have the necessary mechanical properties, a finding that was further corroborated by clinical studies (Figure 20). However, except in the case of two missing middle incisors, juxtaposition of two implants is not recommended for the anterior region for aesthetic reasons (Figure 21). Therefore, in cases where two standard implants cannot be used as treatment for loss of three teeth in the anterior region because of the existing bone, three adjacent narrow-diameter implants should not be applied. In conclusion, although results from clinical studies have partially validated results from the preclinical biomedical studies used to determine implant number and diameter for use in implant supported fixed partial prostheses in functional area, the application of the results for use in aesthetic regions is much more limited due to the distinct differences between functional and aesthetic region requirements.

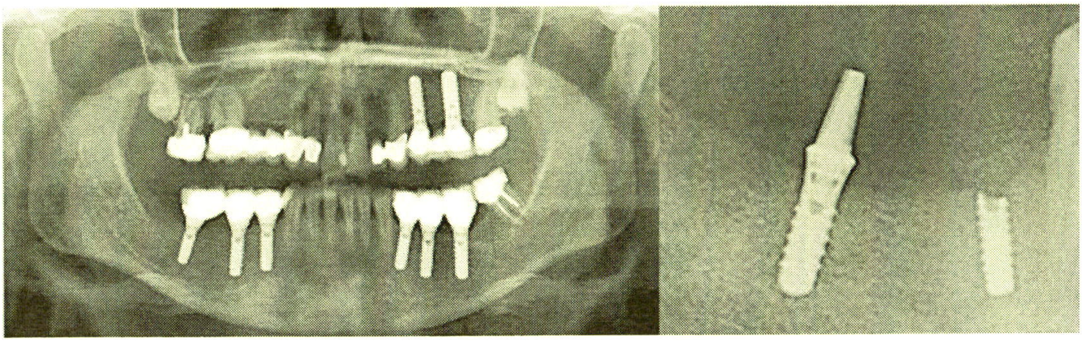

Figure 20. On daily clinical applications use of narrow-diameter implants need further implant support with splinted restorations (a) to avoid mechanical irreversible complications likely to occur with use of them on the regular base (b).

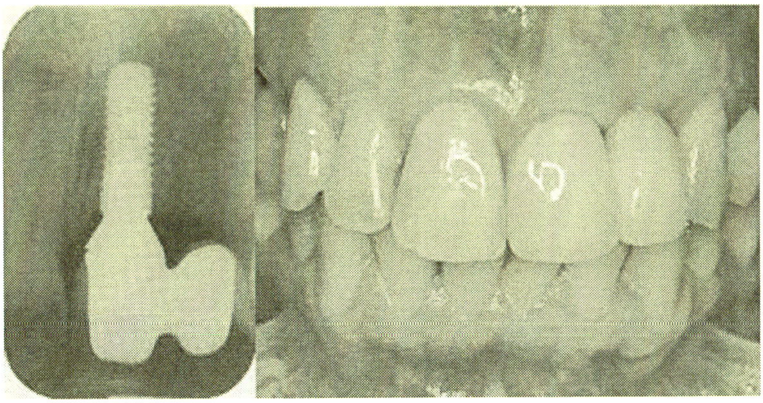

Figure 21. Treatment principles differs between esthetic zone and functional area. In functional area two narrow diameter implants with splinted crowns would be applicable to avoid advanced surgical approaches. However two adjacent implant placement for unilateral missing two teeth in esthetic zone likely to end with esthetic failure (refer Figure17). Placement of an regular diameter with advanced surgical approaches to support a cantileved two unit fixed restorations is highly suggested to fulfill esthetic expectations (a and b).

Determination of the type of abutment to be used in fixed prostheses is an important clinical step. Currently in connection prostheses to the implants are either with use of screw or cement retention (Figure 22). Both screw-retained and cement-retained restorations have been the subject of several *in vitro* preclinical studies. One of the highly evaluated topics in this field, related to function following placement of prostheses, is the evaluation of the forces transmitted to the implant and peri-implant tissues. *In vitro* experimental studies have demonstrated that both types of restorations cannot be passively connected to the implants. However, cement-retained restorations were found to be closer to passive adaptation. Unfortunately, it is not possible for preclinical biomechanical studies to be carried out under clinical conditions. However, when cement-retained and screw-retained fixed partial prostheses were compared, no differences were observed with respect to the tissues surrounding the peri-implant region in clinical studies. It is thought that in *in vitro* experimental studies, when an incompatible screw-retained prosthesis is placed with applied force through its retainer, excessive forces are concentrated on the peri-implant tissues, thereby resulting in biological complications along with negative impact on the bone turnover. This observation, however, has not been verified clinically. It is thought that loosening or breakage of the screws in screw-retained prostheses at an early stage may be due to unusually vigorous tightening of the screws. Thus, it is thought that uneven distribution of forces more likely affects the screw connecting the implant with the prosthesis rather than the bone in the peri-implant region. Similar to previous experiments, results from these *in vitro* experimental studies were not corroborated by clinical studies.

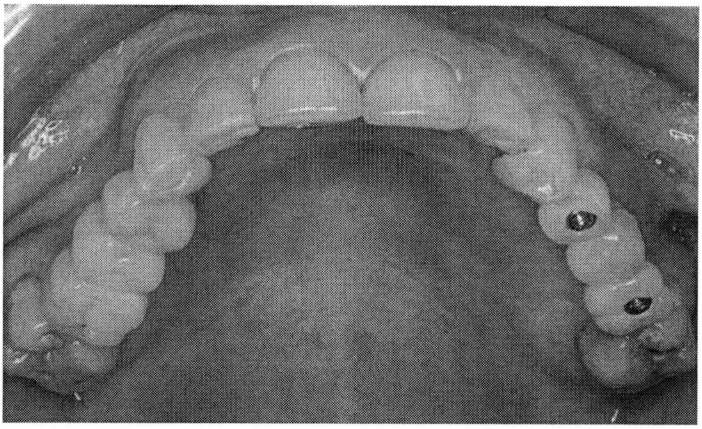

Figure 22. Screw (left)- and cement(right)-retained implant supported fixed prostheses in maxillary arch.

Topics for *in vitro* experimental biomechanical studies have often been cantilever extensions in the design of fixed partial prosthesis superstructures intended for force distribution to the tissues surrounding the peri-implant region. In these studies, it has been demonstrated that concentration of forces specifically in the region near implant surface neighboring the cantilever would result in biological complications. Conversely, results obtained again from clinical studies have clearly showed that two implant supported fixed partial prostheses with one extended occlusal unit have been used with no adverse effects on the success rate or life span of the prosthesis in functional area (Figure 23, 24 and 25).

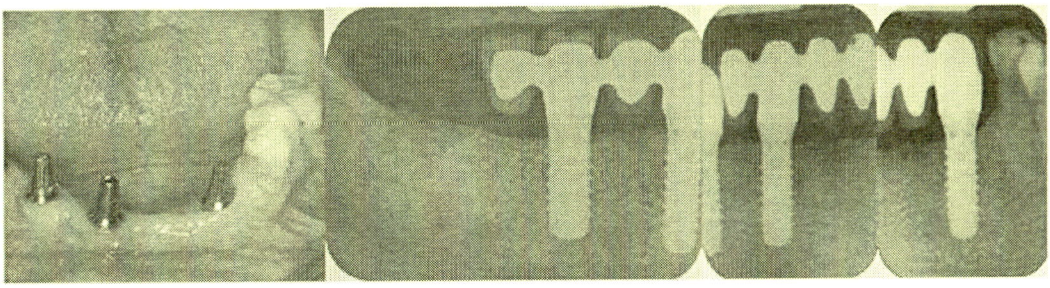

Figure 23. Cantilever designs used in implant supported fixed prostheses to optimize implant number (a) in treatment of mandibular long span edentulism, and their 10-year radiographic follow-up (b to d).

Several *in vitro* experimental studies that have investigated the connection between the implant and natural tooth have largely been refuted due to the hypothesis that biological complications would arise in the pre-implant region as a result of cantilever impact on the implant support (Figure 26). For the functional region, the presence of increased occlusal force has been demonstrated on the implants as a result of support from the tooth and implant on the fixed prostheses with three occlusal members; however, this force was not found to cause biological complications in the peri-implant region. This clinical finding was further supported by preclinical animal studies that were aimed at evaluating effect of increased occlusal forces, which confirmed the lack of biological complications in the peri-implant region.

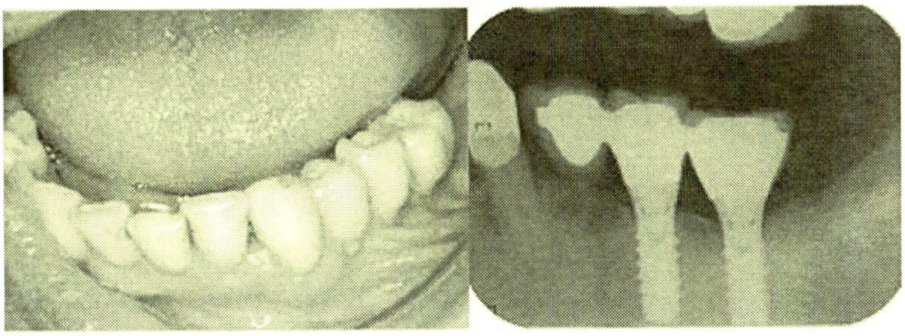

Figure 24. Cantilevered fixed partial prostheses with two adjacent implants placed due to unfavorable position of mental foramen in mandibular posterior region and its 10-year radiographic follow-up (a and b).

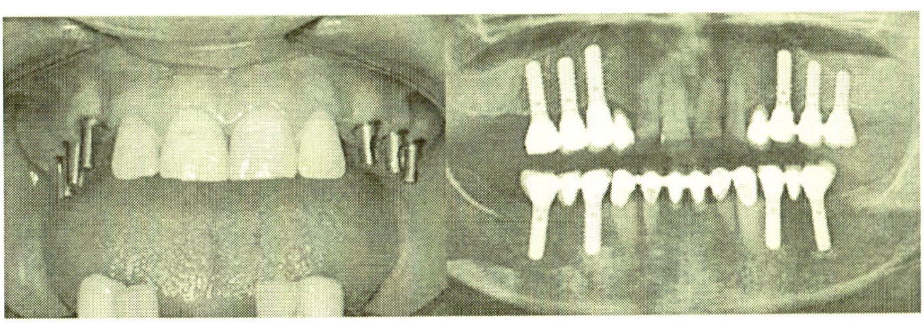

Figure 25. Implant supported fixed partial prostheses into cantilever extensions into canine teeth designed to avoid advanced surgical approaches to enhance the soft/hard tissues in the related region (a) and their 10-year radiographic follow-up (b).

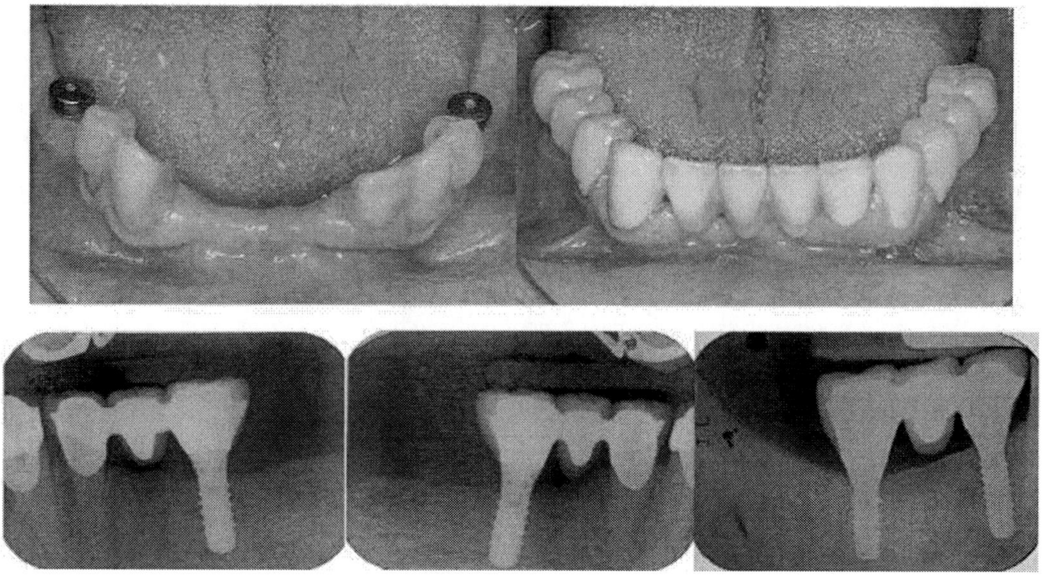

Figure 26. Tooth to implant connection (a and b) results in favorable long-term results (c and d) for cases to restricted three occlusal unit replacement in posterior region with rigid design.

5. Future Directions in Clinical Applications

Principles of daily clinical applications in dental implantology have arisen from evaluation of the results of preclinical studies in accordance to with evidence-based dentistry clinical studies over a long period of time. Because of this, rather than seeking developments that will result in large changes to the practices in the field, focus was given more to improvements in the materials and techniques used. All-ceramic and CAD/CAM technology serve as good examples of this. In preclinical studies on this topic, the requirement for biological tissue integration is minimal; thus, accurate evaluation of results from these studies can more directly lead to changes in clinical applications. For example, reliability of results from an *in vitro* study on the strength of glass ceramic crowns produced directly in the clinic for the restoration of single tooth implants with metal abutment will be much more useful than results from a preclinical study on the effect of single tooth implant supporting two unit restoration with cantilevered extensions on the bone in the pre-implant region.

In the future, we expect to see decreased use of advanced surgical approaches, allowing implant treatments to become more widespread. In this context, the most promising approach is to increase the durability of narrow-diameter implants to allow their placement under unfavorable bone conditions without the need for advanced surgical techniques. In relation to this, the durability of fabricated Ti-Zi alloy implants has been improved, and use of narrow-diameter implants has begun (Figure 27). On the other hand, the effects of changes in force transmission around the peri-implant region are not known due to changes in properties of these materials. Further preclinical experimental studies should be helpful in uncovering these kinds of information. In planning these studies, the differences and similarities between preclinical and clinical studies need to be kept in mind. Preclinical studies conducted this way should produce more realistic results for clinical studies.

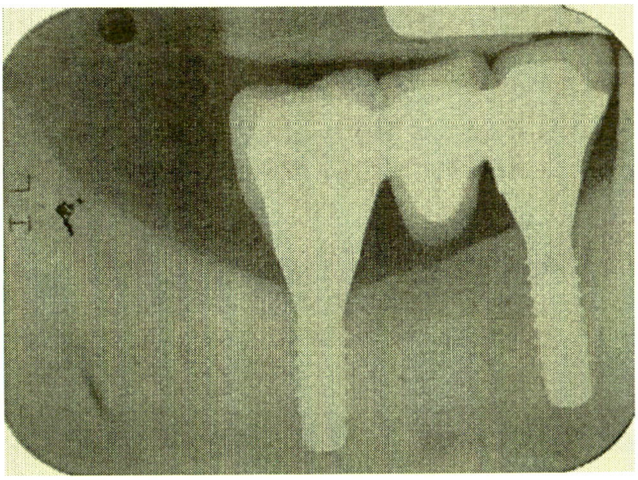

Figure 27. Innovation of new implant material is promising to use of narrow diameter implants on daily practice with the same prosthetic principles applied with regular diameter implants. Need for advanced surgical techniques to house regular diameter implants or increasing number of reduced diameter implants would be beneficial for treatment costs and time loss.

CONCLUDING REMARKS

During the developmental phases of dental implantology, it was believed that *in vitro* experimental biomechanical studies would primarily serve as a guide to clinical applications, and planning of prosthesis would be required to remain within these boundaries. Nevertheless, a better understanding of biological tissues gained through animal studies and long term results from clinical studies aimed evidence-based dentistry applications have helped us understand the limitations of *in vitro* experimental studies. For this reason, efforts should be made to integrate results from animal and clinical studies into experimental studies as much as possible. Thus verified *in vitro* models together with the results obtained from studying various conditions will, in turn, give results with higher validity.

Although efforts to create experimental conditions similar that mimic the physiological medium for dental implants continue to increase, results from these studies will be mostly limited to the functional region. Due to differences in the criteria used to evaluate results from the aesthetic region versus the functional region, there will not be much contribution by preclinical studies towards the planning of implant supported fixed partial prosthesis for the aesthetic region. The topic of utmost importance, in light of information gleaned from preclinical biomechanical studies aimed at clinical application, is the definition and designation of the effects, on the peri-implant tissues, of the forces applied the implant supported fixed partial prostheses under normal conditions in the living tissues. Preclinical studies should be expected to provide information about the normal, increased, and extreme forces for implant supported fixed prostheses and what the effects of these forces are on the tissues occupying the pre-implant region. If progress is made in this topic, preclinical studies can then effectively direct clinical applications. At present, biomechanical-based preclinical studies are limited to serving as the infrastructure for clinical evaluations to be made in clinical applications.

References

Aglietta, M., Siciliano, V.I., Zwahlen, M., Brägger, U., Pjetursson, B.E., Lang, N.P., and Salvi, G.E. (2009) A systematic review of the survival and complication rates of implant supported fixed dental prostheses with cantilever extensions after an observation period of at least 5 years. *Clinical Oral Implants Research*, 20,441-51.

Akça, K., Uysal, S. and Cehreli, M.C. (2006) Implant-tooth-supported fixed partial prostheses: correlations between in vivo occlusal bite forces and marginal bone reactions. *Clinical Oral Implants Research*,17,331-6.

Akça, K., Chang, T.L., Tekdemir, I. and Fanuscu MI. (2006) Biomechanical aspects of initial intraosseous stability and implant design: a quantitative micro-morphometric analysis. *Clinical Oral Implants Research*,17,465-72.

Akça, K. and Cehreli, M.C. (2008) Two-year prospective follow-up of implant/tooth-supported versus freestanding implant-supported fixed partial dentures. *International Journal of Periodontics and Restorative Dentistry*,28,593-9.

Akça, K., Fanuscu, M.I. and Caputo, A.A. (2008) Effect of compromised cortical bone on implant load distribution. *Journal of Prosthodontics*,17,616-20.

Akca, K., Cehreli, M.C. and Uysal, S. (2010) Marginal bone loss and prosthetic maintenance of bar-retained implant-supported overdentures: a prospective study. *International Journal of Oral and Maxillofacial Implants*,25,137-45.

Bilhan, H., Kutay, O., Arat, S., Cekici, A. and Cehreli, M.C. (2010) Astra Tech, Brånemark, and ITI implants in the rehabilitation of partial edentulism: two-year results. *Implant Dentistry*,19,437-46.

Brånemark, P.-I., Zarb, G.A. and Albrektsson, T (eds). (1985) Tissue-Integrated Prostheses: Osseointegration in Clinical Dentistry. Chicago: Quintessence.

Brunette, D.M. (1996) Critical Thinking: Understanding and Evalauting Dental Research. Quintessence Publishing Co..

Buser, D., Wismeijer, D. and Belser, U. (2008) ITI Treatment Guide Volume III: Implant Palcement in Post-Extraction sites, Treatment Options. Quintessence Publishing Co.

Cehreli, M., Duyck, J., De Cooman, M., Puers, R. and Naert, I. (2004) Implant design and interface force transfer: A photoelastic and strain-gauge analysis. *Clinical Oral Implants Research*,15,249-57.

Cehreli, M.C. and Akça, K., Iplikçioğlu, H. and Sahin, S. (2004) Fatigue resistance of implant-abutment junction in an internally notched morse-taper oral implant: influence of abutment design. *Clinical Oral Implants Research*,15,459-65.

Cehreli, M.C. and Akça K. (2004) Narrow-diameter implants as terminal support for occlusal three-unit FPDs: a biomechanical analysis. *International Journal of Periodontics and Restorative Dentistry*,24,513-9.

Cehreli, M.C., Akça, K. and Iplikçioğlu, H. 82004) Force transmission of one- and two-piece morse-taper oral implants: a nonlinear finite element analysis. *Clinical Oral Implants Research*,15,481-9.

Cehreli, M.C. and Akça, K. (2006) Impression techniques and misfit-induced strains on implant-supported superstructures: an in vitro study. *International Journal of Periodontics and Restorative Dentistry*,26,379-85.

Cehreli, M.C., Akkocaoglu, M., Comert, A., Tekdemir, I. and Akca, K. (2007) Bone strains around apically free versus grafted implants in the posterior maxilla of human cadavers. *Medical and Biological Engineering and Computing*,45,395-402.

Cehreli, M.C., Karasoy, D., Kökat, A.M., Akça, K., Eckert, S. (2010) A systematic review of marginal bone loss around implants retaining or supporting overdentures. *International Journal of Oral and Maxillofacial Implants*,25,266-77.

Cehreli, M.C., Karasoy, D., Kökat, A.M., Akça, K., Eckert, S. (2010) Systematic review of prosthetic maintenance requirements for implant-supported overdentures. *International Journal of Oral and Maxillofacial Implants*,25,163-80.

Cordaro, L., Torsello, F., Mirisola Di Torresanto, V. and Rossini, C. (2006) Retrospective evaluation of mandibular incisor replacement with narrow neck implants. *Clinical Oral Implants Research*,17,730-5.

de Oliveira, R.R., Novaes, A.B. Jr, Taba, M. Jr, Papalexiou, V. and Muglia, V.A. (2009) Bone remodeling adjacent to Morse cone-connection implants with platform switch: a fluorescence study in the dog mandible. *International Journal of Oral and Maxillofacial Implants*,24,257-66.

Eser, A., Tonuk, E., Akca, K. and Cehreli, M.C. (2009) Numeric simulation of time-dependent remodeling of bone around loaded oral implants. *International Journal of Oral and Maxillofacial Implants*,24,597-608.

Eser, A., Tonuk, E., Akca, K. and Cehreli, M.C. (2010) Predicting time-dependent remodeling of bone around immediately loaded dental implants with different designs. *Medical Engineering and Physics*,32,22-31.

European Association for Osseointegration First Consensus Conference. Clin Oral Implants Res. 2006; 17 (suppl).

Gallucci, G.O., Doughtie, C.B., Hwang, J.W., Fiorellini, J.P. and Weber, H.P. (2009) Five-year results of fixed implant-supported rehabilitations with distal cantilevers for the edentulous mandible. *Clinical Oral Implants Research*,20,601-7.

Glasziou, P., Del Mar, C. and Salisbury, J. (2009) Evidence-Based Practice Workbook. Blackwell Publishing.

Hackshaw, A., Paul, E. and Davenport, E. (2006) Evidence-Based Dentistry: An Introduction. Blackwell Publishing. 2.

Hoshaw, S.J., Brunski, J.B. and Cochran, G.V.B. (1994) Mechanical loading of Branemark implants affects interfacial bone modelling and remodelling. *International Journal of Oral and Maxillofacial Implants*,9,345-360.

Iplikcioglu, H., Akca, K., Cehreli, M.C. and Sahin, S. (2003) Comparison of non-linear finite element analysis with in vitro strain-gauge measurements on a morse-taper implant. *International Journal of Oral and Maxillofacial Implants*,18,258-265.

Miyata, T., Kobayashi, Y., Araki, H., Motomura, Y. and Shin, K. (1998) The influence of controlled occlusal overload on peri-implant tissue: a histologic study in monkeys. *International Journal of Oral and Maxillofacial Implants*,13,677-83.

Oral Implant Rehabilitation: A state-of-the-art overview of case management. (2008) *Austrian Dental Journal*, 53(suppl 1).

Pessoa, R.S., Vaz, L.G., Marcantonio, E. Jr, Vander Sloten, J., Duyck, J. and Jaecques, S.V. (2010) Biomechanical evaluation of platform switching in different implant protocols: computed tomography-based three-dimensional finite element analysis. *International Journal of Oral and Maxillofacial Implants*,25,911-9.

Pieri, F., Aldini, N.N., Marchetti, C. and Corinaldesi, G. (2011) Influence of implant-abutment interface design on bone and soft tissue levels around immediately placed and restored single-tooth implants: a randomized controlled clinical trial. *International Journal of Oral and Maxillofacial Implants*,26,169-78.

Proceedings of the Second ITI Consensus Conference. (2000) *Clinical Oral Implants Research*,11 (suppl).

Proceedings of the Third ITI Consensus Conference. (2004) *International Journal of Oral and Maxillofacial Implants*,24 (suppl).

Proceedings of the Fourth ITI Consensus Conference. (2009) *International Journal of Oral and Maxillofacial Implants*,24 (suppl).

Romeo, E., Tomasi, C., Finini, I., Casentini, P. and Lops, D. (2009) Implant-supported fixed cantilever prosthesis in partially edentulous jaws: a cohort prospective study. *Clinical Oral Implants Research*,20,1278-85.

Schroeder, A., Sutter, F., Buser, D. and Krekeler, G. (1996) Oral Implantology: Basics, ITI Hollow Cylinder System. New York: Thieme Medical Publishers.

Shunmugasamy, V.C., Gupta, N., Pessoa, R.S., Janal, M.N., Coelho, P.G. (2011) Influence of clinically relevant factors on the immediate biomechanical surrounding for a series of dental implant designs. *Journal of Biomechanical Engineering*,133,031005.

State of the Science on Implant Dentistry. (2007) *International Journal of Oral and Maxillofacial Implants*,22 (suppl).

Tymstra, N., Raghoebar, G.M., Vissink, A. and Meijer, H.J. (2011) Dental implant treatment for two adjacent missing teeth in the maxillary aesthetic zone: a comparative pilot study and test of principle. *Clinical Oral Implants Research*,22,207-13.

Wismeijer, D., Buser, D. and Belser, U. (2008) ITI Treatment Guide Volume II: Loading Protocols in Implant Dentistry, Partially Dentate Patients. Quintessence Publishing Co.

Wismeijer, D., Buser, D. and Belser, U. (2007) ITI Treatment Guide Volume I: Implant Therapy in Esthetic Zone, Single-Tooth Replacements. Quintessence Publishing Co.

Wismeijer, D., Buser, D. and Belser, U. (2010) ITI Treatment Guide Volume IV: Loading Protocols in Implant Dentistry, Edentulous Patients. Quintessence Publishing Co.

Yang, T.C., Maeda, Y. and Gonda, T. (2011) Biomechanical rationale for short implants in splinted restorations: an in vitro study. *International Journal of Prosthodontics*,24,130-2.

In: Biomechanics of Dental Implants
Editor: Murat Çehreli
ISBN 978-1-62100-780-7

Chapter 13

BIOMECHANICS OF POST AND CORE RESTORATIONS

Burak Sadık and Serdar Arıkan

1. INTRODUCTION

Endodontically-treated teeth have been claimed to be weaker and more prone to fracture than vital teeth [1]. It is frequently pronounced that the dentin in endodontically treated teeth is more brittle, due to water loss [2] and loss of collagen cross-linking [3]. Randow and Glantz [4] reported that teeth have a protective feedback mechanism that is lost when the pulp is removed, and this may contribute to tooth fracture. It is the loss of structural integrity associated with the access preparation, rather than changes in the dentin, that lead to a higher occurrence of fractures in endodontically-treated teeth compared with "vital" teeth [5, 6]. One 15-week study involving 11 dentists reported 543 tooth fractures out of which 85.6% were complete cusp fractures, 13.4% incomplete or suspected cusp fractures, and 0.9% root fractures [7]. 1- to 2-year follow-up study of 517 previously restored and mainly vital noncarious fractured posterior teeth, the lingual cusps of mandibular molars, the mesiobuccal and distolingual cusps of maxillary molars, and the buccal cusps of maxillary premolars fractured most frequently [8].

In this chapter, the authors aimed to focus on the biomechanics of post and core restorations and the methods used to assess stress/strain fields around these load bearing devices, as the same method are used in the context of orthodontic and oral implants. It was beyond the scope of this chapter to present traditional post and core information, which can be found in classical text books.

2. THE IMPACT OF POST DESIGN

2.1. Retention and Resistance of Post

The primary purpose of a post is to retain a core in a tooth with extensive loss of coronal tooth structure [9, 10]. Ikram et al. [11] compared the volume of hard tooth tissue lost after caries removal, access cavity preparation, root canal preparation, fiber post space and cast post preparation in carious premolar teeth. They reported that access cavity and post space preparation are procedures that result in the largest loss of hard tooth structure. They also reported that cast post space preparation causes a larger loss of tooth structure than fiber post space preparation.

Post retention refers to the ability of a post to resist vertical dislodging forces. Retention is influenced by the length, diameter and taper of the post, the luting cement used, and whether the post is active or passive [12-14]. Increasing the length and diameter of the post can increase retention. Stress analyses have shown better stress distribution within dentine when longer metal, fiber, or zirconia posts are used [15, 16] (Figure1). Similarly, increased fracture resistance is associated with increased post length [17, 18]. Longer posts provide greater rigidity and less root bending than short posts [19]. An endodontic post should extend beyond the level of alveolar bone to provide better root support [20]. Overall, the optimum post length depends on several factors including root length, crown height, level of bone support, and technique of cementation. Adhesive cements, the presence of ferrule effect, and full coronal restoration may reduce the biomechanical effect of post length on the tooth fracture resistance [20, 21]. Mechanical fatigue is a risk factor for fractures. The following factors influence the risk for fractures [22]:

- Magnitude and frequency of occlusal loads
- Direction of forces
- Dimension and shape of dentin and restorative materials

Torbjörner et al. [22] showed conclusion that lending the prosthesis a favourable occlusal design is more important for the survival of the root canal treated tooth than the type of post used, and the teeth, cement, and restorative materials will be less susceptible to fatigue failures.

A smaller post diameter is recommended to retain more dentine during preparation of post channel, which improves the fracture resistance of dowel-restored teeth [23, 24]. The ability of a tooth to resist fracture is directly related to the amount of remaining dentine around the post [23, 25]. Increased radicular dentinal stresses were observed with increased post diameter [26, 27]. Generally, the larger the post diameter, the less the fracture resistance of a dowel-restored tooth [23].

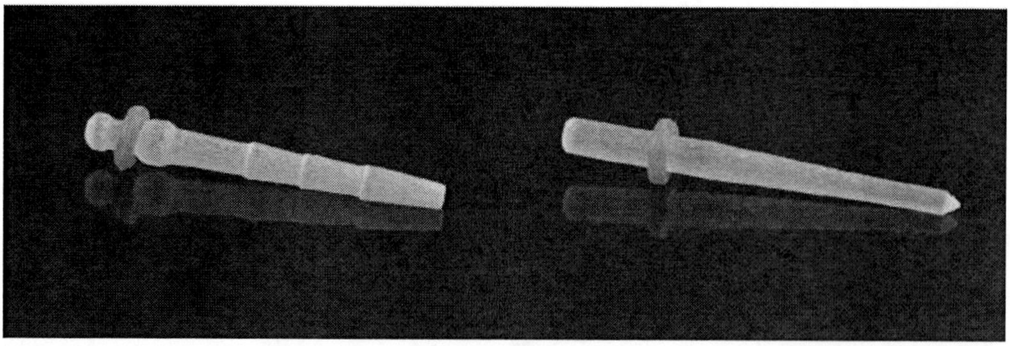

Figure 1. Examples of fiber posts.

Parallel posts are more retentive than tapered posts [13, 28]. Teixeira et al. [29] divided 44 teeth into four groups according to the type of post they would receive; parallel fiberglass posts, double tapered fiber quartz posts, tapered fiberglass posts and two different types of parallel fiberglass posts. Both tapered posts showed lower retention than did the paralel posts. Tapered metal posts cause greater cervical stress concentration than parallel-sided posts [16, 30]. Active posts are more retentive than passive posts [13]. Diameter is less important than the other factors listed [31]. Even though retention can be increased slightly by enlarging the post diameter, the loss of tooth structure weakens the tooth. Therefore, this is not a recommended method to increase retention.

Several studies compared the retention of post systems. Purton and Love [32] reported greater retention with stainless-steel posts than carbon fiber posts. Both were luted with resin cement. Gallo et al. [33] reported that stainless-steel posts luted with zinc phosphate cement had greater retention than a variety of fiber posts luted with resin cement. Qualtrough et al. [34] reported higher retentive values with a quartz fiber post than titanium, glass fiber, or carbon fiber posts. The same resin cement was used for all post systems. Drummond [35] reported no difference in retention between stainless steel and three fiber posts, all with the same resin-luting agent.

Resistance refers to the ability of the post and tooth to withstand lateral and rotational forces. It is influenced by the remaining tooth structure, the length and rigidity of the post, the presence of antirotation features, and the presence of a ferrule. A restoration lacking resistance form is not likely to be a long-term success, regardless of the retentiveness of the post [36, 37]. Salameh et al. [38, 39] showed that endodontically-treated teeth restored with fiber posts and ceramic crowns were more resistant to fracture and had less catastrophic fracture patterns than the ones restored with ceramic crowns without posts.

2.2. Fracture Strength of Posts: Effects of Material Properties

Many mechanical studies were conducted to investigate the effect of post placement and related factors on the fracture resistance of endodontically-treated teeth [23, 40-44]. Extracted teeth, especially incisors and premolars, were used in these studies. Static loading at a constant angle was applied to restored teeth in some studies [42, 45]. However, actual masticatory forces are multidirectional and repeatedly applied on larger areas [19]. In order to mimic such conditions, cyclic loading was also applied in some studies [43, 44, 46]. Most of the in vitro studies comparing fracture strength of post systems used continuous or intermittent loading. In tests of fracture load or failure load, the post/tooth complex is traditionally loaded with a continuous force on a test machine until failure, and the load values are recorded and compared. In recent years, cyclic or intermittent loading has become more popular, because it is thought to be more representative of the forces that occur in vivo [42]. Cyclic loading is continued until failure, or to a specified number of cycles, and the results are reported as the number of cycles to failure, or as the number of failures when cycling loading was stopped. Some of these studies also reported on failure mode. The results of static loading studies are inconclusive, but are slightly more favourable toward metal posts. Cormier et al. [47] studied posts made of stainless-steel, gold, and four commercial brands of fiber posts and found that teeth with stainless-steel posts had the highest fracture load, whereas teeth with one of the quartz fiber post systems had the lowest. Martinez-Insua et al.

[48] reported higher fracture loads with cast gold than carbon fiber. Newman et al. [49] compared stainless-steel posts with three brands of fiber posts and found higher failure loads in teeth with the stainless-steel posts. Sidoli et al. [40] compared cast gold, stainless steel, and carbon fiber and reported that teeth containing metal posts were equivalent and had significantly higher failure loads than teeth containing carbon fiber posts. Akkayan et al. [50] compared titanium posts with glass fiber, quartz fiber, and zirconium. They reported the highest failure loads for the teeth with quartz fiber posts. Ottl et al. [51] reported the highest failure loads with carbon fiber posts, followed by stainless steel and ceramic posts. Zirconium posts had the lowest values. Raygot et al. [52] found no difference between cast gold, stainless-steel, and carbon fiber posts. Similar inconclusive results were reported in studies that used intermittent loading (also called cyclic or dynamic loading). Isidor et al. [36] compared cast gold, titanium, and carbon fiber posts and reported better results with the metal posts than the carbon fiber post. Reid et al. [53] compared titanium posts with three carbon fiber post and one quartz fiber post, and reported no differences. Butz et al. [54] used cyclic loading, followed by continuous loading, to compare cast gold, titanium with a composite core, and zirconium with composite or ceramic cores. The zirconium/composite post and core performed significantly worse fracture load than the other post systems, which were comparable. Three of the studies reported more favorable failure modes with fiber posts than metal posts [47, 49, 50]. Martinez-Insua et al. [48] reported similar results but commented that the fracture loads exceeded those ordinarily found clinically.

3. Biomechanical Testing of Post and Cores

Experimental studies in general have many limitations. The strength of dentine varies according to age, pulpal condition before extraction, and the storing media, which may affect the fracture patterns [19]. Resins and stones are commonly used for mounting teeth during testing [24, 55, 56]. They set by the exothermic reaction, which may affect the dental structures [57].

Photoelastic models have been used to study the effect of post placement, design, and related factors on the patterns of stress distribution within endodontically-treated teeth [58-60]. Photoelasticity, as already presented in this book, is the property of transparent materials to exhibit colourful patterns known as ''isochromatic fringes'' when stressed under polarized light [58]. A transparent double-refraction plastic sheet is used to fabricate specimens for two-dimensional photoelastic stress analysis [60]. Stress concentration areas can be identified according to the sequence of color bands of the fringes. Three-dimensional photoelastic models, made from photoleastic sheets are not suitable for post and core testing and are less frequently used because of the difficulty of their construction [58]. Moreover, photoelastic coatings on posts cannot elucidate distribution of stresses, as stress gradients around the relatively small threads cannot be viewed even on high-resolution images.

Sook-Hyun Nam et al. [61] compared the fracture resistances and the failure patterns of 100 simulated mandibular premolars of a different number of coronal walls (zero to four walls) with or without fiber-reinforced composite (FRC) posts. The fracture resistance was measured at a 45° angle with a crosshead speed of 1 mm/min, and the failure patterns were observed. The photoelastic stress distribution of specimens with or without FRC posts was also evaluated. They found that the FRC post was advantageous in lower premolars, especially with two or more walls in terms of the fracture resistance and stress distribution.

The finite element method (FEM) has been frequently used for stress analysis in many aspects of dentistry. Many studies used the FEM to investigate the effect of post placement and related parameters on the stress picture within dowel-restored teeth [15, 27, 62-68]. Stress analysis is performed by solving differential equations of elements to quantify stresses generated within these elements. FEM offers the advantages of easy simulation of nonhomogenous models and easy changing of parameters like material properties and loading conditions [27]. However, material properties, loading conditions, and boundaries simulated have not so far represented the absolute clinical situation that constitutes a limitation for the application of this method in studies concerning post and core restorations [19]. Soares CJ et al. [69] tested the hypothesis that the stress distribution and bond strength of glass posts to intraradicular dentin is influenced by the mechanical testing methodology. Tapered fiber-glass posts were used with dual-cure resin cement for this study. The teeth were randomly divided into three groups; hourglass-shaped microtensile, rectangular stick-shaped microtensile, and micro-push-out. These tests methods were analyzed by finite element analysis. The authors found that push-out test demonstrated a more homogenous stress distribution by FEA and less variability in mechanical testing.

Strain-gauge analysis is another method that has been also used to a lesser extent in the context of post and core restorations. Application of this technique relies on use of electrical resistance strain gauges bonded on tooth tissue [70]. Santos-Filho et al. [71] investigated ex vivo the effects of different post systems and lengths on the strain and fracture resistance of root-filled teeth. They used bovine incisors and three different post; fiberglass post; prefabricated steel post; cast post and core. Each group was divided into three subgroups ($n = 15$) according to the post length: 5.0 mm; 7.5 mm; 10.0 mm. As a result the cast post and core when the length was 10.0 mm had the highest fracture resistance; however, the fiberglass post was effective with the three post lengths, with higher fracture resistance than metal posts when the length was 5.0 mm.

The inner dentine of the root is usually less mineralized and possesses more water content than the outer dentine [72]. Therefore, the inner dentine has a higher potential for plastic deformation and crack formation. Using experimental and clinical investigations, Kishen et al. [72] examined fractured post and core restored teeth using laser scanning confocal microscopy and scanning electron microscopy and observed numerous microcracks within the inner dentine material adjacent to the endodontic post. They also used FEMs of dentine to relate crack formation and root fractures to tensile stresses generated within dentine. High strains were generated within the inner dentine substance upon tensile loading. They concluded that crack formation and fracture progression in postrestored teeth were initiated from the inner region of dentine [72-74]. Thickness of the inner dentine and factors related to the postdentine interface seem to play a major role in stress distribution and fracture resistance of endodontically-treated teeth. Therefore, the removal of inner dentine during post placement should be minimized as much as possible to maintain adequate fracture resistance of dowel-restored teeth [72].

Cone beam computed tomography (CBCT) is a medical imaging technique that is a complished by using a rotating gantry to which an x-ray source and detector are fixed [75]. CBCT scanners are now finding many uses in dentistry, such as in the fields of endodontics and orthodontics, as well. During a CBCT scan, the scanner rotates around the patient's head, obtaining up to nearly 600 distinct images. The scanning software collects the data and reconstructs it, producing what is termed a *digital volume* composed of three dimensional voxels of anatomical data that can then be manipulated and visualized with specialized software[76]. Estrela et al. [77] verified the influence of intracanal post on apical periodontitis detected by cone-beam computed tomography (CBCT). Consecutive samples of 1020 images were taken from 619 patients. From a total of 1,020 teeth used in this study, apical periodontitis was detected in 397 (38.92%) by periapical radiography and in 614 (60.19%) by CBCT scans. Intracanal posts lengths did not influence apical periodontitis.

A universal testing machine is used to test the tensile stress and compressive strength of materials. Some experimental work has attempted to measure the routine limits of temperature change induced by eating and drinking. Gale et al. [78] indicated that the standart regimen defined is: 35°C (25 s), 15°C (2 s), 35°C (28 s), 45°C (2 s) and provisional estimate of approximately 10 000 cycles per year is suggested. It is difficult to be precise about such events, as eating and drinking are very erratic habits and large variations are expected between occasions, subjects [79] and locations in the mouth [79]. Narmin Mohammadi et al. [80] evaluated the effect of fiber post and cusp coverage on fracture resistance of endodontically-treated maxillary premolars directly restored with composite resin. The specimens were exposed to thermocycling at 5°C to 55°C with 30 seconds execution time and 5 seconds for transfer for 500 cycles. Fracture resistance was evaluated in a universal testing machine (H5 K-S model; Hounsfield Test Equipment, Ltd., Surrey, England). They found that there were no significant differences in fracture resistance between the groups. However, there were statistically significant differences in failure mode between the groups. Albashaireh et al. [81] evaluated the influence of post surface conditioning methods and artificial aging on the retention and microleakage of adhesively luted glass fiber-reinforced composite resin posts. Seventy-two endodontically treated single-rooted teeth were used for this study. Specimens were stored in water at 37°C for 30 days and thermal cycled for 7500 cycles with a dwell time of 30 seconds and a transition time of 6 seconds. Specimens were subjected to 300,000 cycles of mechanical loading parallel to the long axis of the post with a masticatory simulator at 30-N force and 1.6 Hz. For the retentive force test, a screw-driven universal testing machine was used to apply a tensile load parallel to the long axis of the posts at a crosshead speed of 2 mm/min. The force (N) required to dislodge each post was recorded. The retention values of the airborne-particle-abrasion group were significantly higher than those of the acidic-treatment and no-treatment groups. The application of bonding agent on the post surface produced no significant influence on retention. The mean retention values after artificial aging were significantly higher than without artificial aging. Nurit Bittner et al. [82] evaluated the accuracy of fit of milled zirconia posts and cores and to compare the shear strength with other post-and-core systems. Eighty five maxillary central incisors and canines received endodontic treatment were used for this study. Teeth were divided into five groups; cast gold post and core (Au) as control, 1-piece milled zirconia post and core (Zr), prefabricated zirconia post with heat-pressed ceramic core (Zr/Cer), titanium post and composite resin core (Ti), and combined fiber/zirconia post with composite resin core (Fiber/Zr). Specimens were thermal cycled for 2500 cycles, at 5° and 55°C, with a 30-second dwell time and an intermediate time of 12 seconds. After thermal cycling, the specimens were stored in water for 24 hours. Loading was accomplished by securing the specimens in a universal testing machine at a 135-degree angle to simulate the arch position of anterior teeth. Mean loads to failure were highest for the Fiber/Zr group in comparison with custom-made milled zirconia posts and cores. Kerstin Bitter et al. [83] investigated the effects of endodontic treatment, post placement and ceramic restoration type on the fracture resistance of premolars. One hundred and twenty teeth were used for this study. Specimens were stored in water for 21 days at 37 °C. Chewing simulation was performed with 500.000 cycles occlusal load of 49 N in horizontal direction. Specimens were subjected to 5.000 thermal cycles in deionized water from 5–55 °C with a dwelling time of 30 s in each bath and a transfer time of 2 s. The fracture resistance of the buccal cusp was tested using a universal testing machine, and the load was applied using a steel ball (diameter 3.5 mm) in a 30° angle to the long axis of the tooth with a cross head speed of 0.5 mm min^{-1}. Fracture resistance was significantly affected by the restoration type and endodontic treatment/post placement. Mesio-occlusal-distal-inlays with a buccal and palatal wall of 2 mm (MOD) group showed significantly lower fracture resistance compared to partial onlays with palatal cusp coverage group. Compared to non-endodontically treated teeth, root filled teeth revealed significantly lower fracture resistance.

REFERENCES

[1] Schwartz R.S., Robbins J.W. (2004) Post placement and restoration of endodontically treated teeth: a literature review. *Journal of Endodontics*, 30, 289-301.

[2] Helfer A.R., Melnick S., Schilder H. (1972) Determination of moisture content of vital and pulpless teeth. *Oral Surgery Oral Medicine Oral Pathology, Oral Radiology, and Endodontics*,34,661-70.

[3] Rivera E.M., Yamauchi M. (1993) Site comparisons of dentine collagen crosslinks from extracted human teeth. *Archives of Oral Biology*, 38, 541-6.

[4] Randow K., Glantz P. (1986) On cantilever loading of vital and non-vital teeth. *Acta Odontologica Scandinavica*, 44, 271-7.

[5] Reeh E.S., Messer H.H., Dougles W.H. (1989) Reduction in tooth stiffness as a result of endodontic restorative procedures, *Journal of Endodontics*, 15, 512-6.

[6] Peroz I., Blankenstein F., Lange K.P., Naumann M. (2005) Restoring endodontically treated teeth with posts and cores: a review. *Quintessence International*, 36, 737-46.

[7] Bader J.D., Martin J., Shugars D.A. (2001) Incidence rates for complete cusp fracture. *Community Dentistry and Oral Epidemiology*, 29, 346-53.

[8] Bader J.D., Shugars D., Sturdevant J.R. (2004) Consequences of posterior cusp fracture. *General Dentistry*, 52, 128-31.

[9] Robbins J.W. (1990) Guidelines for the restoration of endodontically treated teeth. *The Journal of the American Dental Association*, 120, 558-66.

[10] Goodacre C.J., Spolnik K. (1994) The prosthodontic management of endodontically treated teeth: a literature review. Part I. Success and failure data, treatment concepts. *Journal of Prosthodontists*, 3, 243-50.

[11] Ikram O.H, Patel S., Sauro S., Mannocci F. (2009) Micro-computed tomography of tooth tissue volume changes following endodontic procedures and post space preparation. *International Endodontic Journal*, 42, 1071-6.

[12] Standlee J.P., Caputo A., Collard E.W., Pollack M.H. (1972) Analysis of stress distribution by endodontic posts. *Oral Surg Oral Med Oral Pathology, Oral Radiology, and Endodontics*, 33, 952-60.

[13] Standlee J.P., Caputo A., Hanson E.C. (1978) Retention of endodontic dowels: effects of cement, dowel length, diameter, and design. *The Journal Prosthetic Dentistry*, 39, 401-5.

[14] Felton D.A., Webb E., Kanoy B.E., Dugoni J. (1991) Threaded endodontic dowels: effect of post design on incidence of root fracture. *The Journal Prosthetic Dentistry*, 65, 179-87.

[15] Asmussen E., Peutzfeldt A., Sahafi A., (2005) Finite element analysis of stresses in endodontically treated, dowel-restored teeth. *The Journal Prosthetic Dentistry*,0 94, 321-9.

[16] Davy D.T, Dilley G.L., Krejci R.F. (1981) Determination of stress patterns in root-filled teeth incorporating various dowel designs. *Journal of Dental Research*, 60, 1301-10.

[17] Sokol D.J. (1984) Effective use of core and post concepts. *The Journal Prosthetic Dentistry*, 52, 231-4.

[18] Buttel L., Krastl G., Lorch H., Naumann M., Zitzmann N.U., Weiger R. (2009) Influence of post fit and post length on fracture resistance. *International Endodontic Journal*, 42, 47-53.

[19] Fernandes A., Dessai G. (2001) Factors affecting the fracture resistance of post-core reconstructed teeth: a review. *The International Journal of Prosthodontics*, 14, 355-63.

[20] Leary J.M., Aquilino S., Svare C.W. (1987) An evaluation of post length within the elastic limits of dentine, *The Journal Prosthetic Dentistry*, 57, 277-81.

[21] Nissan J., Dimitry Y., Assif D., (2001) The use of reinforced composite resin cement as compensation for reduced post length. *The Journal Prosthetic Dentistry,* 86, 304-8.

[22] Torbjörner A., Fransson B. (2004) Biomechanical aspects of prosthetic treatment of structurally compromised teeth. *The International Journal of Prosthodontics*, 17, 135-41.

[23] Tjan A.H., Whang S. (1985) Resistance to root fracture of dowel channels with various thicknesses of buccal dentin walls. *The Journal Prosthetic Dentistry*, 53, 496-500.

[24] Sorensen J.A., Engelman M.J. (1990) Effect of post adaptation on fracture resistance of endodontically treated teeth. *The Journal Prosthetic Dentistry*, 64, 419-24.

[25] Assif D., Gorfil C. (1994) Biomechanical considerations in restoring endodontically treated teeth. *The Journal Prosthetic Dentistry,* 71, 565-7.

[26] Mattison G.D. (1982) Photoelastic stress analysis of cast-gold endodontic posts. *The Journal Prosthetic Dentistry,* 48, 407-11.

[27] de Castro Albuquerque R., Polleto L., Fontana R.H., Cimini C.A. (2003) Stress analysis of an upper central incisor restored with different post,. *Journal of Oral Rehabilitation,* 30, 936-43.

[28] Johnson J.K., Sakumura J. (1978) Dowel form and tensile force. *The Journal Prosthetic Dentistry*, 40, 645-9.

[29] Teixeira E.C., Teixeira F.B., Piasick J.R., Thompson J.Y. (2006) An in vitro assessment of prefabricated fiber post systems, *The Journal of the American Dental Association*, 137, 1006-12.

[30] McAndrew R., Jacobsen P.H. (2002) The relationship between crown and post design on root stress—a finite element study. *The European Journal of Prosthodontics Restorative Dentistry*, 10, 9-13.

[31] Nergiz I., Schmage P., Ozcan M., Platzer U. (2002) Effect of length and diameter of tapered posts on the retention. *Journal of Oral Rehabilitation*, 29, 28-34.

[32] Purton D.G., Love R. (1996) Rigidity and retention of carbon fibre versus stainless steel root canal posts. *International Endodontic Journal*, 29, 262-5.

[33] Gallo J.R. 3rd, Miller T., Xu X., Burgess J.O. (2002) In vitro evaluation of the retention of composite fiber and stainless steel posts, *Journal of Prosthodontists*, 11, 25-9.

[34] Qualtrough A.J., Chandler N.P., Purton D.G. (2003) A comparison of the retention of tooth-colored posts. *Quintessence International*, 34, 199-201.

[35] Drummond J.L. (2000) In vitro evaluation of endodontic posts. *American Journal of Dentistry*, 13, 5B-8B.

[36] Isador F., Brondum K., Ravnholt G., (1999) The influence of post length and crown ferrule length on the resistance to cyclic loading of bovine teeth with prefabricated titanium post, *The International Journal of Prosthodontics*, 12, 78-82.

[37] Lambjerg-Hansen H., Asmussen E., (1997) Mechanical properties of endodontic posts. *Journal of Oral Rehabilitation*, 24, 882-7.

[38] Salameh Z., Sorrentino R., Ounsi H.F., Goracci C., Tashkandi E., Tay F.R., Ferrari M. (2007) Effect of different all-ceramic crown system on fracture resistance and failure pattern of endodontically treated maxillary premolars restored with and without glass fiber posts. *Journal of Endodontics*, 33, 848-51.

[39] Salameh Z., Ounsi H.F., Aboushelib M.N., Sadiq W., Ferrari M. (2008) Fracture resistance and failure patterns of endodontically treated mandibular molars with and without glass fiber post in combination with a zirconia-ceramic crown, *Journal of Dentistry*, 36, 513-9.

[40] Sidoli G.E., King P.A., Setchell D.J. (1997) An in vitro evaluation of a carbon fiber-based post and core system. *Journal of Prosthodontists*, 78, 5-9.

[41] Costa LC, Pegoraro L., Bonfante G. (1997) Influence of different metal restorations bonded with resin of fracture resistance of endodontically treated maxillary premolars. *Journal of Prosthodontists*, 77, 365-9.

[42] Drummond J.L., Bapna M., (2003) Static and cyclic loading of fiber-reinforced dental resin. *Dental Materials*, 19, 226-31.

[43] Cohen B.I., Pagnillo M.K., Newman I., Musikant B.L., Deutsch A.S. (2000) Pilot study of the cyclic fatigue characteristics of five endodontic posts with four core materials. *Journal of Oral Rehabilitation*, 27, 83-92.

[44] Mannocci F., Ferrari M., Watson T.F. (2001) Microleakage of endodontically treated teeth restored with fiber posts and composite cores after cyclic loading: a confocal microscopic study. *Journal of Prosthodontists*, 85, 284-91.

[45] Martinez-Gonzalez A., Amigo-Borras V., Fons-Font A., Selva-Otaolaurruchi E., Labaiq-Rueda C. (2001) Response of three types of cast posts and cores to static loading. *Quintessence International*, 32, 552-60.

[46] Isidor F., Brondum K., Ravnholt G. (1999) The influence of post length and crown ferrule length on the resistance to cyclic loading of bovine teeth with prefabricated titanium posts. *The International Journal of Prosthodontics*, 12, 78-82.

[47] Cormier C.J., Burns D., Moon P. (2001) In vitro comparison of the fracture resistance and failure mode of fiber, ceramic, and conventional post systems at various stages of restoration, *Journal of Prosthodontists*,. 10, 26-36.

[48] Martinez-Insua A., da Silva L., Rilo B., Santana U. (1998) Comparison of the fracture resistances of pulpless teeth restored with a cast post and core or carbon-fiber post with a composite core. *The Journal Prosthetic Dentistry*, 80, 527-32.

[49] Newman M.P., Yaman P., Dennison J., Rafter M., Billy E. (2003) Fracture resistance of endodontically treated teeth restored with composite posts. *The Journal Prosthetic Dentistry*, 89, 360-7.

[50] Akkayan B., Gülmez T., (2002) Resistance to fracture of endodontically treated teeth restored with different post systems. *The Journal Prosthetic Dentistry*, 87, 431-7.

[51] Ottl P., Hahn L., Lauer H.C., Fay M. (2002) Fracture characteristics of carbon fibre, ceramic and non-palladium endodontic post systems at monotonously increasing loads. *Journal of Oral Rehabilitation*, 29, 175-83.

[52] Raygot C.G., Chai J., Jameson D.L. (2001) Fracture resistance and primary failure mode of endodontically treated teeth restored with a carbon fiberreinforced resin post system in vitro. *The International Journal of Prosthodontics*, 14, 141-5.

[53] Reid L.C., Kazemi R.B., Meiers J.C. (2003) Effect of fatigue testing on core integrity and post microleakage of teeth restored with different post systems. *Journal of Endodontics*, 29, 125-31.

[54] Butz F., Lennon A., Heydecke G., Strub J.R. (2001) Survival rate and fracture strength of endodontically treated maxillary incisors with moderate defects restored with different post-and-core systems: an in vitro study. *The International Journal of Prosthodontics*, 14, 58-64.

[55] Mendoza D.B., Eakle W.S., Kahl E.A., Ho R., (1997) Root reinforcement with a resin bonded preformed post. *The Journal Prosthetic Dentistry*, 78, 10-4.

[56] King P.A., Setchell D.J. (1990) An in vitro evaluation of a prototype CFRC prefabricated post developed for the restoration of pulpless teeth, *Journal of Oral Rehabilitation*, 17, 599-609.

[57] Sirimai S., Riis D.N., Morgano S.M. (1999) An in vitro study of the fracture resistance and the incidence of vertical root fracture of pulpless teeth restored with six post-and-core systems, *The Journal Prosthetic Dentistry*, 81, 262-9.

[58] Loney R.W., Kotowicz W.E., McDowell G.C. (1990) Three-dimensional photoelastic stress analysis of the ferrule effect in cast post and core, *The Journal Prosthetic Dentistry*, 63, 506-12.

[59] Assif D., Oren E., Marshak B.L., Aviv I. (1989) Photoelastic analysis of stress transfer by endodontically treated teeth to the supporting structure using different restorative techniques. *The Journal Prosthetic Dentistry*, 61, 535-43.

[60] Cohen B.I., Pagnillo M., Musikant B.L, Deutsch A.S. (1999) Comparison of the retentive and photoelastic properties of two prefabricated endodontic post systems. *Journal of Oral Rehabilitation*, 26, 488-94.

[61] Nam SH, Cang H.S., Min K.S., Lee Y., Cho H.W., Bae J.M. (2010) Effect of the number of residual walls on fracture resistances, failure patterns, and photoelasticity of simulated premolars restored with or without fiber-reinforced composite posts, *Journal of Endodontics*, 36, 297-301.

[62] Toparli M. (2003) Stress analysis in a post-restored tooth utilizing the finite element method. *Journal of Oral Rehabilitation*, 30, 470-6.

[63] Pegoretti A., Fambri L., Zappini G., Bianchetti M. (2002) Finite element analysis of a glass fibre reinforced composite endodontic post. *Biomaterials*, 23, 2667-82.

[64] Pierrisnard L., Bohin F., Renault P., Barguins M. (2002) Corono-radicular reconstruction of pulpless teeth: a mechanical study using finite element analysis. *The Journal Prosthetic Dentistry*, 88, 442-8.

[65] Ho M.H., Lee S., Chen H.H., Lee M.C. (1994) Three-dimensional finite element analysis of the effects of posts on stress distribution in dentin. *The Journal Prosthetic Dentistry*, 72, 367-72.

[66] Lewgoy H.R., Youssef M., Matson M.R., Bocangel J.A., Netto C.A., Amore R. (2003) Finite elements study of the Flexi Post and Flexi Flange post systems in a maxillary central incisor. *Pesquisa Odontologica Brasileira*, 17, 132-6.

[67] Lanza A., Aversa R., Rengo S., Apicella D., Apicella A. (2005) 3D FEA of cemented steel, glass and carbon posts in a maxillary incisor. *Dental Materials*, 21, 709-15.

[68] Yang H.S., Lang L.A., Molina A., Felton D.A. (2001) The effects of dowel design and load direction on dowel-and-core restorations. *The Journal Prosthetic Dentistry*, 85, 558-67.

[69] Soares C.J., Santana F.R., Castro C.G., Santos-Filho P.C., Soares P.V., Qian F., Armstrong S.R. (2008) Finite element analysis and bond strength of a glass post to intraradicular dentin: comparison between microtensile and push-out tests. *Dental Materials*, 24, 1405-11.

[70] Santana F. R., Castro C.G., Simamoto-Ju´ nior P. C., Soares P. V., Quagliatto P. S., and Estrela C., Soares C.J. (2011) Influence of post system and remaining coronal tooth tissue on biomechanical behaviour of root filled molar teeth. *International Endodontic Journal*, 44, 386-394.

[71] Santos-Filho P.C., Castro C.G., Silva G.R., Campos R.E., Soares C.J. (2008) Effects of post system and length on the strain and fracture resistance of root filled bovine teeth. *International Endodontic Journal*, 41, 493-501.

[72] Kishen A., Kumar G., Chen N.N. (2004) Stress-strain response in human dentine: rethinking fracture predilection in postcore restored teeth. *Dental Traumatology*, 20, 90-100.

[73] Fan P., Nicholls J.I., Kois J.C. (1995) Load fatigue of five restoration modalities in structurally compromised premolars. *The International Journal of Prosthodontics*, 8, 213-20.

[74] Thorsteinsson T.S., Yaman P., Craig R.G. (1992) Stress analyses of four prefabricated posts. *The Journal Prosthetic Dentistry*, 67, 30-3.

[75] Scarfe W.C., Farman A.G. (2008) What is cone-beam CT and how does it work? *Dental Clinics of North America*, 52, 707-30.

[76] Hatcher D.C. (2010) Operational principles for cone-beam computed tomography. *The Journal of the American Dental Association*, 141, 3S-6S.

[77] Estrela C., Bueno M., Porto O.C., Rodrigues C.D., Pécora J.D. (2009) Influence of intracanal post on apical periodontitis identified by cone-beam computed tomography. *Brazilian Dental Journal*, 20, 370-5.

[78] Gale M.S., Darvell B. (1999) Thermal cycling procedures for laboratory testing of dental restorations. *Journal of Dentistry*, 27, 89-99.

[79] Spierings T.A., Peters M., Bosman F., Plasschaert A.J. (1987) Verification of theoretical modelling of heat transmission in teeth by in vivo experiments. *Journal of Dental Research*, 66, 1336-1339.

[80] Mohammadi N., Kahnamoii M., Yeganeh P.K., Navimipour E.J. (2009) Effect of fiber post and cusp coverage on fracture resistance of endodontically treated maxillary premolars directly restored with composite resin. *Journal of Endodontics*, 35, 1428-32.

[81] Albashaireh Z.S., Ghazal M., Kern M. (2010) Effects of endodontic post surface treatment, dentin conditioning, and artificial aging on the retention of glass fiber-reinforced composite resin posts. *The Journal Prosthetic Dentistry*, 103, 31-9.

[82] Bittner N., Hill T., Randi A. (2010) Evaluation of a one-piece milled zirconia post and core with different post-and-core systems: An in vitro study. *The Journal Prosthetic Dentistry*, 103, 369-79.

[83] Bitter K., Meyer-Lueckel H., Fotiadis N., Blunck U., Neumann K., Kielbassa A.M., Paris S. (2010) Influence of endodontic treatment, post insertion, and ceramic restoration on the fracture resistance of maxillary premolars. *International Endodontic Journal*, 43, 469-77.

In: Biomechanics of Dental Implants
Editor: Murat Çehreli

ISBN 978-1-62100-780-7

Chapter 14

EVIDENCE BASED DENTISTRY HIERARCHY

Steven Eckert

1. INTRODUCTION

Interviews with patients identify a number of different reasons why they seek dental care. Although the words may differ, careful analysis of these conversations generally identify concerns in the three categories of comfort, function, or aesthetics. Some patients identify issues in each of these categories while others express concerns that are specific to only one.

The one component to dental care that is rarely discussed by patients is the long-term service provided by dental restorations or prostheses. Patients appear to expect that dental care will provide a lifetime of service. Clinicians contribute to this misperception by describing temporary, provisional and permanent prostheses. Clearly the term "permanent prosthesis" suggests a prosthetic device that will continue to meet the needs of the patient throughout their life.

Obviously the description of a permanent prosthesis is a misnomer. Clinicians understand that dental prostheses have a finite lifespan but this information is not always conveyed to the patients. Further it appears that the lifespan of dental procedures may be decreasing as there is an increased emphasis placed upon aesthetic results. While functional restorations may rely heavily on materials that have demonstrated long terms of service, aesthetic restorations are often provided with materials that have been released to the marketplace with little or no evidence of long-term clinical performance. Many of the materials used today were created through research and development programs that assessed material physical properties but did not test clinical performance and may not have evaluated loading conditions that simulate those found in the oral cavity. Further it must be recognized that short-term clinical performance, if such studies exist, may not be indicative of long-term performance especially when materials that are subject to fatigue failure are utilized.

Most clinicians understand that the practice of dentistry demands the use of a number of different materials to repair damaged or diseased teeth. The physical demands on dental materials increase dramatically when prostheses are designed to replace multiple missing teeth.

When discussing dental materials, the initial concerns relate to the mechanical properties of the materials. Terms such as stress, force per unit area, and strain, change in length per unit length, are described for virtually every dental material. Characteristics such as elasticity, plasticity, fatigue, hardness, and others may apply to one material and not to another. Thermal properties are critical for cast metals and ceramics materials but may have little influence on other materials. Although physical properties such as these are important these properties alone cannot predict the long-term clinical performance of dental materials and specifically of those dental materials used in the fabrication of dental prostheses.

Considering that dental materials must function within a biologic environment it becomes important to appreciate the effects of materials on the surrounding physical structures. In some instances the stiffness of a material may be beneficial as it protects the integrity of the prosthesis while in other situations a stiff material could prevent stimulatory forces from being transmitted to the underlying supporting structures.

The increasing number of clinical applications for dental materials results in the need for new testing modalities. Whereas dental material studies have existed throughout the era of modern dentistry the field of biomechanics is relatively new. [1]

Biomechanics is defined as: *The scientific study of the role of mechanics in biological systems. The study of biomechanics includes the analysis of motion in animals, the fluid dynamics of blood, and the role of mechanical processes in the development of disease.* Through this definition it becomes clear that biomechanical studies evaluating dynamic conditions should be much more predictive of long-term clinical performance.

When fabricating prostheses clinicians generally trust that the materials themselves will perform adequately in function. There is an additional expectation that the materials will allow favorable tissue responses which are possible only if the materials are truly biocompatible.

Clinical performance can be assessed through long-term clinical studies but these studies, as the name implies, demand tremendous investments of time and expense. All the while such studies would exhibit trial and error approaches to treatment that may not be ethically acceptable. An alternative to this approach is to perform biomechanical tests to ensure that complications are not the result of the biomaterial but are instead unavoidable biologic or technical events. Hence materials that are used for tooth repair or replacement must be evaluated in simulated oral environments to ensure their long-term performance.

Biomechanical testing of biomaterials attempts to achieve the purpose of being an appropriate surrogate for long term clinical testing. In contrast to traditional material testing, biomechanical testing attempts to mimic clinical situations. It is important that the failure patterns exhibited during testing mimics those that have been observed in clinical practice as the replication of clinical complications ensures that the modeling system used for the biomechanical testing is appropriate. Failure to establish appropriate modeling that mimics the clinical situation will create outcomes that may be irrelevant.

The oral environment can be modeled in a number of different ways. Synthetic materials may be used to simulate the physical characteristics of bone, teeth, mucosa or the periodontal ligament. The physical properties of biologic tissues and dental biomaterials must be well understood before models are created. Perhaps the most difficult modeling challenges relate to the variability of oral tissues. Further, despite the efforts of researchers involved in biomechanical research there is recognition that some patients may have physical characteristics that are outside the normal range and in those clinical situations the predicted results may not be attainable.

2. DESCRIPTION OF EBD HIERARCHY

It is obvious that biomechanical testing plays an important role in the prediction of clinical performance of biomaterials. One critical question however is how is biomechanical testing is used to make clinical decisions? This question of clinical decision-making using best available evidence is the hallmark of the evidence-based dentistry.

Clinicians are routinely presented with a variety of materials and techniques that are represented as dramatic improvements over those that existed previously. Since most dentists work in isolated settings their ability to respond to manufacturer's claims must be entirely their own decision. In contrast to colleagues in medical disciplines, where practice committees establish institutional guidelines, dentists must rely upon a combination of their own knowledge, manufacturer claims and published reports.

When considering changes in mode of practice clinicians must weigh the available evidence against their own clinical experiences to ensure that the promise of improved performance can be met. Logically, if the clinician's previous experience was essentially the same as the claims for newer products there is no reason to change practices. This is another guideline of evidence-based dentistry. Clinicians should make informed decisions based upon the best available evidence in comparison to their own clinical experiences. In this situation it is critical that clinicians know what the data demonstrates regarding their own practice experiences. This is, unfortunately, the most critical factor in evidence-based decision-making as so few clinicians truly understand the data from their own practice.

Unfortunately many of the claims for products describe factors they cannot truly be quantified. When a product is described as "easier to use", "less technically demanding", or "more aesthetic" it becomes clear that these subjective factors can neither be confirmed nor denied. For this reason many new products reach the profession without any clear understanding of the benefits of the products.

The Center for Evidence Based Medicine has established a hierarchy for clinical studies that might be used in clinical decision-making. The primary reason for establishing the hierarchy is to determine the potential for studies to be biased. In the hierarchy factors such as randomization, controls, blinding and comparative clinical trials demonstrate designs that are difficult to manipulate in a biased fashion. Although this does not ensure the absolute truth of the study it does provide the reader with a degree of confidence that the study was not intentionally manipulated. Likewise, studies that are near the bottom of the hierarchy are not necessarily biased studies but are instead studies that could potentially be designed to reflect the bias of the investigators.

It is recognized that the desire to avoid bias establishes a hierarchy for evidence-based medicine. Readers may be comforted by the notion that a study is not biased and therefore may be more closely aligned to truth.

Biomechanical studies fall near the bottom of the evidence-based hierarchy. The reason for this is easy to understand as a researcher who is familiar with the technical intricacies of a specific material or device can easily select testing methodology that favors that specific device. Unfortunately the complexity of biomechanical testing often makes it difficult for clinicians/readers to understand when the study was designed to demonstrate superiority of one product over another. In contrast the goal of an appropriately designed study is simply to identify the factors that influence clinical performance. The latter are designed without intentional bias and are therefore more closely aligned with truth.

Considering the potential for bias one must carefully evaluate the value of biomechanical testing. The most beneficial assessments in biomechanics simulate clinical settings. Conversely, testing that fails to simulate clinical settings would be of little value. At present it is unclear whether biomechanical studies should be performed in advance of product release onto the marketplace or following product release as issues relative to performance are identified. Realistically, both situations should be assessed in studies to provide meaningful product improvement and establish guidelines for post market surveillance.

3. Benefits of Biomechanical Testing

Assuming that testing is performed without systematic bias, biomechanical testing will provide valuable benefits to clinicians and researchers alike. Both have an expressed desire to work with biomaterials that achieve the goals of providing comfortable, functional and esthetic restorations. The most favorable restorations will accomplish these tasks while also providing long term service.

Durability is often difficult to test clinically as it is not until time passes that the longevity of a material is known. Simple evaluation of factors such as resistance to wear, ultimate tensile or compressive strength, shade stability, fracture toughness, etc. may provide the clinician with a sense of comfort but these are not always factors that translate into clinical performance. Even the use of simulated clinical performance through the use of devices that will attempt to replicate chewing function in a wet environment with temperature variations that can occur in the oral cavity may still fail to predict the actual clinical outcomes.

In contrast, biomechanical testing that is used to accelerate aging of materials may be the best alternative that is available. Without such testing the clinician becomes the beta tester for the industry resulting in the possibility that clinical performance of new materials or devices may fall well short of anticipated targets.

Ultimately the clinician maintains a responsibility to the patient to provide the best possible service. Defining “best service” is not always easy. In some situations this may be a pure aesthetic decision while in other situations, based upon patient desire, other factors such as durability, comfort or function may be the predominant concerns. Products that have not been tested through biomechanical means may be brought to the market with only a concept behind the product. Unfortunately many concepts that seem perfectly logical fail to perform in the clinical setting. Although biomechanical testing is not perfect it does serve as a necessary first step towards the advancement of dental science.

Conclusion

This chapter has attempted to identify the rationale for biomechanical testing while also identifying the risks when too much emphasis is placed on the results of such testing. Clinicians must carefully evaluate observed failure patterns to ensure that biomechanical testing demonstrates similar failure patterns otherwise the tests may not prove valuable to the reader.

Oxford Centre for Evidence-based Medicine Levels of Evidence (March 2009) (for definitions of terms used see glossary at http://www.cebm.net/?o=1116)

Level	Therapy/Prevention, Aetiology/Harm	Prognosis	Diagnosis	Differential diagnosis/symptom prevalence study	Economic and decision analyses
1a	SR (with homogeneity*) of RCTs	SR (with homogeneity*) of inception cohort studies; CDR † validated in different populations	SR (with homogeneity*) of Level 1 diagnostic studies; CDR † with 1b studies from different clinical centres	SR (with homogeneity*) of prospective cohort studies	SR (with homogeneity*) of Level 1 economic studies
1b	Individual RCT (with narrow Confidence Interval ‡)	Individual inception cohort study with > 80% follow-up; CDR † validated in a single population	Validating** cohort study with good ††† reference standards; or CDR † tested within one clinical centre	Prospective cohort study with good follow-up****	Analysis based on clinically sensible costs or alternatives; systematic review(s) of the evidence; and including multi-way sensitivity analyses
1c	All or none §	All or none case-series	Absolute SpPins and SnNouts ††	All or none case-series	Absolute better-value or worse-value analyses ††††
2a	SR (with homogeneity*) of cohort studies	SR (with homogeneity*) of either retrospective cohort studies or untreated control groups in RCTs	SR (with homogeneity*) of Level >2 diagnostic studies	SR (with homogeneity*) of 2b and better studies	SR (with homogeneity*) of Level >2 economic studies
2b	Individual cohort study (including low quality RCT; e.g., <80% follow-up)	Retrospective cohort study or follow-up of untreated control patients in an RCT; Derivation of CDR† or validated on split-sample §§§ only	Exploratory** cohort study with good ††† reference standards; CDR † after derivation, or validated only on split-sample §§§ or databases	Retrospective cohort study, or poor follow-up	Analysis based on clinically sensible costs or alternatives; limited review(s) of the evidence, or single studies; and including multi-way sensitivity analyses

Table. (Continued)

Level	Therapy/Prevention, Aetiology/Harm	Prognosis	Diagnosis	Differential diagnosis/symptom prevalence study	Economic and decision analyses
2c	"Outcomes" Research; Ecological studies	"Outcomes" Research		Ecological studies	Audit or outcomes research
3a	SR (with homogeneity*) of case-control studies		SR (with homogeneity*) of 3b and better studies	SR (with homogeneity*) of 3b and better studies	SR (with homogeneity*) of 3b and better studies
3b	Individual Case-Control Study		Non-consecutive study; or without consistently applied reference standards	Non-consecutive cohort study, or very limited population	Analysis based on limited alternatives or costs, poor quality estimates of data, but including sensitivity analyses incorporating clinically sensible variations.
4	Case-series (and poor quality cohort and case-control studies §§)	Case-series (and poor quality prognostic cohort studies***)	Case-control study, poor or non-independent reference standard	Case-series or superseded reference standards	Analysis with no sensitivity analysis
5	Expert opinion without explicit critical appraisal, or based on physiology, bench research or "first principles"	Expert opinion without explicit critical appraisal, or based on physiology, bench research or "first principles"	Expert opinion without explicit critical appraisal, or based on physiology, bench research or "first principles"	Expert opinion without explicit critical appraisal, or based on physiology, bench research or "first principles"	Expert opinion without explicit critical appraisal, or based on economic theory or "first principles"

Produced by Bob Phillips, Chris Ball, Dave Sackett, Doug Badenoch, Sharon Straus, Brian Haynes, Martin Dawes since November 1998. Updated by Jeremy Howick March 2009.

NotesUsers can add a minus-sign "-" to denote the level of that fails to provide a conclusive answer because:

EITHER a single result with a wide Confidence IntervalOR a Systematic Review with troublesome heterogeneity. Such evidence is inconclusive, and therefore can only generate Grade D recommendations.

* By homogeneity we mean a systematic review that is free of worrisome variations (heterogeneity) in the directions and degrees of results between individual studies. Not all systematic reviews with statistically significant heterogeneity need be worrisome, and not all worrisome heterogeneity need be statistically significant. As noted above, studies displaying worrisome heterogeneity should be tagged with a "-" at the end of their designated level.

† Clinical Decision Rule. (These are algorithms or scoring systems that lead to a prognostic estimation or a diagnostic category.).

‡ See note above for advice on how to understand, rate and use trials or other studies with wide confidence intervals.

§ Met when all patients died before the Rx became available, but some now survive on it; or when some patients died before the Rx became available, but none now die on it.

§§ By poor quality cohort study we mean one that failed to clearly define comparison groups and/or failed to measure exposures and outcomes in the same (preferably blinded), objective way in both exposed and non-exposed individuals and/or failed to identify or appropriately control known confounders and/or failed to carry out a sufficiently long and complete follow-up of patients. By poor quality case-control study we mean one that failed to clearly define comparison groups and/or failed to measure exposures and outcomes in the same (preferably blinded), objective way in both cases and controls and/or failed to identify or appropriately control known confounders.

§§§ Split-sample validation is achieved by collecting all the information in a single tranche, then artificially dividing this into "derivation" and "validation" samples.

†† An "Absolute SpPin" is a diagnostic finding whose Specificity is so high that a Positive result rules-in the diagnosis. An "Absolute SnNout" is a diagnostic finding whose Sensitivity is so high that a Negative result rules-out the diagnosis.

‡‡ Good, better, bad and worse refer to the comparisons between treatments in terms of their clinical risks and benefits.

††† Good reference standards are independent of the test, and applied blindly or objectively to applied to all patients. Poor reference standards are haphazardly applied, but still independent of the test. Use of a non-independent reference standard (where the 'test' is included in the 'reference', or where the 'testing' affects the 'reference') implies a level 4 study.

†††† Better-value treatments are clearly as good but cheaper, or better at the same or reduced cost. Worse-value treatments are as good and more expensive, or worse and the equally or more expensive.

** Validating studies test the quality of a specific diagnostic test, based on prior evidence. An exploratory study collects information and trawls the data (e.g. using a regression analysis) to find which factors are 'significant'.

*** By poor quality prognostic cohort study we mean one in which sampling was biased in favour of patients who already had the target outcome, or the measurement of outcomes was accomplished in <80% of study patients, or outcomes were determined in an unblinded, non-objective way, or there was no correction for confounding factors.

**** Good follow-up in a differential diagnosis study is >80%, with adequate time for alternative diagnoses to emerge (for example 1-6 months acute, 1 - 5 years chronic).

Grades of Recommendation

A	consistent level 1 studies
B	consistent level 2 or 3 studies *or* extrapolations from level 1 studies
C	level 4 studies *or* extrapolations from level 2 or 3 studies
D	level 5 evidence *or* troublingly inconsistent or inconclusive studies of any level

"Extrapolations" are where data is used in a situation that has potentially clinically important differences than the original study situation.

REFERENCES

[1] Biomechanics. (n.d.). The American Heritage® Science Dictionary. Retrieved January 17, 2010, from Dictionary.com website: http://dictionary.reference.com/browse/biomechanics.

INDEX

A

B

C

D

E

F

G

N

O

P

Q

R

S

W

X

Y

Z